Ultrasound Imaging and Therapy

Edited by

Aaron Fenster

Imaging Research Laboratories, Robarts Research Institute
Department of Medical Biophysics and Department of Medical Imaging
University of Western Ontario

James C. Lacefield

Imaging Research Laboratories, Robarts Research Institute
Department of Electrical and Computer Engineering and Department of Medical Biophysics
University of Western Ontario

CRC Press

Taylor & Francis Group
Boca Raton London New York

CRC Press is an imprint of the
Taylor & Francis Group, an **informa** business

A TAYLOR & FRANCIS BOOK

IMAGING IN MEDICAL DIAGNOSIS AND THERAPY

Series Editors: Andrew Karellas and Bruce R. Thomadsen

Published titles

Quality and Safety in Radiotherapy
Todd Pawlicki, Peter B. Dunscombe,
Arno J. Mundt, and Pierre Scalliet, Editors
ISBN: 978-1-4398-0436-0

Adaptive Radiation Therapy
X. Allen Li, Editor
ISBN: 978-1-4398-1634-9

Quantitative MRI in Cancer
Thomas E. Yankeelov, David R. Pickens,
and Ronald R. Price, Editors
ISBN: 978-1-4398-2057-5

Informatics in Medical Imaging
George C. Kagadis and Steve G. Langer, Editors
ISBN: 978-1-4398-3124-3

Adaptive Motion Compensation in Radiotherapy
Martin J. Murphy, Editor
ISBN: 978-1-4398-2193-0

Image-Guided Radiation Therapy
Daniel J. Bourland, Editor
ISBN: 978-1-4398-0273-1

Targeted Molecular Imaging
Michael J. Welch and William C. Eckelman,
Editors
ISBN: 978-1-4398-4195-0

Proton and Carbon Ion Therapy
C.-M. Charlie Ma and Tony Lomax, Editors
ISBN: 978-1-4398-1607-3

Comprehensive Brachytherapy: Physical and Clinical Aspects
Jack Venselaar, Dimos Baltas, Peter J. Hoskin,
and Ali Soleimani-Meigooni, Editors
ISBN: 978-1-4398-4498-4

Physics of Mammographic Imaging
Mia K. Markey, Editor
ISBN: 978-1-4398-7544-5

Physics of Thermal Therapy: Fundamentals and Clinical Applications
Eduardo Moros, Editor
ISBN: 978-1-4398-4890-6

Emerging Imaging Technologies in Medicine
Mark A. Anastasio and Patrick La Riviere, Editors
ISBN: 978-1-4398-8041-8

Cancer Nanotechnology: Principles and Applications in Radiation Oncology
Sang Hyun Cho and Sunil Krishnan, Editors
ISBN: 978-1-4398-7875-0

Monte Carlo Techniques in Radiation Therapy
Joao Seco and Frank Verhaegen, Editors
ISBN: 978-1-4665-0792-0

Image Processing in Radiation Therapy
Kristy Kay Brock, Editor
ISBN: 978-1-4398-3017-8

Informatics in Radiation Oncology
George Starkschall and R. Alfredo C. Siochi,
Editors
ISBN: 978-1-4398-2582-2

Cone Beam Computed Tomography
Chris C. Shaw, Editor
ISBN: 978-1-4398-4626-1

Tomosynthesis Imaging
Ingrid Reiser and Stephen Glick, Editors
ISBN: 978-1-4398-7870-5

Stereotactic Radiosurgery and Stereotactic Body Radiation Therapy
Stanley H. Benedict, David J. Schlesinger, Steven
J. Goetsch, and Brian D. Kavanagh, Editors
ISBN: 978-1-4398-4197-6

Computer-Aided Detection and Diagnosis in Medical Imaging
Qiang Li and Robert M. Nishikawa, Editors
ISBN: 978-1-4398-7176-8

Ultrasound Imaging and Therapy
Aaron Fenster and James C. Lacefield, Editors
ISBN: 978-1-4398-6628-3

IMAGING IN MEDICAL DIAGNOSIS AND THERAPY

Series Editors: Andrew Karellas and Bruce R. Thomadsen

Forthcoming titles

Handbook of Small Animal Imaging: Preclinical Imaging, Therapy, and Applications
George Kagadis, Nancy L. Ford, George K. Loudos, and Dimitrios Karnabatidis, Editors

Cardiovascular and Neurovascular Imaging: Physics and Technology
Carlo Cavedon and Stephen Rudin, Editors

Physics of PET and SPECT Imaging
Magnus Dahlbom, Editor

Hybrid Imaging in Cardiovascular Medicine
Yi-Hwa Liu and Albert Sinusas, Editors

Scintillation Dosimetry
Sam Beddar and Luc Beaulieu, Editors

CRC Press
Taylor & Francis Group
6000 Broken Sound Parkway NW, Suite 300
Boca Raton, FL 33487-2742

First issued in paperback 2018

© 2015 by Taylor & Francis Group, LLC
CRC Press is an imprint of Taylor & Francis Group, an Informa business

No claim to original U.S. Government works

ISBN 13: 978-1-138-89435-8 (pbk)
ISBN 13: 978-1-4398-6628-3 (hbk)

Visit the Taylor & Francis Web site at
http://www.taylorandfrancis.com

and the CRC Press Web site at
http://www.crcpress.com

Contents

SECTION I   Ultrasound Instrumentation

SECTION II   Diagnostic Ultrasound Imaging

SECTION III Therapeutic and Interventional Ultrasound Imaging

Series Preface

Advances in the science and technology of medical imaging and radiation therapy are more profound and rapid than ever before since their inception over a century ago. Further, the disciplines are increasingly cross-linked as imaging methods become more widely used for planning, guiding, monitoring, and assessing treatments in radiation therapy. Today, the technologies of medical imaging and radiation therapy are so complex and so computer-driven that it is difficult for those (physicians and technologists) responsible for their clinical use to know exactly what is happening at the point of care when a patient is being examined or treated. Medical physicists are well equipped to understand the technologies and their applications, and they assume greater responsibilities in the clinical arena to ensure that what is intended for the patient is actually delivered in a safe and effective manner.

The growing responsibilities of medical physicists in the clinical arenas of medical imaging and radiation therapy are not without their challenges, however. Most medical physicists are knowledgeable in either radiation therapy or medical imaging and expert in one or a small number of areas within their discipline. They sustain their expertise in these areas by reading scientific articles and attending scientific talks at meetings. However, their responsibilities increasingly extend beyond their specific areas of expertise. To meet these responsibilities, medical physicists periodically must refresh their knowledge on the advances in medical imaging and radiation therapy, and they must be prepared to function at the intersection of these two fields. To accomplish these objectives is a challenge.

At the 2007 annual meeting of the American Association of Physicists in Medicine in Minneapolis, this challenge was the topic of conversation during a lunch hosted by Taylor & Francis Group and involving a group of senior medical physicists (Arthur L. Boyer, Joseph O. Deasy, C.-M. Charlie Ma, Todd A. Pawlicki, Ervin B. Podgorsak, Elke Reitzel, Anthony B. Wolbarst, and Ellen D. Yorke). The conclusion of the discussion was that a book series should be launched under the Taylor & Francis Group banner, with each book in the series addressing a rapidly advancing area of medical imaging or radiation therapy of importance to medical physicists. The aim would be for each book to provide medical physicists with the information needed to understand technologies driving rapid advances and their applications to safe and effective delivery of patient care.

Each book in the series is edited by one or more individuals with recognized expertise in the technological area encompassed by the book. The editors are responsible for selecting the authors of individual chapters and ensuring that the chapters are comprehensive and intelligible to someone without such expertise. The enthusiasm of the book editors and chapter authors has been gratifying and reinforces the conclusion of the Minneapolis luncheon that this series addresses a major need of medical physicists.

Imaging in Medical Diagnosis and Therapy would not have been possible without the encouragement and support of the series manager, Luna Han of Taylor & Francis Group. The editors and authors, and most of all I, are indebted to her steady guidance throughout the project.

William Hendee
Founding Series Editor
Rochester, Minnesota

Preface

For the past 50 years, ultrasound imaging has been used extensively for diagnosis of a wide range of diseases. With improvements in image quality and reduction of cost for advanced features, ultrasound imaging is playing an ever-greater role in diagnosis and image-guided interventions. The pace of innovations is increasing, and new improved applications are constantly being described. Many of these have been adopted by clinicians for routine use. This book offers an overview of ultrasound imaging for diagnosis, covering its use in image-guided interventions and ultrasound-based therapy and highlighting the latest advances. It discusses both improvements on current techniques already in clinical use as well as techniques in an advanced state of testing with great potential for adoption into routine clinical use. The scope extends from background on the state of the art in transducers and beam formers for use in 2-D, 3-D, and 4-D ultrasound as well as developments in tissue characterization, Doppler techniques, use of ultrasound contrast agents, ultrasound-guided biopsy and therapy, and use of ultrasound to deliver therapy.

Many books have been written on this subject, but this field is advancing rapidly, with ever-expanding applications. During this last decade, ultrasound imaging has increased its role in image-guided delivery and monitoring of therapy. As a result, increasing numbers of medical physicists, radiation therapy physicists, and biomedical engineers are making use of this technology in their work and research. In addition, more computer scientists have been needed to develop image processing algorithms for diagnostic and interventional applications. Thus, this book has two objectives: (1) to inform the audience on the state of the art of current and developing techniques and (2) to identify trends in the use of ultrasound imaging and the technical and computational problems that need to be solved.

We have aimed the book at individuals working on diagnostic and therapeutic applications. Thus, the audience is quite broad and includes researchers, trainees, academic physicians, technicians, and technologists in research laboratories and diagnostic and therapy departments. It will be of particular importance to researchers and their trainees who are trying to identify areas that require innovative solutions to unsolved problems. In addition, it will be of value to those working in diagnostic and treatment centers with interest in identifying trends and future offerings by vendors. Because many of the applications require computational algorithmic solutions, computer science researchers and trainees will find a useful review of major problems and specifications that should be met.

The book is organized into three main sections. The first chapters deal with advances in the technology, including transducers (2-D, 3-D, and 4-D), beamformers, 3-D imaging systems, and blood velocity estimation systems. The second section deals with diagnostic applications, including elastography, quantitative techniques for therapy monitoring and diagnostic imaging, and ultrasound tomography. The last two chapters address the use of ultrasound in image-guided interventions, for image-guided biopsy and brain imaging.

Editors

Aaron Fenster, PhD, is a founding director of the Imaging Research Laboratories (IRL) at the Robarts Research Institute and a professor in the Department of Medical Biophysics and Department of Medical Imaging at the University of Western Ontario (UWO). He is also the founder and associate director of the Graduate Program in Biomedical Engineering at UWO. Dr. Fenster earned his PhD degree from the Department of Medical Biophysics of the University of Toronto for research under the supervision of Dr. H. E. Johns. His first academic appointment was at the Department of Radiology and Medical Biophysics of the University of Toronto from 1979 to 1987 as a director of the radiological research laboratories of the Department of Radiology.

His research group focuses on the development of 3-D ultrasound imaging with diagnostic and surgical and therapeutic cancer applications. His team developed the world's firsts in 3-D ultrasound imaging of the carotids and prostate, 3-D ultrasound-guided prostate cryosurgery and brachytherapy, 3-D ultrasound-guided prostate and breast biopsy for early diagnosis of cancer, and 3-D ultrasound images of mouse tumors and their vasculature. Among his numerous honors, Dr. Fenster is the recipient of the 2007 Premier's Discovery Award for Innovation and Leadership, the 2008 Hellmuth Prize for Achievement in Research at the UWO, and the Canadian Organization of Medical Physicists 2010 Gold Medal Award. He is also a fellow of the Canadian Academy of Health Sciences.

James C. Lacefield, PhD, is an associate professor jointly appointed to the Department of Electrical and Computer Engineering and the Department of Medical Biophysics at the University of Western Ontario. He is also a faculty member of the Graduate Program in Biomedical Engineering, an associate scientist of the Imaging Research Laboratories at Robarts Research Institute, and a mentor in Western's CIHR Strategic Training Program in Cancer Research and Technology Transfer. Dr. Lacefield earned his PhD in biomedical engineering at Duke University, where he was an NSF/ERC predoctoral fellow in the Center for Emerging Cardiovascular Technologies. He served as a visiting research associate of the Diagnostic Ultrasound Research Laboratory in the Department of Electrical and Computer Engineering at the University of Rochester from 1999 through 2001.

His research activities address physical acoustics and signal-processing aspects of biomedical ultrasound imaging, with an emphasis on applications of ultrasound to cancer and cardiovascular research. Dr. Lacefield is a member of the Acoustical Society of America, the American Society for Engineering Education, the Institute of Electrical and Electronics Engineers, and the Association of Professional Engineers of Ontario.

Contributors

Craig K. Abbey
Department of Psychological and
  Brain Sciences
University of California, Santa
  Barbara
Santa Barbara, California

Paul E. Barbone
Department of Mechanical
  Engineering
Boston University
Boston, Massachusetts

Jeff Bax
Imaging Research Laboratories
Robarts Research Institute
and
Graduate Program in Biomedical
  Engineering
University of Western Ontario
London, Ontario, Canada

Matthew Bayer
Department of Medical Physics
University of Wisconsin
Madison, Wisconsin

Bernard C. Y. Chiu
Department of Electronic Engineering
City University of Hong Kong
Kowloon, Hong Kong

Derek Cool
Imaging Research Laboratories
Robarts Research Institute
and
Biomedical Imaging Research Centre
and
Department of Medical Imaging
University of Western Ontario
London, Ontario, Canada

Joshua R. Doherty
Department of Biomedical
  Engineering
Duke University
Durham, North Carolina

Neb Duric
Karmanos Cancer Institute
Wayne State University
Detroit, Michigan

Aaron Fenster
Imaging Research Laboratories
Robarts Research Institute
and
Biomedical Imaging Research Centre
and
Graduate Program in Biomedical
  Engineering
and
Department of Medical Biophysics
and
Department of Medical Imaging
University of Western Ontario
London, Ontario, Canada

Timothy J. Hall
Department of Medical Physics
and
Department of Biomedical
 Engineering
University of Wisconsin
Madison, Wisconsin

Kullervo Hynynen
Department of Medical Biophysics
University of Toronto
Toronto, Ontario, Canada

Michael F. Insana
Department of Bioengineering
Beckman Institute for Advanced
 Science and Technology
University of Illinois at
 Urbana-Champaign
Urbana, Illinois

Jørgen Arendt Jensen
Center for Fast Ultrasound Imaging
Department of Electrical Engineering
Technical University of Denmark
Lyngby, Denmark

Vaishali Karnik
Imaging Research Laboratories
Robarts Research Institute
and
Graduate Program in Biomedical
 Engineering
University of Western Ontario
London, Ontario, Canada

Cuiping Li
Karmanos Cancer Institute
Wayne State University
Detroit, Michigan

Peter J. Littrup
Karmanos Cancer Institute
Wayne State University
Detroit, Michigan

Nghia Q. Nguyen
Department of Engineering
University of Cambridge
Cambridge, United Kingdom

Kathryn R. Nightingale
Department of Biomedical
 Engineering
Duke University
Durham, North Carolina

Assad A. Oberai
Department of Mechanical, Aerospace
 and Nuclear Engineering
and
Scientific Research Computation
 Center
Rensselaer Polytechnic Institute
Troy, New York

Michael L. Oelze
Bioacoustics Research Laboratory
Department of Electrical and
 Computer Engineering
University of Illinois,
 Urbana-Champaign
Urbana, Illinois

Meaghan A. O'Reilly
Physical Sciences Platform
Sunnybrook Research Institute
Toronto, Ontario, Canada

Mark L. Palmeri
Department of Biomedical
Engineering
Duke University
Durham, North Carolina

Grace Parraga
Imaging Research Laboratories
Robarts Research Institute
and
Graduate Program in Biomedical
Engineering
and
Department of Medical Biophysics
and
Department of Medical Imaging
University of Western Ontario
London, Ontario, Canada

Cesare Romagnoli
Biomedical Imaging Research Centre
and
Department of Medical Imaging
University of Western Ontario
London, Ontario, Canada

Olivier Roy
Karmanos Cancer Institute
Wayne State University
Detroit, Michigan

Steve Schmidt
Karmanos Cancer Institute
Wayne State University
Detroit, Michigan

K. Kirk Shung
Department of Biomedical
Engineering
University of Southern California
Los Angeles, California

Gregg E. Trahey
Department of Biomedical
Engineering
Duke University
Durham, North Carolina

Eranga Ukwatta
Imaging Research Laboratories
Robarts Research Institute
and
Graduate Program in Biomedical
Engineering
University of Western Ontario
London, Ontario, Canada

Aaron Ward
Imaging Research Laboratories
Robarts Research Institute
and
Biomedical Imaging Research Centre
and
Graduate Program in Biomedical
Engineering
and
Department of Medical Biophysics
University of Western Ontario
London, Ontario, Canada

Jesse Yen
Department of Biomedical
Engineering
University of Southern California
Los Angeles, California

Ultrasound Instrumentation

1. Array Transducers and Beamformers

K. Kirk Shung and Jesse Yen

Chapter 1

Ultrasound Imaging and Therapy. Edited by Aaron Fenster and James C. Lacefield © 2015 CRC Press/Taylor & Francis Group, LLC. ISBN: 978-1-4398-6628-3

1.1 Introduction

All ultrasonic imaging or therapeutic systems require an ultrasonic transducer to convert electrical energy into ultrasonic or acoustic energy and vice versa. Ultrasonic transducers come in a variety of shapes and sizes ranging from single-element transducers for mechanical scanning, to linear arrays, to multidimensional arrays for electronic scanning. The most critical component of an ultrasonic transducer is a piezoelectric element.

1.2 Piezoelectric Effect

The phenomenon that a material upon the application of an electrical field changes its physical dimensions and vice versa is known as the piezoelectric effect (pressure-electric effect), discovered by French physicists Pierre and Jacques Curie in 1880. The direct and reverse piezoelectric effects are illustrated in Figure 1.1a and b, respectively, where dashed lines represent the shape of the piezoelectric material before external disturbance. Certain naturally occurring crystals such as quartz and tourmaline are piezoelectric but are not used often today because of their poor piezoelectric properties. A class of materials called ferroelectric materials, which are polycrystalline (Safari and Akdogan 2008), possesses very strong piezoelectric properties following a preparation step called poling. The most popular ferroelectric material is lead zirconate titanate, $Pb(Zr, Ti)O_3$ or PZT, which can be doped to enhance certain properties. For instance, PZT 5H is preferred for imaging systems because of its superior piezoelectric conversion capability, whereas PZT 4 is preferred for therapeutic systems because of its capability of handling heating.

Poling or polarization is conducted by heating a ferroelectric material to a temperature just above the Curie temperature of the material, in which the material loses piezoelectricity. Then the material is cooled slowly in the presence of a strong electric field, typically in the order of 20 kV/cm, applied in the direction in which the piezoelectric effect is required. There are a great variety of ferroelectric materials, including barium titanate ($BaTiO_3$), lead metaniobate ($PbNb_2O_6$), and lithium niobate ($LiNbO_3$).

There are several piezoelectric coefficients frequently specified for piezoelectric materials for assessing their performance. The piezoelectric stress constant (e) is defined as the change in stress per unit change in electric field without strain or while being clamped. It has the unit of newtons per volt-meter or coulombs per square meter. The transmission or piezoelectric strain constant (d) is a measure of the transmission performance of a piezoelectric

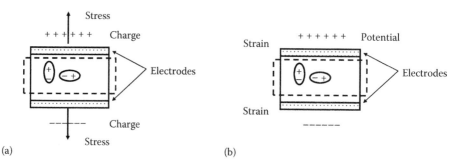

(a)   (b)

FIGURE 1.1 (a) Direct piezoelectric effect where a stress induces a charge separation. (b) Reverse piezoelectric effect where a potential difference across the electrodes induces a strain.

material representing the change in strain per unit change in electric field with a unit of coulombs per newton when there is no stress. By contrast, the receiving constant (g) with a unit of volt-meters per newton is a measure of piezoelectric material performance during reception, representing the change in electric field per unit change in applied stress when there is no current or under open circuit condition. There are two dielectric constants (ε) associated with a piezoelectric material. One is the dielectric constant when there is no stress or free dielectric constant, and the other is the dielectric constant when there is no strain or clamped dielectric constant. It should be noted that these properties are direction dependent because the piezoelectric materials are anisotropic (Shung 2005; Safari and Akdogan 2008).

For crystals such as quartz, the principal axes are defined by the crystalline axes; for example, a plate cut with its surface perpendicular to the x-axis is called an x-cut. The x, y, and z directions are indicated by numbers 1, 2, and 3, respectively. For polarized ferroelectric ceramics, direction 3 is usually used to denote the polarization direction. A piezoelectric strain constant, d_{33}, represents the strain produced in direction 3 by applying an electric field in direction 3. Here it is important to note that the piezoelectric properties of a material depend on boundary conditions and therefore on the shape of the material. For example, the piezoelectric constant of a material in a plate form is different from that in a rod form.

The ability of a piezoelectric material to convert one form of energy into another is measured by its electromechanical coupling coefficient, k, defined as

$$k = \sqrt{\frac{\text{stored mechancial energy}}{\text{total stored energy}}}.$$

Total stored energy includes both mechanical and electrical energy. Therefore,

$$k^2 = \frac{\text{stored mechanical energy}}{\text{total stored energy}}.$$

It should be noted that this quantity is not the efficiency of the transducer. If the transducer is lossless, its efficiency is 100%. However, the electromechanical coupling coefficient is not necessarily 100% because some of the energy is stored as mechanical energy, but the rest may be stored dielectrically in a form of electrical potential energy. The electromechanical coupling coefficient is a measure of the performance of a material as a transducer because only the stored mechanical energy is useful. The piezoelectric constants for a few important piezoelectric materials are listed in Table 1.1.

In addition to PZT, piezoelectric polymers have also been found to be useful in several applications (Brown 2000). One of these polymers is polyvinylidene difluoride (PVDF), which is semicrystalline. After processes such as polymerization, stretching, and poling, a thin sheet of PVDF with a thickness in the order of 6 to 50 μm can be used as a transducer material. The advantages of this material are that it is wideband, flexible, and inexpensive. The disadvantages are that it has a very low transmitting constant, its dielectric loss is large, and the dielectric constant is low. Although PVDF is not an ideal transmitting material, it does possess a fairly high receiving constant. Miniature PVDF hydrophones are commercially available. P(VDF-TrFE) co-polymers have been shown to have a higher electromechanical coupling coefficient.

Chapter 1

Table 1.1 Piezoelectric Material Properties

Piezoelectric Property	PVDF	Quartz (X-cut)	PZT 5H
Transmission coefficient d_{33} (10^{-12} c/n)	15	2.3	583
Receiving constant g_{33} (10^{-2} V-m/n)	14	5.8	191
Electromechanical coupling coefficient, k_t	0.11	0.14	0.55
Clamped dielectric constant	5.0	4.5	1470
Sound velocity (cm/s)	2070	5740	3970
Density (kg/m³)	1760	2650	7450
Curie temperature (°C)	100	573	190

Note: k_t indicates electromechanical coupling coefficient measured with the piezoelectric material in the form of a disc where the radius is much greater than the thickness. d_{33} and g_{33} are constants measured with the response and the excitation all in the 3 directions.

One of the most promising frontiers in transducer technology is the development of piezoelectric composite materials (Smith 1989). A notation of 1-3, 2-2, and so on, has been coined by Newnham et al. (1978) to describe the composite structure. A notation of 1-3 means that one phase of the composite is connected only in one direction whereas the second phase is connected in all three directions. A notation of 2-2 means that both phases are connected in two directions as illustrated in Figure 1.2. These composites, typically in a volume concentration of 20% to 70% PZT, have a lower acoustic impedance (4–25 MRayls) than conventional PZT (34 MRayls), which better matches the acoustic impedance of human skin. The composite material can be made flexible with an adjustable dielectric constant and a higher electromechanical coupling coefficient than the bulk PZT. Higher coupling coefficient and better impedance matching can lead to higher transducer sensitivity and improved image resolution.

More recently, several single-crystal ferroelectric materials such as $Pb(Zn_{1/3}Nb_{2/3})O_3$-$PbTiO_3$ (PZN-PT), $Pb(Mg_{1/3}Nb_{2/3})O_3$-$PbTiO_3$ (PMN-PT), and $Pb(In_{1/2}Nb_{1/2})$-$Pb(Mg_{1/3}Nb_{2/3})$

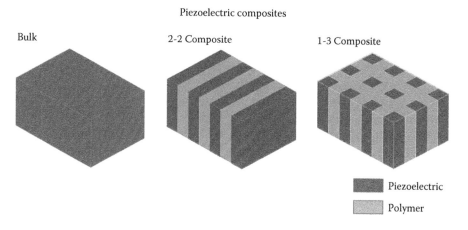

Piezoelectric composites

Bulk   2-2 Composite   1-3 Composite

■ Piezoelectric
▨ Polymer

FIGURE 1.2 Two different configurations of piezoceramic composites: 1-3 composites and 2-2 composites.

Table 1.2   Piezoelectric Properties of Single-Crystal Piezoelectric Materials

Property	PZN-PT	PMN-PT	PIN-PMN-PT
Electromechanical coupling coefficient in the form of a pillar, k_{33}	0.93	0.94	0.94
Curie temperature, °C	140	155	160
Clamped dielectric constant	294	800	700
Acoustic impedance, MRayls	26	30	30

O_3-$PbTiO_3$ (PIN-PMN-PT) with higher electromechanical coupling coefficients than conventional PZT have been developed (Shrout and Fielding 1990; Tian et al. 2007). These materials possess extremely high electromechanical coupling coefficients, which can be as high as 0.9. Table 1.2 lists the piezoelectric properties of these single-crystal materials. It is known that several commercial clinical scanner probes at frequencies from 3 to 7 MHz now are made from single-crystal piezoelectric materials and they exhibit superior bandwidth and sensitivity, which are measures of transducer performance to be discussed in the next section.

1.3   Ultrasonic Transducers

The simplest ultrasonic transducer is a single-element piston transducer shown in Figure 1.3, where panels a and b show a photo and the internal construction of a single-element ultrasonic transducer, respectively. The most important component of such a device is the piezoelectric element. Several factors are involved in choosing a proper piezoelectric material for transmitting and/or receiving the ultrasonic wave. They include stability, piezoelectric properties, and material strength. The surfaces of the element are the electrodes, and the outside electrode is usually grounded to protect the patients from electrical shock. The resonating frequency, f_0, of a disc is determined by its thickness, L, described by the following equation:

$$f_0 = \frac{nc_p}{2L},$$

(1.1)

with the lowest resonant frequency being $n = 1$ and where c_p is the acoustic wave velocity in the transducer material, L is the thickness of the piezoelectric material, and n is an odd integer. In other words, resonance occurs when L is equal to odd multiples of one-half of a wavelength in the piezoelectric material.

The transducer can be treated as a three-port network as shown in Figure 1.4, two being mechanical ports representing the front and back surfaces of the piezoelectric crystal and one being an electrical port representing the electrical connection of the piezoelectric material to the electrical generator (Shung 2005). In Figure 1.4, I, V, F, and u denote current, voltage, force, and medium velocity, respectively. Various sophisticated one-dimensional circuit models exist to model the behavior of the transducer. The most well known are the Mason model, the Redwood model, and the KLM model (Krimholtz et al. 1970; Kino 1987). Commercial software based on these models is available. Among

Chapter 1

(a)   Lithium niobate transducers

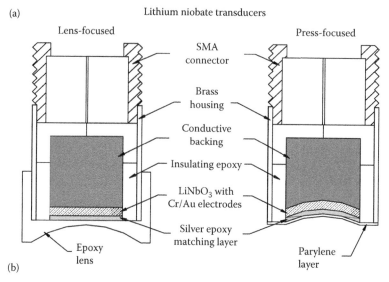

(b)

FIGURE 1.3 (a) Photo and (b) detailed construction of two single-element ultrasonic transducers with one or two matching layers and the backing material. The transducer on the left has a lens whereas the one on the right is self-focusing.

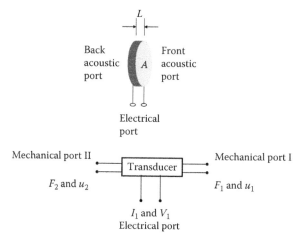

FIGURE 1.4 A piezoceramic disc can be treated as an electromechanical device with two acoustic ports representing the front and rear interfaces between the ceramic and the surrounding media and an electrical port.

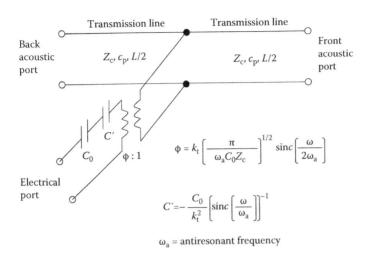

$$\phi = k_t \left[\frac{\pi}{\omega_a C_0 Z_c} \right]^{1/2} sinc \left[\frac{\omega}{2\omega_a} \right]$$

$$C' = -\frac{C_0}{k_t^2} \left[sinc \left[\frac{\omega}{\omega_a} \right] \right]^{-1}$$

ω_a = antiresonant frequency

FIGURE 1.5 The KLM model or equivalent circuit for a single-element transducer.

them, the KLM model is the most popular and is shown in Figure 1.5 for a circular disc with area A and thickness L. This model divides a piezoelectric element into two halves, each represented by a transmission line. It is more physically intuitive. The effects of matching layers and backing material can be readily included. In Figure 1.5, $Z_c = Z_0 A$ is called the radiation impedance, Z_0 is the acoustic impedance of the piezoelectric element, λ_p is the sound velocity in the piezoelectric material, and $C_0 = \epsilon(A/L)$ is the clamped capacitance. The antiresonance frequency, ω_a, is defined as the frequency where the magnitude of the input electrical impedance of the transducer is maximal.

A typical response, including the magnitude and phase of the input electrical impedance for a single-element PZT 5H and the 10 mm diameter circular disc transducer, air loaded and air backed, with a thickness of 0.43λ at 5 MHz, as obtained with the KLM model, is shown in Figure 1.6. The frequency at which electrical impedance is minimal

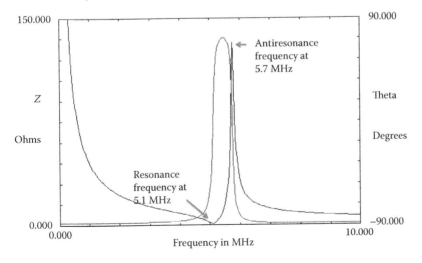

FIGURE 1.6 The magnitude and phase of the input electrical impedance of a circular piston transducer irradiating into air and backed by air.

is defined as the resonance frequency, whereas the frequency at which electrical impedance is maximal is the antiresonance frequency. The vertical scale on the left is the magnitude of electrical impedance in ohm, and on the right is the phase angle in degrees.

1.3.1 Mechanical Matching

When a transducer is excited by an electrical source, it rings at its natural resonant frequency. For continuous wave (CW) applications, the transducers are air backed, allowing as much energy as possible to be irradiated into a forward medium such as water, which has a higher acoustic impedance than air. Because of the mismatch in acoustic impedance between the air and the piezoelectric material, acoustic energy at this interface is reflected into the forward direction. Thus, very little energy is lost out of the back port. The drawback is that this mismatch, which produces the so-called ringing effect for pulse-echo applications, is very undesirable because it lengthens the pulse duration. The pulse duration affects the capability of an imaging system to resolve small objects.

Absorptive backing materials with acoustic impedance similar to that of the piezoelectric material can be used to reduce ringing or to increase bandwidth. The backing material should not only absorb part of the energy from the vibration of the back face but also minimize the mismatch in acoustic impedance. It absorbs as much energy that enters it as possible. It must be noted that the suppression of ringing or shortening of pulse duration is achieved by sacrificing sensitivity because a large portion of the energy is absorbed by the backing material. Various types of backing materials, including tungsten-loaded epoxy and silver-loaded epoxy, have been used with good success.

The performance of a transducer can also be improved with acoustic matching layers mounted in the front. It can be easily shown that for a monochromatic plane wave, 100% transmission occurs for a layer of material of $\lambda_m/4$ thickness and acoustic impedance Z_m, where λ_m is the wavelength in the matching layer material and (Kinsler et al. 2000)

$$Z_m = (Z_p Z_l)^{1/2}. \tag{1.2}$$

In Equation 1.2, Z_p and Z_l are the acoustic impedances of the piezoelectric element and the loading medium, respectively.

For wideband transducers, however, Desilets et al. (1978) showed that for a single matching layer, Equation 1.2 should be modified to

$$Z_m = \left(Z_p Z_l^2 \right)^{1/3}, \tag{1.3}$$

and for two matching layers, the acoustic impedances of the two layers should be as follows:

$$Z_{m1} = \left(Z_p^4 Z_l^3 \right)^{1/7} \tag{1.4}$$

$$Z_{m2} = \left(Z_p Z_l^6 \right)^{1/7}, \tag{1.5}$$

State-of-the-art transducers and arrays that use composites can achieve a bandwidth better than 70% merely with front matching and light backing without losing much sensitivity. The reason for this is that optimal matching enables energy to be transmitted into the forward direction and reduces ringing resulting from reverberation of pulses, thus widening the bandwidth.

1.3.2   Electrical Matching

Maximizing energy transmission and/or bandwidth can also be achieved by matching the electrical characteristics of the transducer to the electrical source and amplifier. Circuit components may be placed between the transducer and the external electrical devices (Desilets et al. 1978). Given that the transducer behaves more like a capacitor at resonance, a shunt inductor may be used to tune out the capacitance. A transformer can be used to match the resistance.

1.4   Transducer Beam Characteristics

The beam characteristics produced by an ultrasonic transducer are far from ideal. The intensity is highest at the center and decreases as a function of the distance from the center. It is possible to calculate the beam profile using the Huygens principle (Kinsler et al. 2000), which states that the resultant wavefront generated by a source of finite aperture can be obtained by considering the source to be composed of an infinite number of point sources. To calculate the beam profile of an ultrasonic transducer, the transducer surface is considered to consist of an infinite number of point sources, each emitting a spherical wave. The summation at a certain point of the spherical wavelets generated by all point sources on the transducer surface yields the field at that point.

Figure 1.7 shows the axial intensity distribution for a 5 MHz transducer of 1 cm diameter. The beam starts to collimate approximately at

$$z_0 = a^2/\lambda = 290 \text{ mm}, \tag{1.6}$$

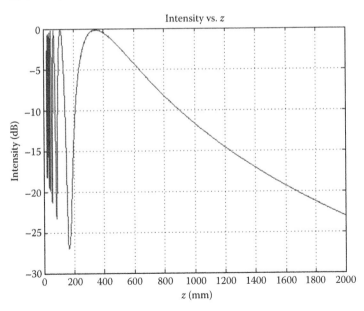

FIGURE 1.7   Axial intensity profile as a function of distance (z) for a disc transducer for CW excitation.

which is called the far field–near field transition point beyond which pressure and intensity decrease as functions of $1/z$ and $1/z^2$, respectively. In Equation 1.6, a is the radius of the disc and λ is the wavelength in the medium where the transducer is immersed.

1.4.1   Lateral Beam Profiles

For a circular aperture of radius a, the angular radiation pattern in the far field is given as follows:

$$H_c(\phi) = H_c(\phi = 0)\frac{2J_1(ka\sin\phi)}{ka\sin\phi},$$ (1.7)

where J_1 is the Bessel function of the first kind of order 1. Equation 1.7 is plotted in Figure 1.8, which shows the angular intensity radiation pattern in the far field of a circular ultrasonic transducer consisting of a main lobe and several side lobes. The number of side lobes and their magnitude relative to that of the main lobe depend on the ratio of transducer aperture size to wavelength and the shape of the piezoelectric element. The first zero occurs at

$$\sin\phi = 0.61(\lambda/a).$$ (1.8)

As the ratio of the aperture size to wavelength becomes larger, ϕ decreases or the beam becomes sharper accompanied by an increase in the number of side lobes. Side lobes are very undesirable in ultrasonic imaging because they produce spurious signals, resulting in artifacts in the image and a reduction in contrast resolution. Therefore, to have a sharper beam by increasing the ratio of transducer aperture size to wavelength, more side lobes are introduced, and z_0 is shifted farther away from the transducer. Consequently, for a particular application, a compromise has to be reached or a lens may be used to shift the focal point closer to the transducer.

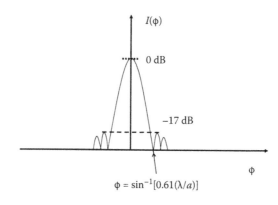

FIGURE 1.8 Angular radiation pattern of a disc transducer.

The angular radiation pattern for a rectangular element is a sinc function in the x-z plane,

$$H_r(\phi_x) = H_r(\phi_x = 0) \frac{\sin\left(kb\dfrac{\sin\phi_x}{2}\right)}{kb\dfrac{\sin\phi_x}{2}}. \tag{1.9}$$

For a rectangular element, which is the basic unit of an array with dimension b in x-direction and h in y-direction, the 3-D directivity function is given as follows:

$$H(\phi_x,\phi_y) = H(\phi_x = 0,\phi_y = 0) \frac{\sin[(kb\sin\phi_x)/2]}{(kb\sin\phi_x)/2} \bullet \frac{\sin[(kh\sin\phi_y)/2]}{(kh\sin\phi_y)/2}, \tag{1.10}$$

where ϕ_x and ϕ_y are angles in the x-z and y-z planes, respectively. The directions of x and y are frequently called elevational and azimuthal directions in the literature. The $(\sin x)/x$ ratio is the sinc function, which is zero when $x = n\pi$, where n is an integer. Therefore, the first zeros for $H(\phi_x,\phi_y)$ are at

$$\phi_x = \sin^{-1}\frac{\lambda}{b}, \quad \phi_y = \sin^{-1}\frac{\lambda}{h}. \tag{1.11}$$

For rectangular elements, the ratio of the magnitude of the main lobe to that of the first side lobe is −13 dB. The far field and the near field transition points on the x-z and y-z planes occur at $b^2/4\lambda$ and $h^2 4\lambda$, respectively.

The beam width d at the focal point of a circular disc transducer of radius a is linearly proportional to the wavelength,

$$d = 2.44 f_\# \lambda, \tag{1.12}$$

where $f_\#$ is the f number defined as the ratio of focal distance to aperture dimension, in this case diameter ($z_0/2a$).

For a rectangular array element, beam widths on the x-z and y-z planes are

$$d_x = 2 f_{\#x} \lambda \quad \text{and} \quad d_y = 2 f_{\#y} \lambda,$$

where $f_{\#x} = z_{0x}/b$ and $f_{\#y} = z_{0y}/h$ are the $f_\#$s on the x-z and y-z planes, respectively.

The depth of focus D_f, that is, the intensity of the beam within −3 dB of the maximal intensity at the focus for a circular aperture and a rectangular aperture within this region, is also found to be linearly related to the wavelength (McKeighen 1998),

$$D_{fc} = 7.2 f_\#^2 \lambda \quad \text{and} \quad D_{fr} = 7.1 f_\#^2 \lambda. \tag{1.13}$$

Chapter 1

From these relationships, it is clear that an increase in frequency that decreases wavelength improves both lateral and axial resolutions by reducing the beam width and the pulse duration if the number of cycles in a pulse is fixed. However, these improvements are achieved at the cost of a shorter depth of focus.

1.4.2 Pulsed Ultrasonic Field

The previous discussion pertains only to CW propagation. Most applications of ultrasound in medicine, however, involve pulsed ultrasound. From the Fourier transform of the pulse and using the principle of superposition, the field characteristics of a transducer transmitting pulses can be readily calculated. When a transducer is pulsed, the radiation pattern and the field characteristics all become much smoother.

1.4.3 Focusing

Better lateral resolution at a certain axial distance can be achieved by acoustic focusing. However, an improvement in the lateral resolution or focusing at a certain range is always accompanied by a loss of resolution in the region beyond the focal zone.

The general principles of focusing are identical to those in optics. Two most often used schemes, a lens and a spherical or bowl type transducer, are illustrated in Figure 1.9a and b, where z_f and D_f are the focal distance and the depth of focus, respectively. The acoustic lens shown in Figure 1.9a is a convex lens, which means that the sound velocity in the lens material is less than the medium into which the beam is launched. The convex lens is preferred in biomedical ultrasonic imaging because it conforms better to the shape of the body curvature. If the sound velocity in the lens material is greater than that in the loading medium, the lens is concave. As illustrated in Figure 1.10, the focal length z_f of a lens is governed by the following equation:

$$z_f = \frac{R_c}{1 - 1/n},$$
(1.14)

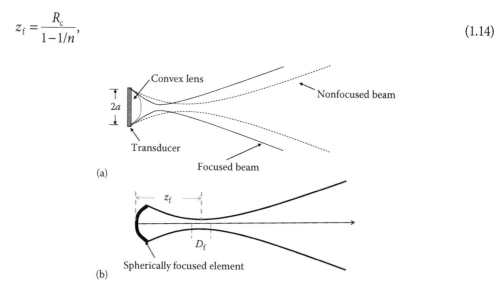

FIGURE 1.9 Two modes of focusing that have been used to focus ultrasonic beams (a) focusing with a lens and (b) self-focusing.

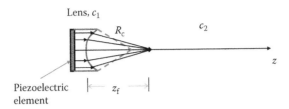

FIGURE 1.10   An ultrasound beam is focused to a point via a lens.

where R_c is the radius of curvature and $n = c_1/c_2$, c_1 being the velocity in the lens and c_2 the velocity in the medium. A popular material for convex lens is an RTV silicon rubber, which has a velocity of 1010 m/s, an acoustic impedance of 1.5 MRayls, and an attenuation of 7 dB/cm-MHz. For a silicon rubber lens in water and a focal distance of 4 cm, R_c can be readily calculated to be 2.12 cm from Equation 1.14. Concave lenses made of polyurethane or polystyrene have also been used. For concave transducers, a suitable filler material is needed to make the transducer face flat. Polyurethane has been shown to fit this need. Ultrasonic imaging is diffraction limited because the beam cannot be properly focused in the region very close to the transducer and beyond the near field and far field transition point. For a circular piston transducer of radius a, $z_0 = a^2/\lambda$. The $f_\#$ is $a/(2\lambda)$, which is determined by the ratio of radius to wavelength. For a ratio of radius to wavelength = 10, $f_\# = 5$. This means that the beam cannot be focused beyond an $f_\#$ of 5. The only way to obtain focusing at a distance greater than this is to either increase the aperture size or decrease the wavelength.

A single-element transducer can be translated or steered mechanically to form an image. Linear translators do not enable movements, resulting in image generation at a rate higher than a few frames per second, although there are sector-scanning devices that steer the transducer within a limited angle at a rate of 30 frames per second. Early real-time ultrasonic imaging devices almost exclusively used this type of transducer, which are called mechanical sector probes. Mechanical sector probes that suffer from poor near field image quality because of reverberations between the transducer and the housing and fixed focusing capability have now been largely replaced by linear arrays.

1.5   Arrays

Arrays are transducer assemblies with more than one element. These elements may be rectangular and arranged in a line called linear array or 1-D array (Figure 1.11a), square and arranged in rows and columns called 2-D array (Figure 1.11b), or ring-shaped and arranged concentrically called annular array (Figures 1.3 through 1.11c).

A linear switched array (sometimes called a linear sequenced or simply a linear array) is operated by applying voltage pulses to groups of elements in succession, as shown in Figure 1.12, where the solid line and the dashed line indicate the first and the second beam, respectively. In this way, the sound beam is moved across the face of the transducer, electronically producing a picture similar to that obtained by scanning a single-element transducer manually. The amplitude of the voltage pulses can be uniform or varied across the aperture, as shown by arrows of varying length in Figure 1.12. Amplitude apodization or variation of the input pulse amplitude across the aperture is

Chapter 1

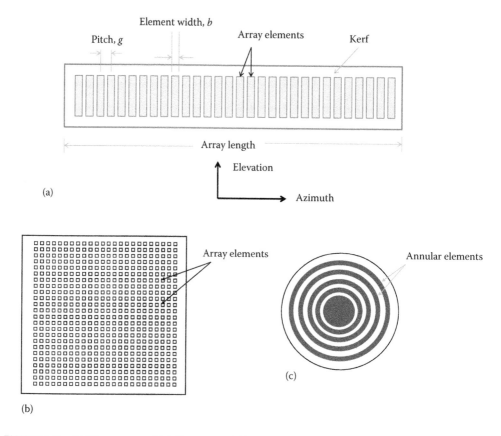

FIGURE 1.11   (a) Linear array, (b) 2-D array, and (c) annular array.

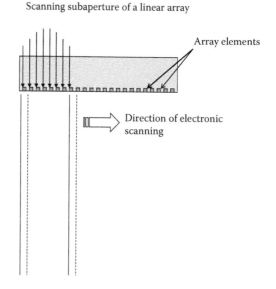

FIGURE 1.12   An image is formed by a linear array by electronically sweeping the beam. A group of elements are fired simultaneously to form one beam.

sometimes used to suppress side lobes at the expense of worsening the lateral resolution. If the electronic sequencing or scanning is repeated fast enough (30 frames per second), a real-time image can be generated. Linear arrays are usually 1 cm wide and 10 to 15 cm long with 128 to 256 elements. Typically, 32 or more elements are fired per group. For the sake of achieving as good a lateral resolution as possible, the irradiating aperture size must be made as large as possible. The aperture size is in turn limited by the requirement of maintaining a large number of scan lines. The basic acoustic stack design of a linear array is similar to a single-element transducer consisting of a backing material, a layer of piezoelectric material sandwiched between two electrodes, and two matching layers. A lens is used to focus the imaging plane in the elevational direction or the slice thickness of the imaging plane. This is a problem of crucial importance in 2-D imaging with 1-D arrays because the slice thickness cannot be controlled throughout the depth of view. The slice thickness is the thinnest only at the focal point of the lens and becomes worse closer to the array or beyond the focal point.

As shown in Figure 1.11a, the space between two elements is called a kerf, and the distance between the centers of two elements is called a pitch. The kerfs may be filled with acoustic isolating material or simply air to minimize acoustic cross talk. The kerfs are often cut into the lens and backing to minimize the acoustic cross talk between adjacent elements through the backing, the lens, and matching layers. The size of a pitch in a linear array ranges from $\lambda/2$ to $3\lambda/2$, where λ is the wavelength in the medium into which ultrasound is launched and is not as critical as in a phased array (Steinberg 1976; Shung 2005).

The linear phased array, although similar in construction, is quite different in operation. A phased array is smaller (1 cm wide and 1–3 cm long) and usually contains fewer elements (96–256). Referring to Figure 1.13a, if the difference in path length between the center element and the element number n is $\Delta r_n = r - r_n$, at a point $P(r,\phi_x)$, the time difference is then

$$\Delta t_n = \frac{\Delta r_n}{c} = \frac{x_n \sin\phi_x}{c} + \frac{x_n^2}{2cr}, \tag{1.15}$$

where the first and the second terms on the right-hand side of the equation indicate the time differences due to steering and focusing, respectively. If the pulse exciting the center element is delayed by a period of Δt_n relative to the pulse exciting element n, the emitted ultrasonic pulses will arrive at point P simultaneously. The ultrasonic beam generated by a phased array can be both focused and steered by properly delaying the signals going to the elements for transmission or arriving at the elements for receiving as illustrated in Figure 1.13b according to Equation 1.15. The radiation pattern in the far field of a linear array is very complicated (Steinberg 1976; Shung 2005). One important feature is the grating lobe. For an irradiating aperture with regularly spaced elements, high side lobes called grating lobes occur at certain angles because of constructive interference. These lobes are related to the wavelength and the pitch by the following equation:

$$\phi_{xg} = \sin^{-1}\left(\frac{m\lambda}{g}\right) \tag{1.16}$$

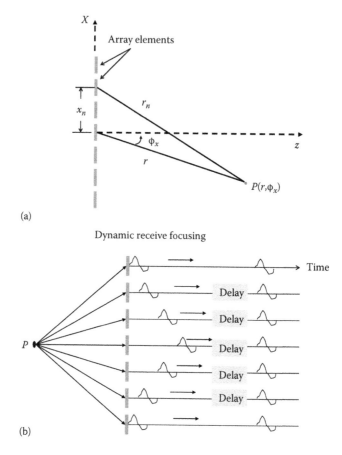

(a)

Dynamic receive focusing

(b)

FIGURE 1.13   (a) The 2-D coordinate system depicting the difference in path length between the center element of a linear array and the *n*th element. (b) Echoes returned from a point scatterer at point *P* can be made to arrive at the same time by appropriately delaying the echoes detected at the elements of a linear array.

where m is an integer = ±1, ±2, …. For the grating lobes to occur at angles greater than 90°, g has to be smaller than $\lambda/2$. When this condition is satisfied, the array is said to be fully sampled. Figure 1.14 shows the radiation pattern of a 5 MHz 32-element array with 1.6λ pitch and element width $b = 1.2\lambda$ for soft kerf filler material, where $\delta = \sin \phi_x$. Here the array length $L_a = 51.2\lambda$, and the acceptance angle of an array is defined as the angle span between the angles where the envelope drops to zero or $\phi_{acceptance} = 2\sin^{-1}(\lambda/b) = 112.9°$. The grating lobes occur at ±38.7°. The first zeros for the main beam occur at $\phi_x = ±1.1°$. The angles where the grating lobes occur is determined by the pitch. The grating lobes move away from the main lobe as the pitch is reduced. The magnitude of the grating lobe relative to the main lobe is determined by the width of element b. The smaller the value for b, the larger the magnitude of grating lobes relative to the main lobe. The width of the main lobe in turn is determined by array length. The greater the length, the smaller the main lobe. There are ways, albeit not perfect, to suppress the grating lobes. These include randomizing the spacings between elements that spread the grating lobe energy in all directions, resulting in a "pedestal" side lobe and subdicing the elements.

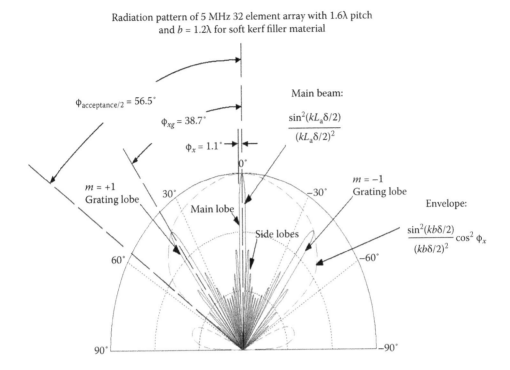

Radiation pattern of 5 MHz 32 element array with 1.6λ pitch and $b = 1.2\lambda$ for soft kerf filler material

$\Phi_{acceptance/2} = 56.5°$

$\phi_{xg} = 38.7°$

$\phi_x = 1.1°$

$0°$

Main beam:

$$\frac{\sin^2(kL_a\delta/2)}{(kL_a\delta/2)^2}$$

$m = +1$
Grating lobe

$m = -1$
Grating lobe

$30°$ $-30°$

Main lobe

Envelope:

Side lobes

$$\frac{\sin^2(kb\delta/2)}{(kb\delta/2)^2}\cos^2\phi_x$$

$60°$ $-60°$

$90°$ $-90°$

FIGURE 1.14 Angular radiation pattern of a linear array with grating lobes.

Phased arrays enable dynamic focusing and beam steering. Dynamic focusing can be achieved in both transmission and reception. However, multiple transmissions of pulses are needed for dynamic focusing during transmission, slowing down the frame rate. Transmission dynamic focusing is usually conducted in discrete zones, whereas receiving dynamic focusing can be conducted in many more zones or almost continuously. After all data are acquired, a composite image is formed, taking only the data from the zones where the beam is focused. To maintain the beam width throughout the depth of view, state-of-the-art scanners also enable variation in aperture size. The aperture size is varied as a function of time to enable proper focus of the beam at different distances. The reduction of aperture size is especially crucial in the near field where the beam cannot be focused for a transducer of a given size.

A photo of a 30 MHz 256-element linear array is shown in Figure 1.15a. The array is connected to a PCB connector via flexible or flex circuits. The PCB connector is then connected to a system termination box called zero insertion force (ZIF) connector via cable as illustrated in Figure 1.15b.

A variation of the linear array is the curved array that enables the formation of a pie-shaped image without resorting to phased array technology, which is more complicated and expensive. Figure 1.15c shows three curved linear arrays and a linear array for comparison. The advantages of the curved linear array are that (1) the beams are always perpendicular to the aperture unlike phased arrays in which the steered beam is affected by the steering angle, losing sensitivity and lateral resolution as the steering angle is increased; (2) the aperture conforms better to the body surface; (3) it produces an image with a wider field of view; and (4) the pitch does not have to be λ/2. There are

Chapter 1

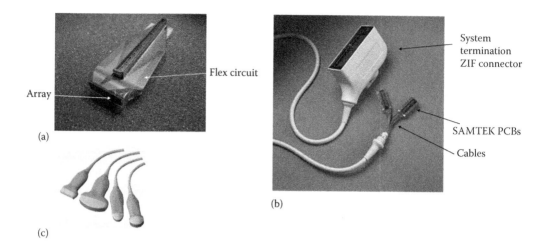

(a)

Flex circuit

Array

System
termination
ZIF connector

SAMTEK PCBs

Cables

(b)

(c)

FIGURE 1.15 (a) Photo of a linear array, (b) interconnected components between a probe and the imaging console, and (c) a variety of probes, including linear array, linear curved array, and tightly curved linear array.

also a couple of disadvantages: (1) larger aperture than phased array and (2) nonuniform scan line density compared with linear array.

Linear arrays can be focused and steered only in one plane, the azimuthal plane. Focusing in the elevation plane perpendicular to the imaging plane, which determines the slice thickness of the imaging plane, is achieved with a lens. This problem may be alleviated by using multidimensional arrays, 1.5-D or 2-D arrays (Shung 2005). A three-row 1.5-D design that is used to provide limited focusing capability in the elevational plane and to reduce slice thickness is shown in Figure 1.16. It is used as an alternative to 2-D arrays. In 1.5-D arrays, the additional elements in the elevation direction increase

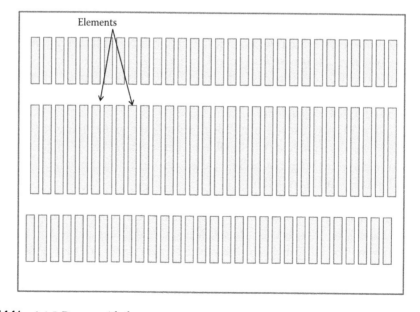

Elements

FIGURE 1.16 A 1.5-D array with three rows.

the number of electronic channels and complexity in array fabrication. Two additional concerns associated with 1.5-D arrays that do not exist in 1-D arrays are grating lobes in the elevational plane as a result of the small number of elements and increased footprint or aperture size.

Commercial 2-D arrays are now available capable of high-speed 3-D or 4-D cardiac imaging (Greenstein et al. 1997; Savord and Soloman 2003). The current 2-D arrays consist of 9212 elements at 2.5 to 3.5 MHz with fewer than a few hundred elements actually wired. A novel beamforming scheme in which beamforming is conducted in groups of elements is used to reduce the total number of electronic channels to a manageable level. The 2-D array suffers from a severe difficulty in electrical interconnection due to the large number of elements and channels, a low signal-to-noise ratio (SNR) due to electrical impedance mismatching, and a small element size. Fiber optics and multilayer architecture are possible solutions to the interconnection problem and array stack design.

The annular arrays shown in Figure 1.11c can also achieve biplane focusing. With appropriate externally controllable delay lines or dynamic focusing, focusing throughout the field of view can be attained. A major disadvantage of annular arrays is that mechanical steering has to be used to generate 2-D images.

1.6 Ultrasound Array Beamforming

1.6.1 Overview

The beamformer can be considered the engine or heart of an ultrasound system. Although the design and performance of a transducer array is paramount, beamformer performance in terms of SNR, number of channels, bit quantization, flexibility, and sampling frequency can affect the beam shape, sensitivity, and thus clinical utility. Besides providing anatomic B-mode imaging, other state-of-the art functions of the ultrasound system such as color, pulsed wave, and power Doppler imaging use the beamformer output as inputs into these functions. Furthermore, advanced algorithms and imaging methods such as the many variations in elasticity imaging and tissue characterization all require beamformed data. A poorly performing beamformer can adversely affect the performance and therefore the diagnostic value of these methods. Figure 1.17 contains a block diagram of a typical ultrasound system showing the relationship between the array transducer, the beamformer, and other modes of ultrasound imaging. This section will present general principles of beamforming, relevant physics, and instrumentation. A description of advanced beamforming methods will also be provided.

1.6.2 Rayleigh–Sommerfeld Diffraction

The primary task of a beamformer is to focus ultrasound energy at a particular point in the field or target. Minimizing energy away from the focus and having the ability to focus at all depths of interest for uniform image quality are also important beamformer tasks. At the conceptual level, focusing an ultrasound beam is analogous to focusing light in optics. In particular, the concept of optical diffraction can be applied to ultrasound because of the wave nature of both light and sound. Generally, diffraction effects must be considered when the object or source and wavelength are of comparable size.

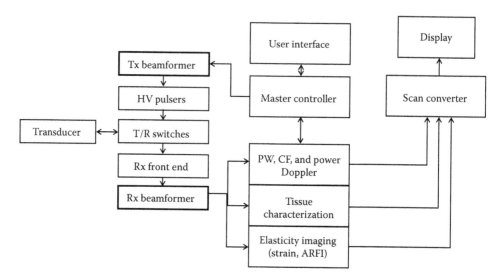

FIGURE 1.17 Ultrasound system diagram. The transmit and receive beamformer sections are emphasized in bold.

The Rayleigh–Sommerfeld diffraction formula (Goodman 1996) describes the field pattern $\phi(x,z)$ given an aperture function along the x_0 direction $a_t(x_0)$ as follows:

$$\phi(x,z) = \frac{1}{j\lambda} \int_{-\infty}^{\infty} a_t(x_0) \frac{e^{jkR}}{R} \frac{z}{R} dx_0. \tag{1.17}$$

The coordinate system is shown in Figure 1.18, $R = \sqrt{(x-x_0)^2 + z^2}$, and k is the wave number and is equal to $2\pi/\lambda$, where λ is the wavelength of the ultrasound. The variable x_0 is the lateral coordinates of the aperture plane located at $z = 0$. As an example, an element having a length D would have an aperture function $a_t(x_0) = \text{rect}\left(\frac{x_0}{D}\right)$. An ideal point source located at the origin would have a function $a_t(x_0) = \delta(x_0)$, where $\delta()$ indicates the Dirac delta function. For the sake of clarity, we confine the aperture function to a 1-D function in the x_0 direction. The equations shown here can be extended to account for 2-D aperture functions by including a y_0 aperture coordinate and a y field coordinate.

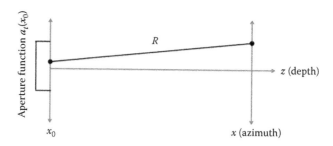

FIGURE 1.18 Coordinate system for the Rayleigh–Sommerfeld diffraction theory.

To gain insight into the Rayleigh–Sommerfeld diffraction formula, two approximations can be made to further simplify Equation 1.17. The first approximation is the Fresnel approximation, which assumes that the lateral field coordinate x is smaller than z. When this is true, the binomial approximation $\sqrt{1+b} \approx 1 + \dfrac{b}{2}$ is used to approximate R as follows:

$$R \approx z + \frac{x_0^2}{2z} - \frac{xx_0}{z} + \frac{x^2}{2z}. \tag{1.18}$$

Replacing R with the approximation, Equation 1.17 can now be rewritten as follows:

$$\phi(x,z) = \frac{1}{j\lambda z} e^{jkz} e^{jk\frac{x^2}{2z}} \int_{-\infty}^{\infty} a_t(x_0) e^{jk\left(\frac{x_0^2}{2z} - \frac{xx_0}{z}\right)} dx_0. \tag{1.19}$$

The Fraunhofer approximation is the second approximation. It assumes that z is much greater than the size of the aperture, or $z \gg x_0$. The quadratic phase term associated with the aperture coordinate, $e^{jk\frac{x_0^2}{2z}}$, is assumed to have a phase of 0 radians; hence, the transmit field can be written as follows:

$$\phi(x,y,z) = \frac{1}{j\lambda z} e^{jkz} e^{jk\frac{x^2}{2z}} \int_{-\infty}^{\infty} a_t(x_0) e^{\frac{-jkxx_0}{z}} dx_0. \tag{1.20}$$

When ignoring the phase terms outside the integral, Equation 1.20 now states that the complex transmitted field is the Fourier transform of the aperture function with a spatial frequency $u_x = \dfrac{x}{\lambda z}$. Although several approximations have been made, this Fourier transform relationship between the apertures is valid for many situations and serves as a useful conceptual tool to understand how the aperture dimensions, depth of interest, and wavelength affects the beam shape and thus image quality. Because of this Fourier transform relationship, it can be easily shown that narrower beams can be achieved when the aperture is larger, the wavelength is shorter, or the depth of interest is closer to the transducer. In certain cases such as transthoracic cardiac imaging, it is not possible to have all three of these attributes so a compromise in image quality is inevitable.

1.6.3   Focusing and the Rayleigh–Sommerfeld Diffraction Formula

Traditional beamforming or focusing can be described as adding a phase shift across the aperture function $a_t(x_0)$. The phase shift for a specific point on the aperture is related by the geometry or distance from a particular point on the aperture to the focal point located at (x_f, z_f). Here it is assumed that $x_f = 0$. Using the distance formula, the distance R_f from the aperture to the focal point $(0, z_f)$ is as follows:

$$R_f = \sqrt{x_0^2 + z_f^2}. \tag{1.21}$$

Applying the binomial approximation to this focusing term gives the following equation:

$$R_f \approx z_f + \frac{x_0^2}{2z_f}.\tag{1.22}$$

The phase shift for focusing being applied is then $-jkR_f$, and the complex transmitted field with focusing $\phi_f(x,z)$ can now be written as follows:

$$\phi_f(x,z) = \frac{1}{j\lambda z} e^{jkz} e^{-jkz_f} e^{jk\frac{x^2}{2z}} \int_{-\infty}^{\infty} a_t(x_0) e^{jk\left(\frac{x_0^2}{2z} - \frac{xx_0}{z} - \frac{x_0^2}{2z_f}\right)} dx_0.\tag{1.23}$$

At the focal depth $z = z_f$, the quadratic phase factors in the integrand cancel in Equation 1.23, and the result is that the focused beam and the aperture function are once again related by the Fourier transform with spatial frequency $u_x = \frac{x}{\lambda z}$. The effects of the aperture size, wavelength, and depth on the beam or pulse-echo field still hold.

The receive field can be determined in the same manner as the transmit field because of reciprocity, and the pulse-echo field is the product of the transmit field and the receive field. The focused pulse-echo field $\psi_f(x,z)$ at z_f then becomes

$$\psi_f(x,z) = \left(\frac{1}{j\lambda z}\right)^2 e^{jk\frac{x^2}{z}} \int_{-\infty}^{\infty} a_t(x_0) e^{\frac{-jkxx_0}{z}} dx_0 \int_{-\infty}^{\infty} a_r(x_0) e^{\frac{-jkxx_0}{z}} dx_0,\tag{1.24}$$

where $a_r(x_0)$ is the receive aperture function. If the receive aperture and the transmit aperture are equal, then the pulse-echo beam is the Fourier transform of the aperture squared.

Figure 1.19 shows the Rayleigh–Sommerfeld integral of a 3.5 MHz, 19.8 mm aperture focused on a 50 mm depth in both transmission and reception. The beam shape is quite close to a sinc squared with first side lobes near −26 dB in magnitude. The result from the Rayleigh–Sommerfeld integral (black) and the approximations (gray) are both shown. There is quite good agreement between the two, particularly around the main lobe. The −6 dB main lobe width is 1.2 mm, which roughly corresponds to $\frac{\lambda z}{D}$.

A limitation of the Rayleigh–Sommerfeld diffraction is that it is based on a single frequency or CW excitation. Nevertheless, the Rayleigh–Sommerfeld diffraction formula, along with the Fresnel and Fraunhofer approximations, can give a reasonable first-order result. The Fourier transform relationship between the aperture and the beam provides insight and intuition of what the beam shape will be. For more in-depth investigation into beamforming with broadband ultrasound, computer simulation programs such as Field II should be used (Jensen and Svendsen 1992).

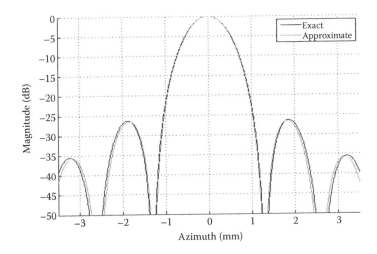

FIGURE 1.19 Transmit–receive CW beam with a 3.5 MHz, 19.8 mm focused aperture at 50 mm depth using the exact Rayleigh–Sommerfeld integral (black) and the small angle approximations (gray).

1.6.4 Ultrasound System Front End

The front end of an ultrasound beamformer consists of the following: high-voltage pulsers, transmit/receive (T/R) switch, low-noise amplifier (LNA), variable gain amplifier (VGA), band-pass filter, antialias filter, and analog-to-digital converter. Transmitters are capable of generating 20–200 Vpp pulses. For low- to mid-range systems, these transmitters are bipolar pulsers, which can send out a programmed integer number of cycles. The center frequency can be set by dividing the clock frequency by an integer number. For example, a system with a 40 MHz clock will divide by 8 to achieve a 5 MHz excitation signal. More sophisticated systems have transmit channels capable of transmitting arbitrary waveforms using a combination of digital-to-analog converters and high-voltage broadband amplifiers. These systems use coded excitation methods to achieve improvements in penetration depth, signal SNR, and image quality. For pulse-echo imaging using the same array, T/R switches protect the sensitive receive electronics from the high-voltage transmit signals. T/R switches can consist of a resistor with parallel-reversed diode bridges, a diode bridge network, or a high-voltage switch. In cases where the system has fewer channels than array elements such as with linear or curvilinear imaging, multiplexers are also included to select the appropriate elements for a given image line.

In recent years, semiconductor companies have had a keen interest in developing chips or chip sets specifically for ultrasound imaging by integrating many of the components. For example, the AD8334 from analog devices combines LNA with VGA. Four channels are provided in a single package. The AFE5804 by Texas Instruments is a "fully integrated" eight-channel chip with LNA, VGA, attenuator, and analog-to-digital converter. Most ultrasound systems may have 32–128 channels so only several of these chips are needed. The number of bits of an analog-to-digital converter can also limit the dynamic range of the system. Normally, 10–14 bit ADCs appear to be the industry standard today.

Chapter 1

1.6.5   Array Beamforming

With array transducers, focusing the ultrasound beam in both transmission and reception is conducted electronically to enable real-time B-mode imaging as well as other imaging modes such as pulsed wave, color, and power Doppler. This section describes the standard delay-and-sum transmit and receive beamformer followed by more sophisticated methods of beamforming for improved image quality and/or frame rate.

Transmit beamforming involves applying time delayed pulses from individual array elements to maximize energy at a particular point in the field as well as minimize energy at all other field points. Assuming a sound speed of $c = 1540$ m/s, time delays are calculated using the distance formula. For a given focus at point (x_f, z_f), the time delay Δt_i required for each element i in an array each having a location x_i can be calculated by

$$t_i = \frac{\sqrt{(x_i - x_f)^2 + z_f^2} - z_f}{c}.$$

(1.25)

The transmit focus is then stepped laterally (x-direction) to create multiple image lines. These image lines are combined to form an image. Figure 1.20 shows a drawing of an array focused on axis at Focus 1 and steered off axis at Focus 2. The triangles indicate the time at which each array element, shown as a square, are electrically excited. A symmetric delay profile is seen for the on-axis focus, and an asymmetric profile is seen in the case of steering to Focus 2. The element at the bottom of the figure is farthest away from Focus 2 and therefore fires first.

Typically, a single transmit focus is applied for each image line formed. However, this leads to nonuniform image quality as the ultrasound beam spreads at depths away from the transmit focus. More uniform image quality can be achieved by multiple transmit foci set at different depths. In addition, improvements in SNR throughout depth can also be achieved by having multiple transmit foci. The tradeoff is frame rate. If N transmit foci are used, the frame rate will be reduced by a factor of N.

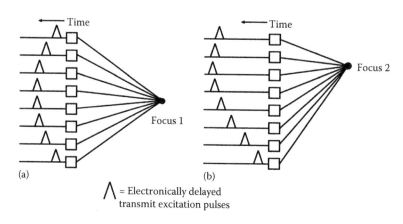

= Electronically delayed
transmit excitation pulses

FIGURE 1.20   Illustration of (a) transmit focus and (b) focusing with steering.

The process of focusing the echoes on reception can be achieved by applying time delays to the digitized echoes using Equation 1.25. Instead of a single or only a few foci, the focus can be dynamically changed as the echoes arrive. Echoes coming from deeper tissue arrive at a later time. Consequently, the receive focus can be dynamically updated. Figure 1.21 shows a plot of the receive delays for a 3.33 MHz phased array probe with 300 μm pitch when the steering angle is set to 0°, or on axis (Figure 1.21a), and when the steering angle is set to 17° (Figure 1.21b). Distances of 25, 35, 45, 55, and 65 mm are shown. As R increases, the difference between the maximum and the minimum delays decreases because the differences in distance among elements decreases with increasing depth.

To compare the effects of different focusing schemes, Figure 1.22 shows simulated images of point targets using a 5 MHz linear array with a 64-element subaperture. Five

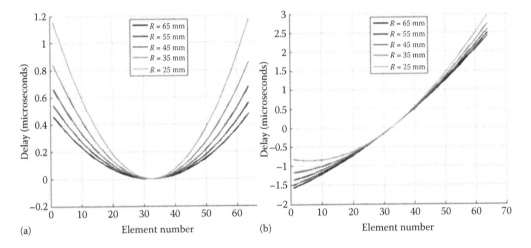

FIGURE 1.21 Dynamic receive focus for (a) on-axis and (b) steering at 17° off-axis.

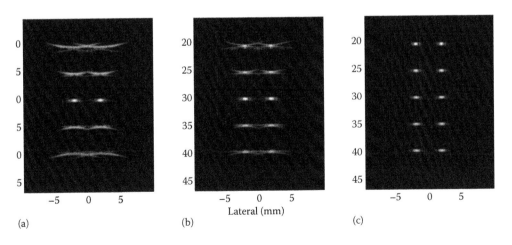

FIGURE 1.22 Simulated images of point targets separated by 5 mm laterally and axially. Point targets are located axially at 20, 25, 30, 35, and 40 mm depths. From left to right, images are formed with (a) single transmit and single receive focus at 30 mm, (b) single transmit and dynamic receive focus, and (c) dynamic transmit and dynamic receive.

Chapter 1

pairs of point targets were separated 5 mm laterally and axially. The pairs of point targets are located at depths of 20, 25, 30, 35, and 40 mm. Figure 1.22a shows a simulated B-mode image where only a single transmit and receive focus at 30 mm depth is applied. Consequently, the two point targets at 30 mm are the most easily identifiable, and the pair of targets at 20 and 40 mm are the least distinguishable because they are farthest away from the focus. Figure 1.22b uses a single transmit focus with dynamic receive focus. The pair at 30 mm are still the most distinguishable, and the targets away from the transmit focus are more easily seen. Finally, with dynamic transmit and receive focus, the point targets at all depths are clearly seen (Figure 1.22c). All images are displayed on a 50 dB dynamic range.

1.6.6 Analog Beamforming

Before the advent of the digital electronics, beamforming in the analog domain was achieved using delay lines. Delay lines consisted of coaxial cables, inductor capacitor circuits, or charge-coupled devices. Figure 1.23 shows a block diagram of an analog beamformer. The front end has not been included to improve clarity. After the analog delay line, the signals are also summed in the analog domain. The digitization of the echo occurs after summation.

To implement delay lines, coaxial cables can be cut to the length corresponding to the desired delay given the propagation speed of the coaxial cable. However, coaxial cables can become quite long and unwieldy for a typical range of delays for diagnostic ultrasound. An alternative is to use a series of inductors and capacitors. Generally, longer delay lines have lower bandwidth. Cascading shorter delay lines will also have a similar effect. Depending on the imaging application, it may be difficult to satisfy both delay and bandwidth requirements. Switches or multiplexers can be used for the selection of appropriate delays. For a given channel, two or more switches could be used, where one used one switch to select a coarse delay. After the signal passed through the coarse delay, a finer delay was selected using a second switch. Often, these switches or multiplexers have glitches because of the injection charge. The glitches, if not properly processed, can lead to artifacts in the image in the form of lines in the lateral direction. The advantages of analog beamforming are that they can be less power consuming, and one may be able to design them in a much more compact manner if delay values can be kept small. This is especially true when one considers the amount of power consumed and the area required for digitization and summation using digital beamformers. The limitation of the analog beamformer is the implementation of the delay values if the delays are large (>1 µs).

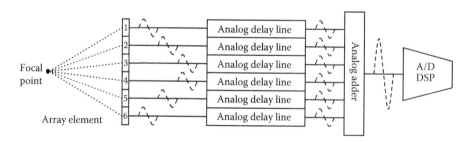

FIGURE 1.23 Block diagram of an analog beamformer.

1.6.7   Digital Beamforming

With advances in digital technology, the beamforming process is now performed digitally. After analog amplification and filtering, the echoes from each element are digitized with an analog-to-digital converter (A/D) with sampling frequencies in the range of 4 to 10 times higher than the center frequency of the transducer (Figure 1.24). After digitization, the echoes are delayed, processed, and then summed all in the digital domain.

Coarse time delays are applied by shifting the digitized samples. Phase rotation or finer delays using finite impulse response (FIR) filters can also be applied. Digital beamforming enables greater flexibility and improvement in image quality compared with analog beamforming. Assuming the sampling frequency is adequate, no bandwidth limitations are imposed as they are in the analog beamformer. Additional capability such as parallel beamforming is more straightforward to implement. Adaptive beamforming techniques such as phase aberration correction are more feasible. Array or channel data processing techniques such as amplitude and phase apodization are simpler. Array or channel data can be collected and/or processed to improve image quality. The generalized coherence factor (Li and Li 2000), phase coherence factor (Camacho et al. 2009), short lag spatial coherence (Lediju et al. 2011), and dual apodization with cross correlation (Seo and Yen 2008) are recent examples. These approaches are not feasible or possible with purely analog beamforming. Digital beamforming is more computationally intensive, requiring analog-to-digital conversion of 64 to 128 channels and therefore consumes more power. This burden decreases because of Moore's law and continuous innovation in digital signal processors, field programmable gate arrays, central processing units, and graphical processing units.

1.6.8   Hybrid Beamforming

A hybrid beamformer uses both analog and digital domains for beamforming. Further analysis of Equation 1.25 shows that clusters of adjacent elements have a relatively small range of delays. Analog delay or beamforming can be used in a practical manner for each cluster of elements. The signals at the output of each miniature analog beamformer is then digitized, delayed, and summed to create the beamformed signal. Hybrid beamformers require fewer digitizers than fully digital beamformers while still maintaining some of the flexibility that comes with a digital beamformer. The point at which to switch to digital is up to the discretion of the system developer who must weigh the

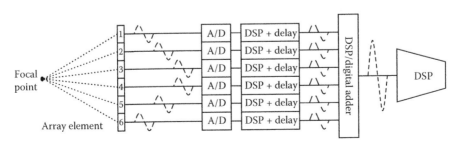

FIGURE 1.24   Block diagram of a digital beamformer.

advantages and disadvantages of analog and digital beamforming against the design requirements of the system.

The hybrid beamformers are particularly useful for 2-D arrays. For 2-D arrays, creating $N \times N$ digital channels is not feasible, not to mention the cable complexity particularly when N is approximately 32–128. For 2-D arrays and 3-D imaging, a combination hybrid beamforming approach has been taken where a small cluster (4×4 or 8×8), for example, is beamformed in the analog domain first. Analog beamforming can happen by using networks of phase shifters (Savord and Soloman 2003). Because the cluster of elements are physically close together, the differences in time delays across the cluster are typically small. The analog beamformed signals from this cluster are summed, sent down the coaxial cable, and then digitized on an ultrasound system with more typical 128 channels.

1.6.9   Beamformer Performance Evaluation

Common evaluation methods of beamformer performance include the measurement of spatial resolution, side lobe levels, and SNR. Lateral resolution is determined by frequency (or wavelength λ), depth, and array aperture size. As the depth z increases, the lateral resolution worsens, which can also cause a nonuniform appearance. To minimize this, a constant f number, defined as the ratio z/D, can be used. With a constant f number, the aperture D grows with increasing z to maintain a constant lateral resolution. The theoretical −6 dB resolution is approximately $\dfrac{\lambda z}{D}$.

Side lobes are sources of off-axis energy that limit contrast, particularly of anechoic targets such as fluid-filled cysts, blood vessels, and heart chambers. Because the beam and the aperture are roughly related by Fourier transform pairs, a hard edge boxcar weighting gives rise to ringing or side lobes. With transmit and receive focus, the pulse-echo beam has approximately a sinc-squared shape, indicating that the first side lobe level will be approximately −26 dB down from the main lobe. Apodization is a method by which the suppression of these side lobes can be achieved. Typical apodization functions include a Hanning, Hamming, or Gaussian amplitude weighting across the aperture. These weightings smooth out the aperture function, lowering or suppressing side lobes. As a tradeoff, the main lobe width is widened, worsening the lateral resolution. Judicious selection between the lateral resolution and the contrast must be chosen based on the clinical application.

SNR or depth of penetration can be evaluated using a speckle-generating tissue-mimicking phantom having a known uniform attenuation (Ustuner and Holley 2003). The ultrasound system is first configured to collect data with the transmitters turned off to obtain a noise measurement. The mean of the noise can then be calculated as a function of depth. Next, the phantom is imaged with the transmitters turned on to obtain the signal. Likewise, the mean of the signal can be determined, and an SNR can be calculated as a function of depth. The depth at which the SNR drops less than 6 dB is the penetration depth.

1.6.10   Synthetic Aperture

The standard delay-and-sum beamformer requires a significant amount of front end transmit and receive hardware as well as back-end processing. For a phased array, 64 to 96 elements are used, indicating that 64 to 96 transmit and receive channels are required. For smaller, compact, and low-cost systems, a synthetic aperture approach may be used where either or both of the transmit and receive apertures are synthesized over multiple firings. A different element or set of elements is selected for each firing. Echoes from each firing are digitized and stored, and the data from all firings can then be combined to coherently form the image.

A major disadvantage with synthetic aperture is that transducer or target motion can blur the images as data from the overall aperture are synthesized (Nock and Trahey 1992; Trahey and Nock 1992). Synthetic aperture methods will normally require some sort of motion estimation and compensation methods to minimize blurring effects. At the very least, an analysis or simulation of image degradation should be performed to ensure that tissue motion will not be a major source of image degradation. Another disadvantage is that synthetic aperture methods may also incur a penalty in frame rate depending on the specific scheme. More firings may be required to form one image compared with standard beamforming and image formation methods. Lastly, SNR can be low depending on the number of transmitters and receivers used. Careful SNR analysis is required to weigh the benefits and limitations of a given synthetic aperture scheme. The rules of thumb are as follows: (1) SNR increases linearly with the number of transmitters used, and (2) SNR increases by the square root of the number of receivers (Lockwood et al. 1998). These rules of thumb assume linearity and that the noise from different receive channels and firings are uncorrelated.

The primary forms of synthetic aperture include monostatic synthetic aperture (Ylitalo and Ermert 1994), synthetic transmit aperture, and synthetic receive aperture. These are shown schematically in Figure 1.25, where the active elements used in either transmission or reception are shown in gray. Unused elements are shown in white.

Using the monostatic approach requires only one transmitter and one receiver. Element 1 is used in transmission and reception in the first firing, element 2 is used in transmission and reception in the second firing, and so on. Data from all firings

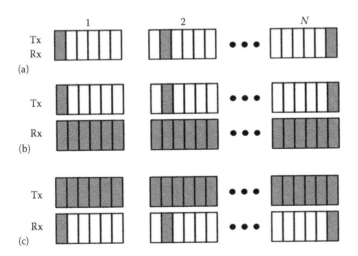

FIGURE 1.25 Common synthetic aperture imaging schemes for an N element array: (a) monostatic approach, (b) synthetic transmit aperture, and (c) synthetic receive aperture. The first two firings and the last Nth firing are shown.

Chapter 1

can then be used to dynamically focus in both transmission and reception. Monostatic approaches have the advantage of requiring only a signal transmitter and receiver. However, the SNR will likely suffer.

In a synthetic transmit aperture (Jensen et al. 2006), only one element or a few elements are used in transmission, and all elements are used in reception. Over several firings, the transmit aperture is synthesized. With sufficient processing, dynamic transmit and receive focus can be achieved yielding a high-quality image.

The STA approach requires a system with many active receive channels and is more likely to suffer from low SNR. However, receive channels are typically the more complex and costly portion of the beamformer compared with transmit beamforming. Minimizing the number of receive channels is thus desirable if cost is a concern. A synthetic receive aperture can be implemented where a full transmit beamformer is used and only one receive element or a few receive elements are active. In this case, the transmit beamformer still has a fixed focus, but dynamic receive focus can be implemented with synthetic receive aperture. The SRA approach will likely not have SNR problems because a full transmit aperture is used, but it does not enable dynamic transmit focus throughout the field of view.

1.6.11 Coded Excitation

The intensity limits imposed by the Food and Drug Administration prevents the use of intensities and mechanical indices of higher values. Coded excitation provides a means of circumventing these limits in a clever manner. In one form of coded excitations, chirps or linear frequency modulation (FM) waveforms can be used to send out higher total energy but of longer duration (Misaridis and Jensen 2005). Although this alone may reduce axial resolution, compression of the chirp can be achieved through match filtering of the echoes. Range side lobes analogous to lateral side lobes may result. However, similar with lateral aperture apodization, these can be suppressed with judicious weighting schemes of the transmitted waveform, the matched filter, or both. The use of coded excitation generally requires an arbitrary waveform generator with digital-to-analog converter followed by a high-voltage amplifier. Additional matched filtering or other digital signal processing is required by the receive beamformer. An increase in SNR by 12 to 20 dB can be achieved depending on the chirp length. Figure 1.26 shows examples of emitted chirps and the result after they have been match filtered. The top row shows a chirp with a boxcar type of weighting (Figure 1.26a). The filtered result (Figure 1.26b) is a condensed pulse with relatively high-range lobes, which are reminiscent of the lateral side lobes seen in Figure 1.19. Using a Hanning weighting to taper the chirp (Figure 1.26c) results in a smoother but longer duration pulse (Figure 1.26d). Using the Hanning weighting here is analogous to using Hanning apodization to suppress lateral side lobes.

Besides FM, phase modulation using binary or phase-modulated codes such as Golay and Barker codes have also been explored (Chiao and Hao 2005). Golay codes require two firings, and motion artifacts may result depending on the amount and direction of probe and tissue motion. Golay codes enable the complete suppression of range side lobes. This reduces frame rate by a factor of 2. Barker codes do not offer complete suppression of range side lobes, but there is no penalty in the frame rate. An increase in SNR by a factor of 4 can be achieved.

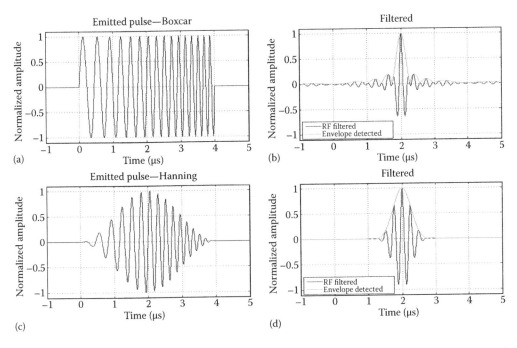

FIGURE 1.26 Examples of chirp coded excitation with a boxcar weighting. (a) The emitted pulse and (b) the matched filter results are shown in the top row. (c) The emitted pulse and (d) the filtered result using Hanning weighting are shown in the bottom row.

1.6.12 Phase Aberration Correction

Most beamformers assume a sound speed of 1540 m/s for focusing. However, it is well known that the sound speed can vary depending on the type of tissue, whether it is muscle, fat, or blood, for example. These variations in sound speed cause aberration effects, leading to lower sensitivity, worse spatial resolution, and contrast. The degradation in contrast is caused by higher side lobe levels. These degradations are akin to distortions seen in astronomy images from Earth-based telescopes because of the atmosphere. Figure 1.27 shows the concept of aberration and its detrimental effect on delay-and-sum focusing. The time at which the signals arrive from focal point 1 have been shifted in an

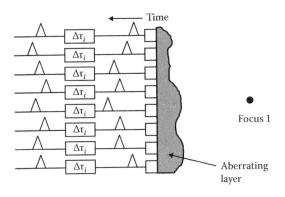

FIGURE 1.27 Effects of an aberrating layer on focusing.

Chapter 1

unknown amount. Even after applying delays based on the distance formula, the echoes will no longer sum coherently.

The use of various algorithms such as speckle brightness (Nock et al. 1989), nearest-neighbor cross correlation, and least squares (Liu and Waag 1994) have been used to improve image quality. The adaptive algorithms work by estimating the time arrival errors among elements. A new transmit delay profile is calculated to compensate for these estimated errors, and a second transmit beam with the new delay profile is emitted. Time arrival errors can again be estimated and compensated for using the resulting echoes from the second firing. Image quality normally improves with up to 2 to 3 iterations, and additional iterations yield diminishing returns. These algorithms can be computationally intensive because of the number of time of arrival estimates. These adaptive methods also require multiple iterations that reduces frame rate. Figure 1.28 shows an example of possible improvements using phase aberration correction with one iteration. Figure 1.28a is an image of a 3 mm diameter cyst in a tissue-mimicking phantom with no aberrator. Figure 1.28b is an image of the same cyst with a 10 mm thick slab of pork to mimic an aberrator. Note that the cyst is barely visible and appears at a deeper depth of 40 mm. After just one iteration of the nearest-neighbor cross-correlation algorithm (Flax and O'Donnell 1988), the cyst becomes more visible.

1.6.13 Parallel Processing

For standard ultrasound image formation, one transmit event creates one image line, which can limit the frame rate based on the depth of interest times the number of image lines. For example, assuming a sound speed of 1540 m/s, a scan depth of 15 cm, and 128 image lines, a maximum frame rate of 40 frames per second is possible. Using multiple transmit foci will result in more uniform image quality but will further reduce frame rate. Using two transmit foci will reduce frame rate by a factor of 2. To improve frame rate, particularly for 3-D imaging, parallel processing methods, where multiple image lines are created for a single transmit event, can be implemented. The transmit beam is purposely made broad by defocusing or no focusing. Focusing on reception can be conducted at multiple field locations laterally within the transmit beam to create more than one image line. For 2-D imaging, 2:1 or 4:1 parallel processing can be conducted to speed up the frame rate by a factor of 2 or 4. For 3-D imaging, 16:1 parallel processing (Smith et al.

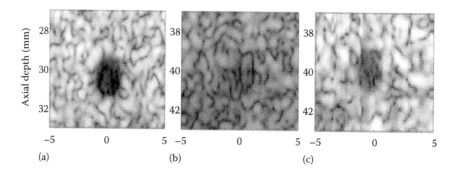

FIGURE 1.28 Images of an anechoic cyst in a quality assurance phantom with (a) no aberration, (b) phase aberration induced using pork tissue, and (c) one iteration of correction using the nearest neighbor cross-correlation algorithm. (Images courtesy of J. Shin.)

1991; von Ramm et al. 1991) can be achieved where a 4 × 4 arrangement of receive beams fit inside an enlarged transmit beam. Figure 1.29 illustrates the concept of parallel beamforming using a 3.5 MHz phased array with 64 elements. The transmit beam has been created using the central 32 elements, whereas the two receive beams use all 64 elements.

A natural extension of parallel processing is plane wave or flash imaging (Montaldo et al. 2009), where all of the array elements fire simultaneously to launch a very broad beam or plane wave into the tissue. Receive beamforming is conducted to form all image lines. Thus, one entire image can be formed by one transmit event, leading to frame rates of more than 5000 frames/s. Figure 1.30 shows a simulation of a 192 element, 7.5 MHz array with λ pitch used in a plane wave imaging mode. The black line shows the broad transmit beam roughly corresponding to the size of the aperture as well as the image width. The receive beams are focused at 30 mm depth and can be spaced roughly every 200 µm. Only 10 receive beams spaced every 2.2 mm are shown here for the sake of clarity.

A disadvantage of both parallel processing and plane wave imaging is that only receive focusing is conducted; thus, image quality is worse compared with images formed using transmit focus. For plane wave imaging, image quality can be recovered by using coherent plane wave compounding (Montaldo et al. 2009). With this technique, plane waves steered at several (typically 5°–30°) angles are sent into the tissue. Echoes from the multitude of plane waves are coherently summed in a process reminiscent of synthetic aperture. High frame rates (>150 frames/s) can still be maintained. The implementation of plane wave compounding requires storage of all channel data from various angles.

1.6.14   Advanced Beamforming Methods

Continuous improvement and innovation in various aspects of beamforming are ongoing. Limited diffraction beams enable constant beam width throughout depth (Lu 1998). Minimum variance beamforming is an adaptive beamforming that seeks to determine optimal amplitude and phase apodizations by minimizing variance while constraining the beamformer power (Synnevag et al. 2005). Minimizing variance can reduce contributions

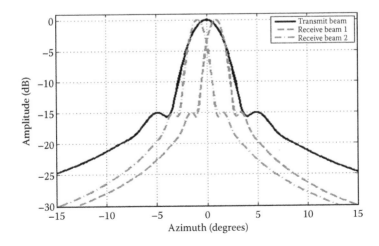

FIGURE 1.29   Transmit (black) and receive beams (gray dashed) used for 2:1 parallel processing. The transmit beam is purposely widened so that two receive beams can fit within one transmit beam.

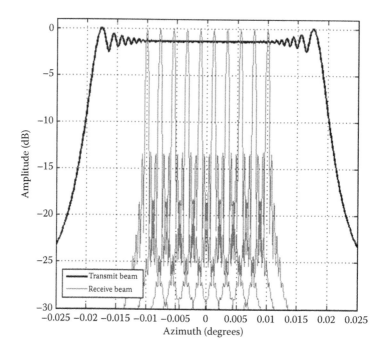

FIGURE 1.30 Simulated beam patterns for a plane wave transmit (black). Multiple receive beams (gray) are created with delay-and-sum focusing.

from off-axis scatterers. Delta-sigma beamformers that use single-bit ADCs at high sample rates have also been explored (Freeman 1999). These beamformers may enable simplicity for systems with high channel counts. Spatially matched filtering techniques analogous to coded excitation have also been explored for focusing of ultrasound beams (Kim et al. 2006). Also of particular note is the development of high-frequency beamformers for linear and annular arrays, which require much higher sampling rates and frame rates.

References

Brown LF. Design considerations for piezoelectric polymer ultrasound transducers. *IEEE Trans Ultrason Ferroelectr Freq Control* 2000:47;1377–1396.

Camacho J, Parilla M, Fritsch C. Phase coherence imaging. *IEEE Trans Ultrason Ferroelectr Freq Control* 2009:56;958–974.

Chiao RY, Hao X. Coded excitation for diagnostic ultrasound: A system developer's perspective. *IEEE Trans Ultrason Ferroelectr Freq Control* 2005:52;160–170.

Desilets CS, Fraser JD, Kino GS. The design of efficient broad-band piezoelectric transducers. *IEEE Trans Sonics Ultrason* 1978:25;115–125.

Greenstein M, Lum P, Yoshida H, Seyed-Bolorforosh MS. A 2.5 MHz array with Z-axis electrically conductive backing. *IEEE Trans Ultrason Ferroelectr Freq Control* 1997:44;970–982.

Flax SW, O'Donnell M. Phase-aberration correction using signals from point reflectors and diffuse scatterers: Basic principles. *IEEE Trans Ultrason Ferroelectr Freq Control* 1988:35;758–767.

Freeman SR. Delta-sigma oversampled ultrasound beamformer with dynamic delays. *IEEE Trans Ultrason Ferroelectr Freq Control* 1999:46;320–332.

Goodman JW. *Introduction to Fourier Optics*, 2nd ed. New York: McGraw-Hill, 1996.

Jensen JA, Svendsen JB. Calculation of pressure fields from arbitrarily shaped, apodized, and excited ultrasound transducers. *IEEE Trans Ultrason Ferroelectr Freq Control* 1992:39;262–267.

Jensen JA, Nikolov SI, Gammelmark KL, Pedersen MH. Synthetic aperture ultrasound imaging. *Ultrasonics* 2006:44;e5–e15.

Kim KS, Liu J, Insana MF. Efficient array beam forming by spatial filtering for ultrasound B-mode imaging. *J Acoust Soc Am* 2006:120;852–861.

Kino GS. *Acoustic Waves*. Englewood Cliffs, NJ: Prentice-Hall, 1987.

Kinsler LE, Frey AR, Coppens AB, Sanders JV. *Fundamentals of Acoustics*, 4th ed. New York: John Wiley, 2000.

Krimholtz R, Leedom D, Matthaei G. New equivalent circuits for elementary piezoelectric transducers. *Electronics Lett* 1970;6:398–399.

Lediju MA, Trahey GE, Byram BC, Dahl JJ. Short-lag spatial coherence of backscattered echoes: Imaging characteristics. *IEEE Trans Ultrason Ferroelectr Freq Control* 2011:58;1377–1387.

Liu DL, Waag RC. Time-shift compensation of ultrasonic pulse focus degradation using least-mean-square error estimates of arrival time. *J Acoust Soc Am* 1994;95;542–555.

Li P-C, Li M-L. Adaptive imaging using the generalized coherence factor. *IEEE Trans Ultrason Ferroelectr Freq Control* 2000:50;93–110.

Lockwood GR, Talman JR, Brunke SS. Real-time 3-D ultrasound imaging using sparse synthetic aperture beamforming. *IEEE Trans Ultrason Ferroelectr Freq Control* 1998:45;980–988.

Lu J. Experimental study of high frame rate imaging with limited diffraction beams. *IEEE Trans Ultrason Ferroelectr Freq Control* 1998:45;84–97.

McKeighen RE. Design guidelines for medical ultrasonic arrays. *Proc. SPIE Vo. 3341 Ultrasonic Transducer Engineering* 1998;2–18.

Misaridis T, Jensen JA. Use of modulated excitation signals in medical ultrasound. Part I: Basic concepts and expected benefits. *IEEE Trans Ultrason Ferroelectr Freq Control* 2005:52;177–191.

Montaldo G, Tanter M, Berfocc J, Benech N, Fink M. Coherent plane-wave compounding for very high frame rate ultrasonography and transient elastography. *IEEE Trans Ultrason Ferroelectr Freq Control* 2009:56;489–506.

Newnham RE, Skinner DP, Cross LE. Connectivity and piezoelectric–pyroelectric composites. *Mater Res Bull* 1978;13:525–536.

Nock LF, Trahey GE. Synthetic receive aperture imaging with phase correction for motion and for tissue inhomogeneities. Part I: Basic principles. *IEEE Trans Ultrason Ferroelectr Freq Control* 1992;39;489–500.

Nock L, Trahey GE, Smith SW. Phase aberration correction in medical ultrasound using speckle brightness as a quality factor. *J Acoust Soc Am* 1989;85;1819–1833.

Safari A, Akdogan EK. *Piezoelectric and Acoustic Materials for Transducer Applications*. New York: Springer, 2008.

Savord B, Soloman R. Fully sampled matrix transducers for real-time 3D ultrasonic imaging. *Proc IEEE Ultrason Symp* 2003;945–953.

Seo C, Yen JT. Sidelobe suppression in ultrasound imaging using dual apodization with cross-correlation. *IEEE Trans Ultrason Ferroelectr Freq Control* 2008:55;2198–2210.

Shrout TR, Fielding J Jr. Relaxor ferroelectric materials. *Proc IEEE Ultrason Symp* 1990;711–720.

Shung KK. *Diagnostic Ultrasound: Imaging and Blood Flow Measurements*. Boca Raton, FL: CRC Press, 2005.

Smith SW, Pavy HG, von Ramm OT. High speed ultrasound volumetric imaging system. Part I: Transducer design and beam steering. *IEEE Trans Ultrason Ferroelectr Freq Control* 1991;38;100–108.

Smith WA. The role of piezocomposites in ultrasonic transducers. *Proc IEEE Ultrason Symp* 1989;755–766.

Steinberg BD. *Principles of Aperture and Array System Design*. New York: John Wiley, 1976.

Synnevag J-F, Austeng A, Holm S. Minimum variance adaptive beamforming applied to medical ultrasound imaging. *Proc IEEE Ultrason Symp* 2005;199–1202.

Tian J, Han PD, Huang XL, Pan HX, Carroll JF, Payne DA. Improved stability for piezoelectric crystal growth in the lead indium niobate-lead magnesium niobate-lead titanate system. *Appl Phys Lett* 2007;91:222903;1–3.

Trahey GE, Nock LF. Synthetic receive aperture imaging with phase correction for motion and for tissue inhomogeneities. Part II: Effects of and correction for motion. *IEEE Trans Ultrason Ferroelectr Freq Control* 1992;39;496–501.

Ustuner KF, Holley GL. Ultrasound imaging system performance assessment. *AAPM Annual Meeting* 2003.

von Ramm OT, Smith SW, Pavy HT. High speed ultrasound volumetric imaging system. Part II: Parallel processing and display. *IEEE Trans Ultrason Ferroelectr Freq Control* 1991;38;109–115.

Ylitalo JT, Ernest H. Ultrasound synthetic aperture imaging: Monostatic approach. *IEEE Transactions on Ultasonic, Ferroelectrics, and Frequency Control* 1994;41;333–339.

Chapter 1

2. Three-Dimensional Ultrasound Imaging

Aaron Fenster, Grace Parraga, Bernard C. Y. Chiu, and Eranga Ukwatta

Chapter 2

Ultrasound Imaging and Therapy. Edited by Aaron Fenster and James C. Lacefield © 2015 CRC Press/Taylor & Francis Group, LLC. ISBN: 978-1-4398-6628-3

2.1 Introduction: Two-Dimensional Medical Imaging

Two-dimensional (2-D) x-ray imaging has been the basis for forming images of the human anatomy since the discovery of x-rays at the end of the nineteenth century. Because 2-D x-ray imaging provides a projection of the three-dimensional (3-D) anatomy onto a 2-D image plane, information of an organ or pathology necessary for diagnosis or treatment is often obscured by overlying structures. Computed tomography (CT), developed in the early 1970s, revolutionized diagnostic radiology by providing physicians with 3-D images of anatomical structures, reconstructed from a set of contiguous tomographic 2-D images. The development of 3-D imaging techniques has continued with the development of 3-D magnetic resonance imaging (MRI), positron emission tomography (PET), and multislice and cone beam CT imaging. These imaging modalities have stimulated the development of a wide variety of clinical applications using advanced 3-D image analysis and visualization techniques.

Although applications of 3-D imaging with CT, PET, and MRI have advanced rapidly, ultrasound (US) imaging has been extended to 3-D imaging more slowly (Elliott 2008). Currently, the majority of US-based diagnostic procedures are still performed using 2-D imaging making use of 1-D arrays, and the development of 3-D US imaging making use of 1-D and 2-D arrays has been growing. Three-dimensional US techniques have been increasingly used in diagnosis, minimally invasive image-guided interventions, and intraoperative use of imaging (Downey et al. 2000; Boctor et al. 2008; Hummel et al. 2008). Advances in 3-D US imaging technology have resulted in high-quality 3-D images of complex anatomical structures and pathology, which are used in diagnosis of disease and to guide interventional and surgical procedures. Increasingly, researchers and commercial companies are incorporating advanced 3-D visualizations into US systems as well as integrating 3-D US imaging into biopsy and therapy procedures (Chin et al. 1996, 1998; Smith et al. 2001; Wei et al. 2004; Carson and Fenster 2009).

This chapter reviews the different methods for obtaining and visualizing 3-D US images using 1-D arrays and describes the use of 3-D US imaging in applications as an example of its potential use in a new way.

2.2 Benefits of 3-D US Imaging

Conventional 2-D US imaging systems making use of 1-D arrays enable users to manipulate the handheld US transducer freely over the body to generate images of organs and pathology. However, the ability to freely manipulate the transducer also results in the following disadvantages, which 3-D US imaging attempts to overcome:

- Freely manipulating the conventional US transducer over the anatomy to generate 2-D US images requires that users mentally integrate many 2-D images to form an impression of the anatomy and pathology in 3-D. In cases of complex anatomy or pathology, this approach leads to longer procedures and may result in variability in diagnosis and guidance in interventional procedures.
- The 2-D US image is difficult to relocate to an exact location and orientation in the body at a later time because the conventional 2-D US imaging transducer is held and manipulated manually. The manual manipulation of a 2-D US image is suboptimal

because monitoring the progression and regression of disease in response to therapy requires imaging of the same location (plane) of the anatomy.

- Conventional 2-D US imaging does not permit viewing planes parallel to the skin—often called *C-mode*. This approach is, at times, suboptimal because diagnostic and interventional procedures sometimes require an arbitrary selection of the image plane for optimally viewing the pathology and guiding the interventional procedure.
- Diagnostic procedures, therapy/surgery planning, and therapy monitoring often require accurate lesion volume measurements. Because conventional 2-D US imaging only provides a cross section of the lesion, measurements of organ or lesion volume is variable and at times inaccurate.

Research investigators have experimented with the production of 3-D US images in the 1970s, and the first commercial system became available in 1989 by Kretz. For the past two decades, many researchers and commercial companies have developed 3-D US imaging systems and used them in a wide variety of applications (Nelson and Pretorius 1992; Greenleaf et al. 1993; King et al. 1993; Rankin et al. 1993; Fenster and Downey 1996; Sklansky 2003; Peralta et al. 2006). Although 3-D imaging techniques based on CT and MRI have progressed rapidly and entered routine clinical use almost immediately, the development of 3-D US systems and their routine use has been slow. The slow integration of 3-D US into routine use may be attributed to the requirement that US images, whether 2-D, 3-D, or 4-D, should be provided to the physician in real time, requiring significant computational speed for acquiring, reconstructing, and viewing 3-D US information on inexpensive systems. The recent availability of low-cost computer technology with significant computational power and advanced visualization techniques based on computer hardware have now made 3-D US imaging a viable technology. This has led to the use of 3-D US in a wide range of applications, as well as the availability of 3-D imaging and viewing on most of the major US systems.

The following sections review the various approaches used in the generation of 3-D US images based on 1-D arrays. An emphasis is placed on the geometric accuracy of the generated 3-D images as well as the use of this technology in interventional and quantitative monitoring applications.

2.3 Use of 1-D Arrays for 3-D US Imaging

Researchers and US companies have used a variety of approaches to produce 3-D US images based on 1-D arrays. The most successful approaches have been based on mechanical and freehand scanning techniques. The use of 2-D arrays for the generation of 3-D US images are described in a separate chapter. Because 1-D arrays produce 2-D US images, the use of these transducers to produce 3-D US images requires that the relative positions and orientations of the 2-D images within the 3-D image volume are determined. Thus, the production of a 3-D US image without any distortions requires that three factors be optimized:

- The relative locations and orientations of the acquired 2-D US images with respect to one another must be accurately determined to avoid geometric distortions in the generation of the 3-D US image. Geometric errors will lead to errors of measurement and guidance.

Chapter 2

- The scanning apparatus must be simple and convenient to use to be easily added to the clinical examination or interventional procedure.
- The acquisition of the 2-D US images needed to construct the 3-D US image must be either rapid (i.e., real time or near real time) or gated to avoid image artifacts due to involuntary, respiratory, or cardiac motion.

For the past two decades, various approaches have been developed to produce 3-D US images; however, current systems make use of one of the following 3-D US imaging approaches: mechanical scanning, freehand scanning with position sensing, and freehand scanning without position sensing.

2.3.1 Mechanical 3-D US Scanning Systems

Mechanical 3-D US scanning systems make use of motorized mechanisms to translate, tilt, or rotate a conventional 2-D US transducer with a 1-D array. To construct the 3-D US image, a sequential series of 2-D US images is rapidly acquired by a computer as the 2-D US transducer is moved mechanically. The relative position and orientation of the acquired 2-D US images are accurately and precisely determined because the scanning geometry in mechanical 3-D US systems is predefined and precisely controlled by a motor and monitored by position or rotary encoders.

The acquired 2-D US images and their relative positions and orientation are then used to construct the 3-D US images as the 2-D images are acquired (i.e., real-time) using novel computational algorithms. These mechanical 3-D scanning systems enable the user to adjust the angular or spatial interval between the acquired 2-D images so that the spatial interval between the acquired 2-D images can be optimized to minimize the scanning time while adequately sampling the volume being investigated (Smith and Fenster 2000).

Currently, the most common approach used to produce 3-D US images involves mechanical scanning mechanisms used to translate or rotate the conventional US transducer. The mechanical 3-D US scanning approach has been developed primarily using two distinct approaches: integrated 3-D US transducers that house the scanning mechanism within the transducer housing and external mechanical fixtures that hold the housing of conventional transducers generating 2-D US images. Both approaches have been successfully used for a variety of clinical applications.

Most US manufacturers now offer integrated 3-D US transducers that are based on a mechanically swept transducer or "wobbler." These systems make use of a 1-D US array, which is wobbled or swept back and forth inside the 3-D transducer housing. As the 1-D US array is wobbled, the 2-D US images that are generated are used in the 3-D US image construction. Because the 3-D transducer housing must contain the 1-D array and enable the array to wobble over an angle sufficient to generate a 3-D image, these transducers are larger than conventional 2-D US transducers but are easier to use than 3-D US systems using external fixtures with conventional 2-D US transducers. Although these types of 3-D US transducers are convenient to use, they require a special US machine that can control the 3-D scanning and reconstruct the acquired 2-D images into a 3-D image. The external mechanical 3-D scanning fixtures, which are discussed in the following sections, are generally bulkier than integrated transducers. They can be used with any US manufacturer's transducer, obviating the need to purchase a special 3-D US machine. In addition, the use of an external fixture to generate 3-D US images can take advantage of improvements in the US machine (e.g., image compounding, contrast agent imaging) and flow information (e.g., Doppler imaging) without any changes in the scanning mechanism.

Both integrated 3-D transducers and external fixture approaches used in mechanical 3-D US scanning enable short imaging times ranging from approximately 3 to 0.2 volumes/s, high-quality 3-D images including B-mode and Doppler, and real-time 3-D reconstruction providing the 3-D image to the physician as it is being acquired. However, mechanical 3-D scanners can be bulky, and their weight makes them sometimes inconvenient to use. Figure 2.1 shows three basic types of mechanical scanners that are being used: tilt scanners, linear scanners, and rotational scanners.

2.3.2   Wobbling or Tilting Mechanical 3–D US Scanners

Figure 2.1a shows a schematic diagram of the mechanical tilt or wobble approach of a conventional 1-D array US transducer about an axis parallel to the face of the transducer, and Figure 2.1b shows the tilting axis away from the face of the transducer. The latter approach is typically used in integrated 3-D scanning mechanisms. In both scanning approaches, the 2-D US images that are generated are arranged as a fan of images with a user-selected angular spacing, for example, 1.0°. In both the integrated 3-D US transducer and the external fixture approaches, the housing of the probe remains fixed on the skin of the patient, whereas the US transducer is wobbled. The time required to acquire a 3-D US image depends on the 2-D US image update rate and the number of 2-D images needed to generate the 3-D image. The 2-D US image update rate depends on the US machine settings (e.g., depth setting and number of focal zones), and the number of acquired 2-D US images is determined by the chosen angular separation between the acquired 2-D images needed to yield a desired image quality and the total scan angle needed to cover the desired anatomy. Typically, these parameters can be adjusted to optimize scanning time, image quality, and size of the volume imaged (Gilja et al.

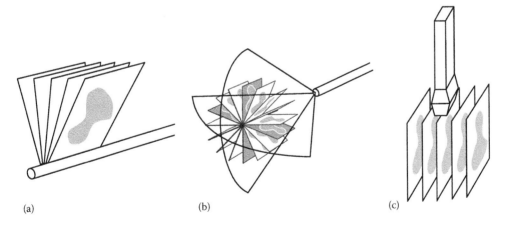

(a)   (b)   (c)

FIGURE 2.1   Schematic diagram of 3-D US mechanical scanning methods. (a) Side-firing transrectal US transducer being mechanically rotated. The acquired images have equal angular spacing. The same approach is used in a mechanically wobbled transducer. (b) A rotational scanning mechanism using an end-firing transducer, typically used in 3-D US-guided prostate biopsy. The acquired images have equal angular spacing. (c) Linear mechanical scanning mechanism. The acquired images have equal spacing.

1994; Delabays et al. 1995; Downey et al. 1995a,b; Fenster et al. 1995; Bax et al. 2008). The most common integrated 3-D transducers using the wobbling technique are used for abdominal and obstetrical imaging (Goncalves et al. 2005; Peralta et al. 2006; Kurjak et al. 2007).

Because the parameters to acquire 3-D US images can be predefined when using the mechanical 3-D US approach, 3-D image construction can occur while the 2-D images are acquired. However, this scanning approach will result in a 3-D image resolution that will not be isotropic. The resolution in the 3-D US image will degrade in the axial direction of the 3-D US images because of the increasing US beam spread in the lateral and elevational directions of the acquired 2-D US images as the distance in the axial direction is increased. In addition, because the geometry of the acquired 2-D images is fanlike, the distance between the acquired US images increases with increasing axial distance. The increased axial distance results in a decrease in the spatial sampling and spatial resolution of the reconstructed 3-D image in the elevational direction of the acquired 2-D US images (Blake et al. 2000).

2.3.3 Linear Mechanical 3-D Scanners

Linear scanner mechanisms use a motorized drive to move the conventional 2-D transducer across the skin of the patient. The 2-D transducer can be fixed to be perpendicular to the surface of the skin or at an angle for acquiring Doppler images. The spacing between the acquired 2-D images is constant but can be selected by the user so that they are parallel and uniformly spaced (Figure 2.1c). The velocity of the transducer as it is being scanned across the skin can be adjusted so that the temporal sampling interval can match the 2-D US frame rate for the US machine and result in the required spatial sampling interval, which should be at least half of the elevational resolution of the transducer (Smith and Fenster 2000).

As with the wobbling or tilting mechanical scanning approach, the predefined and regular geometry of the acquired 2-D US images enables a 3-D image to be reconstructed while the 2-D US images are being acquired. Because the 3-D US image is produced from a series of parallel conventional 2-D US images, its resolution will not be isotropic. The resolution of the constructed 3-D US image will be the same as the original 2-D US images in the direction parallel to the acquired 2-D US images. However, the resolution of the constructed 3-D image will be equal (if spatial sampling is appropriate) to the elevational resolution of the acquired 2-D US images in the direction of the 3-D scanning. The resolution of the 3-D US image will be poorest in the 3-D scanning direction because the elevational resolution is typically poorer than the in-plane resolution of the acquired 2-D US images. Thus, a transducer with good elevational resolution should be used for optimal results (Fenster et al. 2001).

Linear scanning has been successfully implemented in many vascular B-mode and Doppler imaging applications, particularly for carotid arteries (see Section 2.5) (Pretorius et al. 1992; Picot et al. 1993; Downey and Fenster 1995a,b; Fenster et al. 1995; Landry and Fenster 2002; Ainsworth et al. 2005; Landry et al. 2005, 2007; Krasinski et al. 2009) and tumor vascularization (King et al. 1991; Bamber et al. 1992; Carson et al. 1993; Downey and Fenster 1995a). Figure 2.2 shows two examples of linearly scanned 3-D US images of the carotid arteries made with an external fixture.

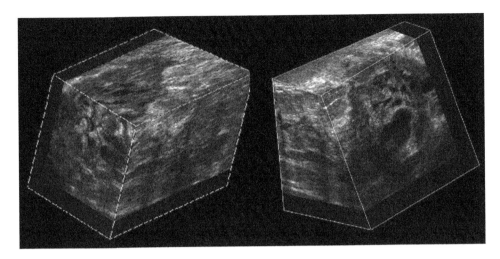

FIGURE 2.2   Two examples of 3-D US images of the carotid arteries obtained with the mechanical linear 3-D scanning approach. The 3-D US images are displayed using the cube-view approach and have been sliced to reveal the details of the atherosclerotic plaque in the carotid arteries in transverse and longitudinal views.

2.3.4   Endocavity Rotational 3-D Scanners

The endocavity rotational 3-D scanning approach uses an external fixture or internal mechanism to rotate an endocavity transducer (e.g., a transrectal ultrasound [TRUS] probe, see Figure 2.1a) about its long axis. For endocavity transducers using an end-firing transducer, which are typically used for prostate biopsy, the set of acquired 2-D images will be arranged as a fan (Figure 2.1b), intersecting in the center of the 3-D US image, resulting in an image as shown in Figure 2.3. For endocavity transducers using a side-firing 1-D array, which are typically used in prostate brachytherapy, the acquired images will also be arranged as a fan but intersect at the axis of rotation of the transducer (see Figure 2.1a). For 3-D TRUS imaging of the prostate, a side-firing transducer is typically rotated from 80° to 110°, and an end-firing transducer is typically rotated by 180° (Tong et al. 1996, 1998; Bax et al. 2008). Figure 2.3 shows an endocavity scanner with end-firing transducer used to image the prostate (Downey and Fenster 1995a; Fenster et al. 1995; Tong et al. 1996) and guide 3-D US biopsy and therapy (Downey et al. 1995a; Onik et al. 1996; Chin et al. 1998, 1999; Wei et al. 2004, 2005; Bax et al. 2008).

A motorized and a manual mechanism with encoders to record the rotation angle have been used to rotate an end-firing endocavity transducer array by (at least) 180° on a fixed axis that perpendicularly bisects the transducer array. In this approach, the resolution of the 3-D image will not be isotropic. Because the spatial sampling is highest near the rotation axis of the transducer and the poorest away from the axis of rotation of the transducer, the resolution of the 3-D US image will degrade as the distance from the rotational axis of the transducer is increased. Similarly, the axial and elevational resolution will decrease as the distance from the transducer is increased. The combination of these effects will cause the 3-D US image resolution to vary—best near the transducer and the rotational axis and poorest away from the transducer and rotational axis.

Chapter 2

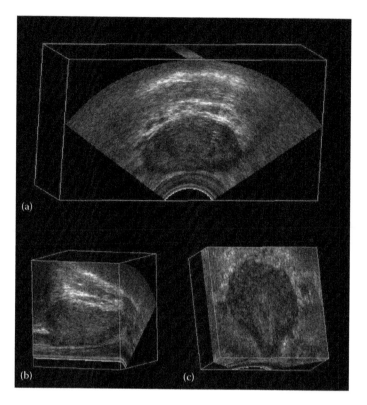

FIGURE 2.3   A 3-D US image of the prostate acquired using a side-firing endocavity rotational 3-D scanning approach (rotation of a transrectal US transducer). The transducer was rotated around its long axis, whereas 3-D US images were acquired and reconstructed. The 3-D US image is displayed using the cube view approach and has been sliced to reveal: (a) transverse view, (b) sagittal view, and (c) coronal view, not possible using conventional 2-D US techniques.

Because the axis of rotation will be in the center of the 3-D US image, 3-D rotational scanning with an end-firing transducer is most sensitive to the motion of the transducer and patient. Because the acquired 2-D images intersect along the rotational axis of the transducer at the center of the 3-D image, any motion during the scan will cause a mismatch in the acquired planes, resulting in the production of artifacts in the center of the 3-D US image. Artifacts in the center of the 3-D US image will also occur if the axis of rotation is not accurately determined; however, proper calibrations can remove this source of potential error. Although scanning the prostate and the uterus using a handheld 3-D rotational can produce excellent 3-D images (see Figure 2.3), for optimal results in long procedures, such as prostate brachytherapy and biopsy (Tong et al. 1996; Bax et al. 2008; Cool et al. 2008), the transducer and its assembly should be mounted onto a fixture, such as a stabilizer in prostate brachytherapy.

2.3.5   Freehand Scanning with Position Sensing

Many researchers have attempted to overcome the bulkiness of mechanical 3-D US scanning mechanisms by developing 3-D scanning techniques that do not require a mechanical scanning device. Because the construction of a 3-D US image from a sequence of 2-D US images requires that the position and orientation of the transducer must be tracked, a freehand scan of a conventional transducer requires a method to track the positions and orientations of the transducer as it is being moved. Several methods have been developed to accomplish this task. All methods require a sensor to be mounted on the transducer to enable the measurement of the conventional 2-D transducer's position and orientation as it is moved over the body. In this way, 2-D images are acquired while the 2-D transducer is moved freehand together with the transducer's relative position and orientation. The images and position/orientation information is then used to construct the 3-D image (Pagoulatos et al. 2001).

The locations and orientations of the acquired 2-D images are not predefined because the user can move the 2-D transducer freehand without a predefined and reproducible motion. To avoid 3-D construction errors or large gaps between acquired 2-D images, the operator must move the transducer over the anatomy at an appropriate speed to ensure that the spatial sampling is proper and to ensure that the set of 2-D images does not have any significant gaps. For the past two decades, several approaches for freehand scanning have been developed: tracked 3-D US with articulated arms, freehand 3-D US with acoustic sensing, freehand 3-D US with magnetic field sensing, and image-based sensing (speckle decorrelation). The method used most commonly is the magnetic field sensing approach, with several companies providing the sensing technology: Ascension—Bird sensor (Boctor et al. 2008), Polhemus—Fastrack sensor (Treece et al. 2001), and Northern Digital—Aurora sensor (Hummel et al. 2008).

2.3.6 Freehand 3-D Scanning with Magnetic Field Sensors

The most successful freehand 3-D US scanning approach made use of magnetic field sensors in many diagnostic applications, such as echocardiography, obstetrics, and vascular imaging (Raab et al. 1979; Detmer et al. 1994; Hodges et al. 1994; Pretorius and Nelson 1994; Nelson and Pretorius 1995; Riccabona et al. 1995; Hughes et al. 1996; Gilja et al. 1997; Leotta et al. 1997; Treece et al. 2001; Boctor et al. 2008; Hsu et al. 2008; Hummel et al. 2008). As with the mechanical scanning approach for 3-D US, magnetic sensor-based freehand scanning also makes use of conventional 1-D US transducers generating 2-D US images. Tracking of the transducer as it is moved over the body is achieved with a small receiver, which is mounted on the transducer and contains three orthogonal coils enabling six-degree-of-freedom sensing. The small receiver is used to sense the strength of the magnetic field in three orthogonal directions, which is generated by a time-varying 3-D magnetic field transmitter placed near the patient. By continuously measuring the strength of the three components of the local magnetic field, the position and orientation of the transducer is calculated as each 2-D image is acquired and used in the 3-D construction algorithm.

Magnetic field sensors are small and unobtrusive devices; hence, they enable the transducer to be tracked without the need for bulky mechanical devices and without the need to keep a clear line of sight as required by optical tracking methods. However, the magnetic field sensing tracking approach is sensitive to electromagnetic interference, which can compromise the accuracy of the tracking. Geometric tracking errors can also occur,

Chapter 2

leading to distortions in the 3-D US image if ferrous (or highly conductive) metals are located nearby. Thus, metal hospital beds in procedure or surgical rooms can cause significant distortions. However, modern magnetic field sensors have been produced to be less susceptible to these sources of error, particularly those that use a magnetic transmitter placed between the bed and the patient. To produce excellent images without distortions, the position of the sensor relative to the transmitter must be calibrated accurately and precisely using one of the numerous calibration techniques that have been developed (Lindseth et al. 2003; Leotta 2004; Dandekar et al. 2005; Gee et al. 2005; Gooding et al. 2005; Mercier et al. 2005; Poon and Rohling 2005; Rousseau et al. 2005; Hsu et al. 2008).

2.3.7 Freehand 3-D US Scanning without Position Sensing

A 3-D US imaging tracked using a mechanical or magnetic field can produce spatially accurate 3-D images. If the 3-D US image is not to be used for distance, area, or volume measurements, the requirements for spatial accuracy are not needed. In this case, 3-D US images can be easily produced without any position sensing or control by a mechanical mechanism. In this approach, the conventional transducer is moved over the body manually in a freehand and steady motion that is predefined. As the user moves the conventional transducer, either in a linear manner over the body or by tilting, a set of 2-D US images is acquired at regular temporal intervals. By assuming that the transducer was moved in a predefined linear or angular velocity, the spatial or angular separations between the acquired 2-D images are calculated and constructed to form a 3-D US image (Downey and Fenster 1995b). However, this approach must not be used for measurements because it does not guarantee that the 3-D US image is geometrically accurate.

2.4 Three-Dimensional US Image Visualization

For the past two decades, many approaches have been developed for the visualization and image processing of 3-D images. The visualization and the manipulation of computer algorithms of 3-D CT and MRI images are currently commonly used by physicians and researchers in many clinical and research applications. However, the visualization of 3-D US images poses significant challenges primarily because 3-D US images suffer from shadowing, poor tissue–tissue contrast, and image speckle. Many visualization approaches have been attempted because the visualization technique of the 3-D US image plays an important role in the ability of a physician to obtain the required information efficiently. The segmentation of organ boundaries is difficult because the field of view is limited and tissue–tissue contrast is poor in 3-D US images. Thus, surface rendering techniques have not generally been successful. However, techniques making use of B-mode data without segmentation have been more successful and are in common use today. Two of the most frequently used techniques for visualizing 3-D US images are multiplanar reformatting (MPR) and volume rendering (VR).

2.4.1 Multiplanar Reformatting

The MPR technique is the most commonly used for 3-D US visualization. In this technique, 2-D US planes are extracted from the 3-D US images and displayed to the user with 3-D spatial cues. The user can view the desired anatomy by choosing which 2-D images to extract from the 3-D US image and manipulate them interactively by causing the 2-D planes to move.

Three MPR techniques are commonly used to display 3-D US images. Figure 2.4a illustrates the crossed-plane approach, in which multiple planes are presented in a view that shows their correct relative orientations. These planes are typically orthogonal and intersect with one another. The user can select any of the planes and move them in a parallel or oblique manner to any other plane to reveal the desired views. A second approach to display 3-D US images makes use of the cube-view approach, which is illustrated in Figure 2.4b. In this approach, the 3-D US image is displayed as a polyhedron (sometimes as a cube) with the extracted set of 2-D US images texture-mapped onto the faces of the polyhedron. The user can select any face of the polyhedron and move it (parallel or obliquely) to any other plane while the appropriate 2-D US image is extracted in real time and texture-mapped on the new face. The appearance of a "solid" polyhedron provides users with 3-D image-based cues, which relates the manipulated plane to the other planes (Fenster and Downey 2000, 2001). In a third approach, three orthogonal planes are extracted from the 3-D US image and displayed together with 3-D cues, such as color-coded lines on each extracted plane to designate its intersection with the other planes. These lines can be moved to extract and display the desired planes.

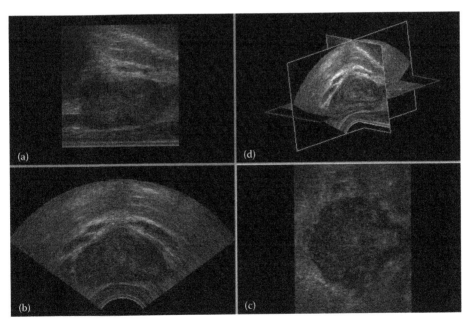

FIGURE 2.4   The 3-D US of the prostate displayed using (a through c) three orthogonal planes from a 3-D image displayed using the MPR approach and (d) the crossed-plane approaches. The 3-D US images were acquired using a side-firing transducer using the mechanical rotation approach.

Chapter 2

2.4.2 Volume Rendering Techniques

Three-dimensional CT and MRI images are frequently viewed using VR techniques. High-quality volume-rendered images require sufficient contrast between different organs or efficient segmentation of organs. CT and MR can provide good contrast between organs; however, 3-D US images do not produce sufficient tissue–tissue contrast needed for VR rendering. However, US imaging does produce excellent contrast between tissue and fluids (i.e., blood and amniotic fluid). Thus, the VR approaches are used extensively by 3-D US system manufacturers to view 3-D US fetal images (Lee 2003; Sklansky 2004) and 4-D US cardiac images (Devore and Polanko 2005; Goncalves et al. 2005; Deng and Rodeck 2006).

VR software is based on ray-casting techniques, which are used to project a 2-D array of lines (rays) through a 3-D image (Levoy 1990; Fruhauf 1996; Mroz et al. 2000). The display of the VR image is then based on the techniques used to determine the intersection of volume elements (voxels) with each ray, which are weighted, summed, and colored or shaded in several ways to produce various effects. Although many VR techniques have been developed, three of the most common approaches used to view 3-D and 4-D US images are maximum intensity projection, translucency rendering, and surface enhancement.

Because the VR techniques project 3-D information onto a 2-D plane together with surface characteristic cues, many VR techniques are not well suited for viewing the details of soft tissues in 3-D US B-mode images. Rather, VR techniques are best suited for viewing anatomical surfaces that are distinguishable in 3-D US B-mode images, including limbs and fetal face surrounded by amniotic fluid (see Figure 2.5) (Pretorius and Nelson 1995; Nelson et al. 1996), tissue–blood interfaces such as endocardiac surfaces and inner vascular surfaces, and structures where the B-mode clutter has been removed from power or color Doppler 3-D images (Downey and Fenster 1995b).

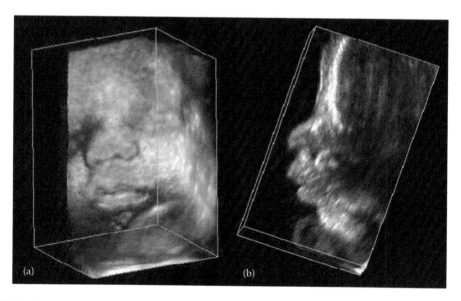

FIGURE 2.5 Two 3-D US images that have been volume rendered. (a) 3-D US image of a fetal face; (b) 3-D US image of the vasculature in a kidney obtained using freehand 3-D Doppler scanning.

2.5 Three-Dimensional Carotid US Imaging

As an example of the use of 3-D US, we describe a quickly developing application using 3-D US imaging in this section. In this application, a 3-D US is used to analyze, quantify, and monitor carotid atherosclerosis. Atherosclerosis is an inflammatory disease in which the inner layer of the arteries progressively accumulates low-density lipoproteins and macrophages over a period of several decades forming plaque (Lusis 2000). The carotid arteries are major sites for developing atherosclerosis in patients. Unstable plaque may suddenly rupture, forming a thrombus causing an embolism, which may ultimately lead to an ischemic stroke by blocking the oxygenated blood supply to parts of the brain. High-quality 3-D US images of carotid arteries are relatively easy to acquire because carotid arteries are superficial structures.

During the last two decades, 2-D US intima-media thickness (IMT) measurement (see Figure 2.6a) has been investigated as a surrogate end point of vascular outcomes for monitoring carotid atherosclerosis in subjects during medical interventions (Lorenz et al. 2006). Although common carotid artery (CCA) IMT is a reproducible measurement that has been shown to significantly correlate with the risk factors of stroke in large clinical trials (O'Leary and Bots 2010), it is not sensitive to the changes in plaque burden, which is a stronger predictor of cardiovascular events (Fosse et al. 2006; Johnsen and Mathiesen 2009; Inaba et al. 2011).

There are new therapies developed to regress plaque and inflammation, but they lack intermediate end points to use as efficacious targets. To study such treatments, it is necessary to directly measure changes in plaque morphology and/or composition. The 3-D US is showing promise in quantifying carotid atherosclerosis for monitoring disease progression and regression in clinical trials.

2.5.1 Three-Dimensional US Image Acquisition and Visualization

Three-dimensional carotid US images can be acquired using a linear scanning approach (see Section 2.3.1) with a conventional vascular US transducer (Fenster et al. 2004). For a 3-D carotid US exam, a distance of at least 4 cm around the bifurcation, encompassing portions of the common carotid artery (CCA), internal carotid artery (ICA), and external carotid artery (ECA), is scanned by translating the US transducer along the patient's neck. There are two methods to translate the US transducer: (1) by using a motorized

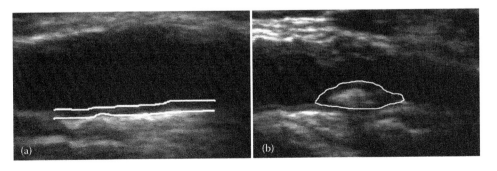

FIGURE 2.6 (a) Manual delineation of the IMT measurement of CCA from a longitudinal 2-D US image. (b) Manual delineation of the plaque boundary for TPA measurement from a longitudinal 2-D US image.

Chapter 2

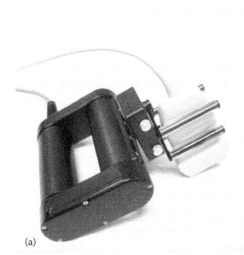
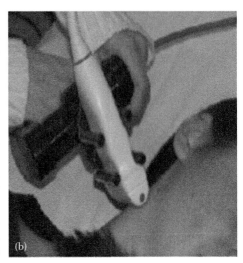

(a)
(b)

FIGURE 2.7 (a) Mechanically assisted 3-D US tranducer probe. (b) Acquiring a 3-D US image by translating the transducer along the patient's neck at least up to a distance of 4 cm encompasing portions of CCA, ICA, and ECA.

controller (Section 2.3.1) and (2) by using a freehand approach with magnetic tracking (Section 2.3.5). Here we describe the acquisition using a motorized controller in brief, and the readers are referred to Fenster et al. (2004, 2006) and Parraga et al. (2011) for more technical details. Figure 2.7a and b shows the motorized scanner used to acquire 3-D carotid US images.

With the utility of the multiplanar cube view (Section 2.4.1), an operator is able to pan any arbitrary plane for a better visualization of the vessel wall boundaries (Fenster et al. 2006). Figure 2.8a and b shows the transverse and longitudinal views of a 3-D carotid US image of a patient with moderate stenosis. Figure 2.8c and d shows the transverse and longitudinal views of a 3-D carotid US image of a patient with a carotid plaque ulceration. The ability to visualize the transverse and longitudinal views side by side greatly aids the operator to estimate or delineate the wall thickening or plaque burden. Notice that the image resolution is anisotropic. The image resolution in the longitudinal direction is poorer than the transverse direction because it depends on the elevational resolution of the transducer, which is poorer than the axial and lateral resolution of the transducer.

2.5.2 Three-Dimensional US Imaging Phenotypes

Recently, carotid US phenotypes have been developed for monitoring carotid atherosclerosis, that is, total plaque area (TPA) (Johnsen and Mathiesen 2009), total plaque volume (TPV) (Ainsworth et al. 2005), vessel wall volume (VWV) (Egger et al. 2007), and vessel-wall-plus-plaque-thickness (VWT) maps (Chiu et al. 2008a,b). The hypothesis is that the area and the volumetric measurements are more reflective of the plaque burden so that they are more sensitive and may provide complementary information to IMT (Johnsen and Mathiesen 2009; Inaba et al. 2011). With these phenotypes, an effect of therapy with fewer subjects in shorter clinical trials with imaging end points may be possible.

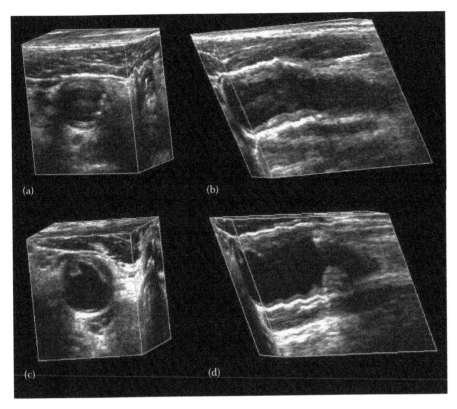

FIGURE 2.8 Multiplaner views of the 3-D US images. (a) Transverse view of a 3-D US image of carotid artery for a subject with moderate stenosis. (b) Longitudinal view of a 3-D US image of carotid artery for a subject with moderate stenosis. (c) Transverse view of a 3-D US image of carotid artery for a subject with ulceration. (d) Longitudinal view of a 3-D US image of carotid artery for a subject with ulceration.

TPV is a direct volumetric measurement of plaque burden in the left and right carotid arteries. TPV measurements encompass complex plaque morphologies and geometries. TPV measurements can be obtained by the manual segmentation of the plaque (Landry et al. 2005, 2007; Fenster et al. 2006). Initially, the observer defines the medial axis of the artery in longitudinal view. After familiarizing with the orientation and geometry of the plaque, the observer outlines the plaque boundary on transverse slices with an interslice distance of 1 mm. Figure 2.9a and b shows the manual segmentations of the vessel wall and lumen overlaid on a 3-D US image and the reconstruced surface for the computation of TPV.

The VWV measurement is the volume enclosed between the outer wall and the lumen boundaries for the CCA, ICA, and less commonly ECA. For this calculation, the outer wall and the lumen boundaries are required to be delineated. Figure 2.9c and d shows manual segmentations of the outer wall and the lumen boundaries overlaid on a 3-D US image and the surfaces of the outer wall and lumen for computing VWV. The manual segmentation of the boundaries is performed on transverse slices with an interslice distance of 1 mm. VWV measurements have a higher reproducibility than TPV measurements because the outer wall and lumen boundaries are relatively easier to delineate than plaque boundaries (Egger et al. 2007).

Chapter 2

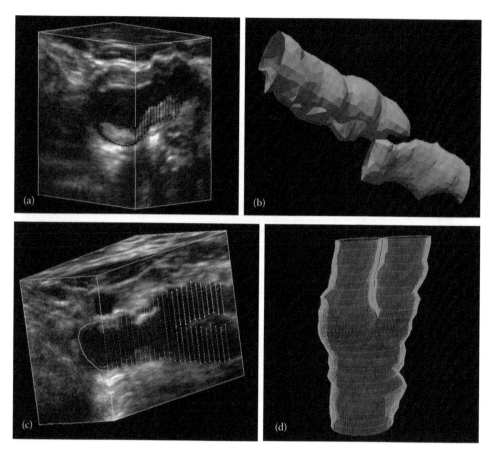

FIGURE 2.9 (a) Manual segmentation of the plaque boundaries of the 3-D US image used for TPV measurement. (b) Reconstructed plaque surfaces from the manual segmentations. (c) Manual segmentation of the CCA, ICA, and ECA lumen and outer wall boundaries from 3-D US images. (d) Reconstructed surfaces for the lumen and outer wall boundaries from the manual segmentations.

The measurements discussed previosuly are global measurements of plaque growth and/or wall thickening. Chiu et al. (2008b) proposed VWT maps and VWT change maps to quantify and visualize local changes in plaque morphology on a point-by-point basis using the segmented surfaces for the lumen and outer wall of the carotid arteries. The identification of the locations of change in plaque burden may assist to develop treatment strategies for patients. To facilitate the visualization and interpretation of these maps for clinicians, Chiu et al. (2008a) proposed a technique to flatten the 3-D VWT maps and the VWT change maps to 2-D (see Figure 2.10a and b).

2.5.3 Segmentation of the Carotid Arteries

The manual segmentation of the outer wall, lumen, and plaque boundaries is time consuming and tedious. Therefore, developing semiautomated techniques for segmentation and analysis is important for translating the use of 3-D carotid US phenotypes to clinical use. Here we provide a brief review of the segmentation techniques to delineate outer wall, lumen, and plaque boundaries from 3-D US images.

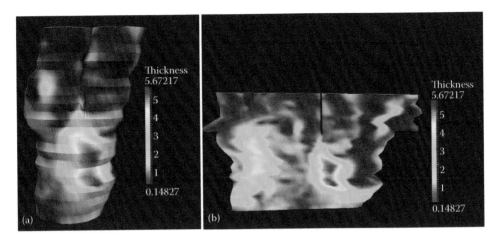

FIGURE 2.10 (a) Vessel wall thickness map for a patient with moderate stenosis. Manual segmentations of the lumen and outerwall boundaries were used to generate the thickness maps. The thickness is indicated in millimeters. (b) Corresponding flattened thickness map for better visualization.

Gill et al. (2000) proposed a semiautomated method based on a dynamic balloon model to only segment the lumen boundary from 3-D US images, but validated using a single image. After placing a seed mesh within the lumen, it is inflated to find the approximate boundary using an inflation force. The mesh is then refined using edge-based forces proximal to the lumen boundary. The algorithm yielded submillimeter value as the mean separation between the manual and the algorithm surface. However, the algorithm needed to be validated using multiple images to adequately evaluate its robustness and usefulness in practice. Zahalka and Fenster (2001) proposed a lumen only segmentation method using a deformable model with a stopping term based on the image gradient, which was validated using only a phantom study. Ukwatta et al. (2011b) proposed a combined 2-D segmentation method to segment the outer wall and the lumen boundaries of the carotid arteries from 3-D US images. Initially, an operator provides four anchor points on each boundary on each transverse slice to be segmented. After initialization, two level set segmentations with different energies were used to segment the outer wall and lumen boundaries. The outer wall was initially segmented so that the resulting segmentation of the outer wall was used to constrain the lumen segmentation from leaking in regions with shadows. The algorithm was evaluated with 231 2-D carotid US slices extracted from 21 3-D US images. The algorithm segmentations were compared with the manual segmentations, as shown in Figure 2.11. The algorithm yielded dice coefficients (>92%), submillimeter errors for the mean absolute distance (MAD) and maximum absolute distance (MAXD), and a comparable coefficient of variation (COV; 5.1% vs. 3.9%) to the manual segmentations for the outer wall and lumen boundaries, respectively (see Table 2.1 for more details). The minimum detectable difference of the algorithm in computing the VWV was comparable (64.2 vs. 50.3 mm³) with manual segmentations. The authors also reported on a coupled level set segmentation method to segment both boundaries simultaneously (Ukwatta et al. 2011a) for a more robust segmentation.

Buchanan et al. (2011) reported on a semiautomated plaque estimated method to outline the plaque boundaries from 3-D US images, thereby computing TPVs. For

Chapter 2

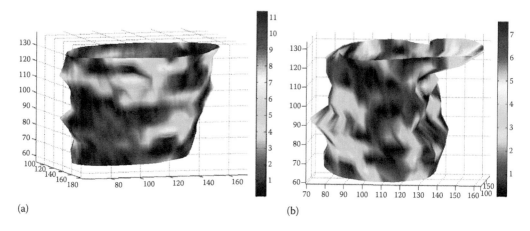

FIGURE 2.11   An example comparison of algorithm segmentations (Ukwatta et al. 2011b) to manual segmentations. The difference in thickness between the algorithm and the manual segmentations is color coded in the units of pixels: (a) outer wall segmentation of CCA and (b) lumen segmentation of CCA.

Table 2.1   Overall CCA Segmentation Results for the Segmentation Algorithm (Ukwatta et al. 2011b) for 231 2-D US Images Extracted from 21 3-D US Images

Metric	Outer Wall Segmentation	Lumen Segmentation
Volume error (%)	2.4 ± 1.9	5.8 ± 2.7
Dice coefficient (%)	95.6 ± 1.5	92.8 ± 3.1
MAD (mm)	0.2 ± 0.1	0.3 ± 0.1
MAXD (mm)	0.6 ± 0.3	0.6 ± 0.4

Note: MAD, mean absoute distance errror; MAXD, maximum absolute distance error.

the initialization, the observer defines the beginning and end points of the plaque in the long axis view followed by two contours in the transverse view to identify the regions of greatest change in shape. The algorithm was evaluated with 22 plaques from 10 3-D US images. The algorithm plaque volume and the manual plaque volumes were significantly correlated ($r = 0.99$, $p < 0.001$) while maintaining a mean COV of ≈12.0%.

2.5.4   Clinical Trials

Clinical trials using 3-D carotid US have focused on investigations about the viability and reproducibility of a 3-D US measurement and the effect of drugs or lifestyle changes. Landry et al. (2004) studied the variability in TPV measurements with 40 subjects. The intraobserver and interobserver reliabilities of the measurement were high (94% and 93.2%). The COV of the TPV measurement decreases with the increased size of plaque (range, 37.43–604.1 mm^3 in size) up to ≈4%. Egger et al. (2007) compared the COV and the intraclass correlation coefficient (ICC) of VWV and TPV measurements for 10 subjects. COV (4.6% vs. 22.7%) and ICC (0.95 vs. 0.85) measurements indicated a higher intraobserver variability and interscan variability in TPV than VWV. Egger et al. (2008) investigated VWT maps to evaluate changes in carotid atherosclerosis in five subjects from 3-D US images. The plaque volume thickness maps showed visual changes in vessel wall thickness and plaque for subjects treated with statin therapy and no visual changes for control subjects.

Ainsworth et al. (2005) studied the effects of atorvastatin in 38 subjects with carotid stenosis >60%, where subjects were randomly assigned to either the statin or the placebo in a 3-month trial. They observed a significant difference between the two groups of 16.81 ± 74.10 mm^3 progression in the placebo group and 90.25 ± 85.12 mm^3 regression in the atorvastatin group.

Shai et al. (2010) studied the effects of dietary weight loss intervention on regression of carotid atherosclerosis using 3-D US VWV and IMT measurements in 140 subjects during a 2-year period, where participants were randomized to low-fat, Mediterranean, and low-carbohydrate diets. The study found that diet-mediated weight loss for a 2-year period induced the significant regression of carotid VWV, although the effect is similar in low-fat, Mediterranean, or low-carbohydrate diets.

2.6   Discussion and Conclusions

Three-dimensional US imaging is being used extensively in obstetrics, cardiology, and image guidance of interventional procedures, such as prostate brachytherapy. Current mechanically integrated 3-D transducers making use of 1-D arrays can provide near real-time 3-D imaging, which are used in obstetrical imaging, whereas 3-D and 4-D imaging based on 2-D arrays are used in cardiology to view the dynamic motion of the heart and its valves. Current 3-D US research and development is focused on establishing its utility in additional clinical applications, particularly in image guidance for interventional techniques. In addition, much effort is focused on the development of image processing techniques, such as the segmentation of organs and lesions from 3-D US images and the multimodality registration of 3-D US images to 3-D CT and MR images. Improved software image processing tools are promising to make 3-D US a routine tool on US machines for use in a variety of applications. The following sections are some of the possible improvements in 3-D US imaging, which may accelerate its use in routine clinical procedures.

2.6.1   Further Analysis of 3-D Carotid US Images

In addition to TPV and VWV quantification, plaque composition and inflammation are also strong indicators of plaque vulnerability (Yuan et al. 2008). Therefore, volumetric measurements alone are not adequate to stratify plaque vulnerability to rupture and cause ischemic stroke. Although the ability of MR imaging for plaque composition (Saam et al. 2005; Clarke et al. 2006) and inflammation analysis (Kerwin et al. 2006) has been validated, the ability of 3-D US to identify individual plaque components such as fibrous, necrotic core, loose connective tissue, and calcifications is still unclear. Therefore, future work lies in investigating the capability of 3-D US imaging to identify plaque components, which could be validated using digital histopathology images of endarterectomy specimens and in vivo MR imaging.

Chapter 2

2.6.2 Improved Visualization Tools

Three-dimensional US imaging systems make use of interactive visualization tools, enabling the physician to manipulate the 3-D US image. However, physicians often find the interactive tools complicated and difficult to use during busy clinical procedures. In addition, the use of a mouse or a track ball to interact with the 3-D image is difficult in a surgical environment. For 3-D US to become more widely accepted for interventional guidance applications, intuitive and efficient tools are required to manipulate the 3-D US image and to display the result with the appropriate background using the appropriate interaction method. The use of voice control in 3-D US images is possible, but it is difficult to be implemented robustly in a busy and at times noisy surgical environment. Perhaps the gesture control of the 3-D image may be more appropriate, but a velar line of sight and easily recognized gestures will be required.

2.6.3 Quantification of Volume Based on 3-D US Segmentation

Many interventional procedures and therapy planning techniques require an accurate quantification of organ and lesion volumes. Thus, segmentation techniques are used to delineate the structure of interest through either manual or algorithmic approaches. Although automated computer segmentation algorithms have been developed for use with 3-D US images, they are not yet sufficiently robust to be used in routine clinical procedures and often require the user to correct segmentation errors. Because 3-D US images suffer from shadowing, image speckle, and poor tissue contrast, semiautomated segmentation approaches have been more successful. These approaches typically require the user to identify the organ or pathology to be segmented (e.g., plaque, prostate, and tumor), and then a computer algorithm performs the segmentation automatically. Although this approach is easier than manual segmentation, it still requires user interaction. Thus, improvements in the manual initialization of semiautomated algorithms and in segmentation algorithms making them robust, accurate, and reproducible would be highly welcomed.

2.6.4 Improvements in the Use of 3-D US for Interventional Procedures

The ability to produce 3-D US images easily and with compact systems has stimulated many investigators to integrate 3-D US in interventional procedures, such as biopsy, therapy, and surgery. For example, the 3-D US-guided prostate biopsy approach improves the physician's ability to accurately guide the biopsy needle to selected targets and record the biopsy location in 3-D (Bax et al. 2008). Using the 3-D TRUS image, the physician is able to observe the patient's prostate in views currently not possible in 2-D procedures. However, significant work and testing is still required to enable physicians to integrate 3-D US imaging efficiently into the interventional procedure. Some required improvements are better 3-D visualization tools, enabling the integration of 3-D US with images from other modalities, registration tools for multimodality registration, efficient and robust 3-D US-based segmentation tools, and 3-D visualization tools to help the physician guide tools within the body.

Acknowledgments

The authors gratefully acknowledge the financial support of the Canadian Institutes of Health Research, the Ontario Institute for Cancer Research, the Ontario Research Fund, the National Science and Engineering Research Council, and the Canada Research Chair program.

References

Ainsworth, C.D., C.C. Blake, A. Tamayo, V. Beletsky, A. Fenster, and J.D. Spence. 2005. "3D ultrasound measurement of change in carotid plaque volume; A tool for rapid evaluation of new therapies." *Stroke* no. 35:1904–1909.

Bamber, J.C., R.J. Eckersley, P. Hubregtse, N.L. Bush, D.S. Bell, and D.C. Crawford. 1992. "Data processing for 3-D ultrasound visualization of tumour anatomy and blood flow." *SPIE* no. 1808:651–663.

Bax, J., D. Cool, L. Gardi, K. Knight, D. Smith, J. Montreuil, S. Sherebrin, C. Romagnoli, and A. Fenster. 2008. "Mechanically assisted 3D ultrasound guided prostate biopsy system." *Med Phys* no. 35 (12):5397–5410.

Blake, C.C., T.L. Elliot, P.J. Slomka, D.B. Downey, and A. Fenster. 2000. "Variability and accuracy of measurements of prostate brachytherapy seed position in vitro using three-dimensional ultrasound: An intra- and inter-observer study." *Med Phys* no. 27 (12):2788–2795.

Boctor, E.M., M.A. Choti, E.C. Burdette, and R.J. Webster III. 2008. "Three-dimensional ultrasound-guided robotic needle placement: An experimental evaluation." *Int J Med Robot* no. 4 (2):180–191.

Buchanan, D., I. Gyacskov, E. Ukwatta, T. Lindenmaiera, A. Fenster, and G. Parraga. 2011. "Semi-automated segmentation of carotid artery total plaque volume from three dimensional ultrasound carotid imaging." In *SPIE Medical Imaging*. San Diego.

Carson, P.L., and A. Fenster. 2009. "Anniversary paper: Evolution of ultrasound physics and the role of medical physicists and the AAPM and its journal in that evolution." *Med Phys* no. 36 (2):411–428.

Carson, P.L., X. Li, J. Pallister, A. Moskalik, J.M. Rubin, and J.B. Fowlkes. 1993. "Approximate quantification of detected fractional blood volume and perfusion from 3-D color flow and Doppler power signal imaging." In *1993 Ultrasonics Symposium Proceedings*, 1023–1026. Piscataway, NJ: IEEE.

Chin, J.L., D.B. Downey, T.L. Elliot, S. Tong, C.A. McLean, M. Fortier, and A. Fenster. 1999. "Three dimensional transrectal ultrasound imaging of the prostate: Clinical validation." *Can J Urol* no. 6 (2):720–726.

Chin, J.L., D.B. Downey, M. Mulligan, and A. Fenster. 1998. "Three-dimensional transrectal ultrasound guided cryoablation for localized prostate cancer in nonsurgical candidates: A feasibility study and report of early results." *J Urol* no. 159 (3):910–914.

Chin, J.L., D.B. Downey, G. Onik, and A. Fenster. 1996. "Three-dimensional prostate ultrasound and its application to cryosurgery." *Tech Urol* no. 2 (4):187–193.

Chiu, B., M. Egger, J.D. Spence, G. Parraga, and A. Fenster. 2008a. "Quantification of carotid vessel wall and plaque thickness change using 3D ultrasound images." *Med Phys* no. 35 (8):3691–3710.

Chiu, B., M. Egger, J. Spence, G. Parraga, and A. Fenster. 2008b. "Development of 3D ultrasound techniques for carotid artery disease assessment and monitoring." *Int J Comput Assist Radiol Surg* no. 3 (1):1–10.

Clarke, S.E., V. Beletsky, R.R. Hammond, R.A. Hegele, and B.K. Rutt. 2006. "Validation of automatically classified magnetic resonance images for carotid plaque compositional analysis." *Stroke* no. 37 (1):93–97.

Cool, D., S. Sherebrin, J. Izawa, J. Chin, and A. Fenster. 2008. "Design and evaluation of a 3D transrectal ultrasound prostate biopsy system." *Med Phys* no. 35 (10):4695–4707.

Dandekar, S., Y. Li, J. Molloy, and J. Hossack. 2005. "A phantom with reduced complexity for spatial 3-D ultrasound calibration." *Ultrasound Med Biol* no. 31 (8):1083–1093.

Delabays, A., N.G. Pandian, Q.L. Cao, L. Sugeng, G. Marx, A. Ludomirski, and S.L. Schwartz. 1995. "Transthoracic real-time three-dimensional echocardiography using a fan-like scanning approach for data acquisition: Methods, strengths, problems, and initial clinical experience." *Echocardiography* no. 12 (1):49–59.

Deng, J., and C.H. Rodeck. 2006. "Current applications of fetal cardiac imaging technology." *Curr Opin Obstet Gynecol* no. 18 (2):177–184.

Detmer, P.R., G. Bashein, T. Hodges, K.W. Beach, E.P. Filer, D.H. Burns, and D.E. Strandness, Jr. 1994. "3D ultrasonic image feature localization based on magnetic scanhead tracking: In vitro calibration and validation." *Ultrasound Med Biol* no. 20 (9):923–936.

Chapter 2

Devore, G.R., and B. Polanko. 2005. "Tomographic ultrasound imaging of the fetal heart: A new technique for identifying normal and abnormal cardiac anatomy." *J Ultrasound Med* no. 24 (12):1685–1696.

Downey, D.B., J.L. Chin, and A. Fenster. 1995. "Three-dimensional US-guided cryosurgery." *Radiology* no. 197(P):539.

Downey, D.B., and A. Fenster. 1995a. "Three-dimensional power Doppler detection of prostate cancer [letter]." no. 165 (3):741.

Downey, D.B., and A. Fenster. 1995b. "Vascular imaging with a three-dimensional power Doppler system." *AJR Am J Roentgenol* no. 165 (3):665–668.

Downey, D.B., A. Fenster, and J.C. Williams. 2000. "Clinical utility of three-dimensional US." *Radiographics* no. 20 (2):559–571.

Downey, D.B., D.A. Nicolle, and A. Fenster. 1995a. "Three-dimensional orbital ultrasonography." *Can J Ophthalmol* no. 30 (7):395–398.

Downey, D.B., D.A. Nicolle, and A. Fenster. 1995b. "Three-dimensional ultrasound of the eye." *Adm Radiol J* no. 14:46–50.

Egger, M., B. Chiu, J. Spence, A. Fenster, and G. Parraga. 2008. "Mapping spatial and temporal changes in carotid atherosclerosis from three-dimensional ultrasound images." *Ultrasound Med Biol* no. 34 (1):64–72.

Egger, M., J.D. Spence, A. Fenster, and G. Parraga. 2007. "Validation of 3D ultrasound vessel wall volume: An imaging phenotype of carotid atherosclerosis." *Ultrasound Med Biol* no. 33 (6):905–914.

Elliott, S.T. 2008. "Volume ultrasound: The next big thing?" *Br J Radiol* no. 81 (961):8–9.

Fenster, A., C. Blake, I. Gyacskov, A. Landry, and J.D. Spence. 2006. "3D ultrasound analysis of carotid plaque volume and surface morphology." *Ultrasonics* no. 44 (Supplement 1):e153–e157.

Fenster, A., and D. Downey. 2001. "Three-dimensional ultrasound imaging." *Proceedings SPIE* no. 4549:1–10.

Fenster, A., and D.B. Downey. 1996. "Three-dimensional ultrasound imaging: A review." *IEEE Eng Med Biol* no. 15:41–51.

Fenster, A., and D.B. Downey. 2000. "Three-dimensional ultrasound imaging." *Annu Rev Biomed Eng* no. 02:457–475.

Fenster, A., D.B. Downey, and H.N. Cardinal. 2001. "Three-dimensional ultrasound imaging." *Phys Med Biol* no. 46 (5):R67–99.

Fenster, A., A. Landry, D.B. Downey, R.A. Hegele, and J.D. Spence. 2004. "3D ultrasound imaging of the carotid arteries." *Curr Drug Targets Cardiovasc Haematol Disord* no. 4 (2):161–175.

Fenster, A., S. Tong, S. Sherebrin, D.B. Downey, and R.N. Rankin. 1995. "Three-dimensional ultrasound imaging." *SPIE Physics of Medical Imaging* no. 2432:176–184.

Fosse, E., S.H. Johnsen, E. Stensland-Bugge, O. Joakimsen, E.B. Mathiesen, E. Arnesen, and I. Njolstad. 2006. "Repeated visual and computer-assisted carotid plaque characterization in a longitudinal population-based ultrasound study: The Tromso study." *Ultrasound Med Biol* no. 32 (1):3–11.

Fruhauf, T. 1996. "Raycasting vector fields." *Visualization '96. Proceedings*:115–120.

Gee, A.H., N.E. Houghton, G.M. Treece, and R.W. Prager. 2005. "A mechanical instrument for 3D ultrasound probe calibration." *Ultrasound Med Biol* no. 31 (4):505–518.

Gilja, O.H., P.R. Detmer, J.M. Jong, D.F. Leotta, X.N. Li, K.W. Beach, R. Martin, and D.E. Strandness, Jr. 1997. "Intragastric distribution and gastric emptying assessed by three-dimensional ultrasonography." *Gastroenterology* no. 113 (1):38–49.

Gilja, O.H., N. Thune, K. Matre, T. Hausken, S. Odegaard, and A. Berstad. 1994. "In vitro evaluation of three-dimensional ultrasonography in volume estimation of abdominal organs." *Ultrasound Med Biol* no. 20 (2):157–165.

Gill, J.D., H.M. Ladak, D.A. Steinman, and A. Fenster. 2000. "Accuracy and variability assessment of a semiautomatic technique for segmentation of the carotid arteries from three-dimensional ultrasound images." *Med Phys* no. 27 (6):1333–1342.

Goncalves, L., J.K. Nien, J. Espinoza, J.P. Kusanovic, J. Lee, B. Swope, and E. Soto. 2005. "Two-dimensional (2D) versus three- and four-dimensional (3D/4D) us in obstetrical practice: Does the new technology add anything?" *Am J Obstet Gynecol* no. 193 (6):S150–S150.

Gooding, M.J., S.H. Kennedy, and J.A. Noble. 2005. "Temporal calibration of freehand three-dimensional ultrasound using image alignment." *Ultrasound Med Biol* no. 31 (7):919–927.

Greenleaf, J.F., M. Belohlavek, T.C. Gerber, D.A. Foley, and J.B. Seward. 1993. "Multidimensional visualization in echocardiography: An introduction [see comments]." *Mayo Clin Proc* no. 68 (3):213–220.

Hodges, T.C., P.R. Detmer, D.H. Burns, K.W. Beach, and D.E. Strandness, Jr. 1994. "Ultrasonic three-dimensional reconstruction: In vitro and in vivo volume and area measurement." *Ultrasound Med Biol* no. 20 (8):719–729.

Hsu, P.W., R.W. Prager, A.H. Gee, and G.M. Treece. 2008. "Real-time freehand 3D ultrasound calibration." *Ultrasound Med Biol* no. 34 (2):239–251.

Hughes, S.W., T.J. D'Arcy, D.J. Maxwell, W. Chiu, A. Milner, J.E. Saunders, and R.J. Sheppard. 1996. "Volume estimation from multiplanar 2D ultrasound images using a remote electromagnetic position and orientation sensor." *Ultrasound Med Biol* no. 22 (5):561–572.

Hummel, J., M. Figl, M. Bax, H. Bergmann, and W. Birkfellner. 2008. "2D/3D registration of endoscopic ultrasound to CT volume data." *Phys Med Biol* no. 53 (16):4303–4316.

Inaba, Y., J.A. Chen, and S.R. Bergmann. 2011. "Carotid plaque, compared with carotid intima-media thickness, more accurately predicts coronary artery disease events: A meta-analysis." *Atherosclerosis*.

Johnsen, S.H., and E.B. Mathiesen. 2009. "Carotid plaque compared with intima-media thickness as a predictor of coronary and cerebrovascular disease." *Curr Cardio Rep* no. 11 (1):21–27.

Kerwin, W.S., K.D. O'Brien, M.S. Ferguson, N. Polissar, T.S. Hatsukami, and C. Yuan. 2006. "Inflammation in carotid atherosclerotic plaque: A dynamic contrast-enhanced MR imaging study." *Radiology* no. 241 (2):459.

King, D.L., A.S. Gopal, P.M. Sapin, K.M. Schroder, and A.N. Demaria. 1993. "Three-dimensional echocardiography." *Am J Card Imaging* no. 7 (3):209–220.

King, D.L., D.L. King, Jr, and M.Y. Shao. 1991. "Evaluation of in vitro measurement accuracy of a three-dimensional ultrasound scanner." *J Ultrasound Med* no. 10 (2):77–82.

Krasinski, A., B. Chiu, J.D. Spence, A. Fenster, and G. Parraga. 2009. "Three-dimensional ultrasound quantification of intensive statin treatment of carotid atherosclerosis." *Ultrasound Med Biol* no. 35 (11):1763–1772.

Kurjak, A., B. Miskovic, W. Andonotopo, M. Stanojevic, G. Azumendi, and H. Vrcic. 2007. "How useful is 3D and 4D ultrasound in perinatal medicine?" *J Perinat Med* no. 35 (1):10–27.

Landry, A., and A. Fenster. 2002. "Theoretical and experimental quantification of carotid plaque volume measurements made by 3D ultrasound using test phantoms." *Med Phys*.

Landry, A., J.D. Spence, and A. Fenster. 2004. "Measurement of carotid plaque volume by 3-dimensional ultrasound." *Stroke* no. 35 (4):864–869.

Landry, A., J.D. Spence, and A. Fenster. 2005. "Quantification of carotid plaque volume measurements using 3D ultrasound imaging." *Ultrasound Med Biol* no. 31 (6):751–762.

Landry, A., C. Ainsworth, C. Blake, J.D. Spence, and A. Fenster. 2007. "Manual planimetric measurement of carotid plaque volume using three-dimensional ultrasound imaging." *Med Phys* no. 34 (4):1496–1505.

Lee, W. 2003. "3D fetal ultrasonography." *Clin Obstet Gynecol* no. 46 (4):850–867.

Leotta, D.F. 2004. "An efficient calibration method for freehand 3-D ultrasound imaging systems." *Ultrasound Med Biol* no. 30 (7):999–1008.

Leotta, D.F., P.R. Detmer, and R.W. Martin. 1997. "Performance of a miniature magnetic position sensor for three-dimensional ultrasound imaging." *Ultrasound Med Biol* no. 23 (4):597–609.

Levoy, M. 1990. "Volume rendering, a hybrid ray tracer for rendering polygon and volume data." *IEEE Comput Graph Appl* no. 10:33–40.

Lindseth, F., G.A. Tangen, T. Lango, and J. Bang. 2003. "Probe calibration for freehand 3-D ultrasound." *Ultrasound Med Biol* no. 29 (11):1607–1623.

Lorenz, M.W., S. von Kegler, H. Steinmetz, H.S. Markus, and M. Sitzer. 2006. "Carotid intima-media thickening indicates a higher vascular risk across a wide age range: Prospective data from the Carotid Atherosclerosis Progression Study (CAPS)." *Stroke* no. 37 (1):87–92.

Lusis, A.J. 2000. "Atherosclerosis." *Nature* no. 407 (6801):233–241.

Mercier, L., T. Lango, F. Lindseth, and D.L. Collins. 2005. "A review of calibration techniques for freehand 3-D ultrasound systems." *Ultrasound Med Biol* no. 31 (4):449–471.

Mroz, L., R. Wegenkittl, and E. Groller. 2000. "Mastering interactive surface rendering for Java-based diagnostic applications." *Visualization 2000, Proceedings*:437–440.

Nelson, T.R., and D.H. Pretorius. 1992. "Three-dimensional ultrasound of fetal surface features." *Ultrasound Obstet Gynecol* no. 2:166–174.

Nelson, T.R., and D.H. Pretorius. 1995. "Visualization of the fetal thoracic skeleton with three-dimensional sonography: A preliminary report." *AJR Am J Roentgenol* no. 164 (6):1485–1488.

Nelson, T.R., D.H. Pretorius, M. Sklansky, and S. Hagen-Ansert. 1996. "Three-dimensional echocardiographic evaluation of fetal heart anatomy and function: Acquisition, analysis, and display." *J Ultrasound Med* no. 15 (1):1–9 quiz 11-2.

O'Leary, D.H., and M.L. Bots. 2010. "Imaging of atherosclerosis: Carotid intima–media thickness." *Eur Heart J* no. 31 (14):1682–1689.

Onik, G.M., D.B. Downey, and A. Fenster. 1996. "Three-dimensional sonographically monitored cryosurgery in a prostate phantom." *J Ultrasound Med* no. 15 (3):267–270.

Pagoulatos, N., D.R. Haynor, and Y. Kim. 2001. "A fast calibration method for 3-D tracking of ultrasound images using a spatial localizer." *Ultrasound Med Biol* no. 27 (9):1219–1229.

Parraga, G., A. Fenster, A. Krasinski, B. Chiu, M. Egger, and J.D. Spence. 2011. "3D carotid ultrasound imaging." In *Atheroclerosis Disease Management*, edited by J.S. Suri, 325–350. Springer.

Peralta, C.F., P. Cavoretto, B. Csapo, O. Falcon, and K.H. Nicolaides. 2006. "Lung and heart volumes by three-dimensional ultrasound in normal fetuses at 12–32 weeks gestation." *Ultrasound Obstet Gynecol* no. 27 (2):128–133.

Picot, P.A., D.W. Rickey, R. Mitchell, R.N. Rankin, and A. Fenster. 1993. "Three-dimensional colour Doppler imaging." *Ultrasound Med Biol* no. 19 (2):95–104.

Poon, T.C., and R.N. Rohling. 2005. "Comparison of calibration methods for spatial tracking of a 3-D ultrasound probe." *Ultrasound Med Biol* no. 31 (8):1095–1108.

Pretorius, D.H., and T.R. Nelson. 1994. "Prenatal visualization of cranial sutures and fontanelles with three-dimensional ultrasonography." *J Ultrasound Med* no. 13 (11):871–876.

Pretorius, D.H., and T.R. Nelson. 1995. "Fetal face visualization using three-dimensional ultrasonography." *J Ultrasound Med* no. 14 (5):349–356.

Pretorius, D.H., T.R. Nelson, and J.S. Jaffe. 1992. "Three-dimensional sonographic analysis based on color flow Doppler and gray scale image data: A preliminary report." *J Ultrasound Med* no. 11 (5):225–232.

Raab, F.H., E.B. Blood, T.O. Steiner, and H.R. Jones. 1979. "Magnetic position and orientation tracking system." *IEEE Trans Aerosp Electron Syst* no. AES-15:709–717.

Rankin, R.N., A. Fenster, D.B. Downey, P.L. Munk, M.F. Levin, and A.D. Vellet. 1993. "Three-dimensional sonographic reconstruction: Techniques and diagnostic applications." *AJR Am J Roentgenol* no. 161 (4):695–702.

Riccabona, M., T.R. Nelson, D.H. Pretorius, and T.E. Davidson. 1995. "Distance and volume measurement using three-dimensional ultrasonography." *J Ultrasound Med* no. 14 (12):881–886.

Rousseau, F., P. Hellier, and C. Barillot. 2005. "Confhusius: A robust and fully automatic calibration method for 3D freehand ultrasound." *Med Image Anal* no. 9 (1):25–38.

Saam, T., M.S. Ferguson, V.L. Yarnykh, N. Takaya, D. Xu, N.L. Polissar, T.S. Hatsukami, and C. Yuan. 2005. "Quantitative evaluation of carotid plaque composition by in vivo MRI." *Arterioscler Thromb Vasc Biol* no. 25 (1):234–239.

Shai, I., J.D. Spence, D. Schwarzfuchs, Y. Henkin, G. Parraga, A. Rudich, A. Fenster, C. Mallett, N. Liel-Cohen, A. Tirosh, and others. 2010. "Dietary intervention to reverse carotid atherosclerosis." *Circulation* no. 121 (10):1200.

Sklansky, M. 2003. "New dimensions and directions in fetal cardiology." *Curr Opin Pediatr* no. 15 (5):463–471.

Sklansky, M. 2004. "Specialty review issue: Fetal cardiology—Introduction." *Pediatric Cardiology* no. 25 (3):189–190.

Smith, W.L., and A. Fenster. 2000. "Optimum scan spacing for three-dimensional ultrasound by Speckle Statistics." *Ultrasound Med Biol* no. 26 (4):551–562.

Smith, W.L., K. Surry, G. Mills, D. Downey, and A. Fenster. 2001. "Three-dimensional ultrasound-guided core needle breast biopsy." *Ultrasound Med Biol* no. 27 (8):1025–1034.

Tong, S., H.N. Cardinal, R.F. McLoughlin, D.B. Downey, and A. Fenster. 1998. "Intra- and inter-observer variability and reliability of prostate volume measurement via two-dimensional and three-dimensional ultrasound imaging." *Ultrasound Med Biol* no. 24 (5):673–681.

Tong, S., D.B. Downey, H.N. Cardinal, and A. Fenster. 1996. "A three-dimensional ultrasound prostate imaging system." *Ultrasound Med Biol* no. 22 (6):735–746.

Treece, G., R. Prager, A. Gee, and L. Berman. 2001. "3D ultrasound measurement of large organ volume." *Med Image Anal* no. 5 (1):41–54.

Ukwatta, E., J. Awad, A.D. Ward, D. Buchanan, G. Parraga, and A. Fenster. 2011a. "Coupled level set approach to segment carotid arteries from 3D ultrasound images." In *Biomedical Imaging: From Nano to Macro, 2011 IEEE International Symposium on*.

Ukwatta, E., J. Awad, A.D. Ward, D. Buchanan, J. Samarabandu, G. Parraga, and A. Fenster. 2011b. "Three-dimensional ultrasound of carotid atherosclerosis: Semiautomated segmentation using a level set-based method." *Med Phys* no. 38:2479.

Wei, Z., L. Gardi, D.B. Downey, and A. Fenster. 2005. "Oblique needle segmentation and tracking for 3D TRUS guided prostate brachytherapy." *Med Phys* no. 32 (9):2928–2941.

Wei, Z., G. Wan, L. Gardi, G. Mills, D. Downey, and A. Fenster. 2004. "Robot-assisted 3D-TRUS guided prostate brachytherapy: System integration and validation." *Med Phys* no. 31 (3):539–548.

Yuan, C., M. Oikawa, Z. Miller, and T. Hatsukami. 2008. "MRI of carotid atherosclerosis." *J Nucl Cardiol* no. 15 (2):266–275.

Zahalka, A., and A. Fenster. 2001. "An automated segmentation method for three-dimensional carotid ultrasound images." *Phys Med Biol* no. 46 (4):1321–1342.

Chapter 2

3. Ultrasound Velocity Imaging

Jørgen Arendt Jensen

3.1 Introduction: Blood Velocity Estimation Systems

This chapter gives an introduction to velocity estimation in medical ultrasound, which is primarily used in cardiovascular imaging, but it can also be applied in, for example, strain imaging and tissue motion estimation. This chapter gives a brief introduction to the human circulation to establish the requirements on blood velocity estimation systems. A simple model for the interaction between ultrasound and point scatterers is derived in Section 3.4 based on the information about scattering from blood in Section 3.3. The model is used in the derivation of velocity estimators based on the signal's frequency content (Section 3.6), phase shift (Section 3.7),

Ultrasound Imaging and Therapy. Edited by Aaron Fenster and James C. Lacefield © 2015 CRC Press/Taylor & Francis Group, LLC. ISBN: 978-1-4398-6628-3

Chapter 3

and time shift (Section 3.8). These sections should give a brief introduction to the function of the most prominent methods used in current commercial ultrasound scanners. Newer and more experimental techniques are described in the sections on vector velocity imaging, synthetic aperture (SA) imaging, and other applications in Sections 3.9 to 3.12.

This chapter is necessarily brief in covering the many topics in this field, and the reader is referred to the more in-depth treatment of the topics by Jensen [1], Evans et al. [2], Evans and McDicken [3], Szabo [4], and Cobbold [5], along with the references given in the text.

3.2   Human Circulation

The major arteries and veins in the human circulation are shown in Figure 3.1, and the dimensions are indicated in Table 3.1 [1,6,7]. Their diameters span from centimeters to

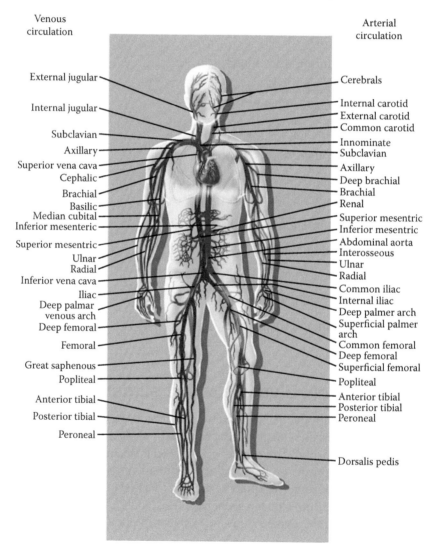

FIGURE 3.1   Major arteries and veins in the body. (Reprinted with permission from Abbott Laboratories, Illinois, USA.)

Table 3.1 Typical Dimensions and Flows of Vessels in the Human Vascular System

Vessel	Internal Diameter (cm)	Wall Thickness (cm)	Length (cm)	Young's Modulus (N/m² × 10⁵)
Ascending aorta	1.0–2.4	0.05–0.08	5	3–6
Descending aorta	0.8–1.8	0.05–0.08	20	3–6
Abdominal aorta	0.5–1.2	0.04–0.06	15	9–11
Femoral artery	0.2–0.8	0.02–0.06	10	9–12
Carotid artery	0.2–0.8	0.02–0.04	10–20	7–11
Arteriole	0.001–0.008	0.002	0.1–0.2	
Capillary	0.0004–0.0008	0.0001	0.02–0.1	
Inferior vena cava	0.6–1.5	0.01–0.02	20–40	0.4–1.0

Vessel	Peak Velocity (cm/s)	Mean Velocity (cm/s)	Reynolds Number (peak)	Pulse Propagation Velocity (cm/s)
Ascending aorta	20–290	10–40	4500	400–600
Descending aorta	25–250	10–40	3400	400–600
Abdominal aorta	50–60	8–20	1250	700–600
Femoral artery	100–120	10–15	1000	800–1030
Common carotid artery (range)	68–171	19–59		600–1100
Common carotid artery (mean)	108	39		600–1100
Arteriole	0.5–1.0		0.09	
Capillary	0.02–0.17		0.001	
Inferior vena cava	15–40		700	100–700

Source: Data from C. G. Caro et al., Mechanics of the circulation. In *MTP International Review of Science Physiology. Series 1, Volume 1: Cardiovascular Physiology*, 15–48. Butterworth, London, 1974; and D. W. Holdsworth et al., *Physiological Measurement*, 20:219–240, 1999. Table adapted from J. A. Jensen, *Estimation of Blood Velocities Using Ultrasound: A Signal Processing Approach*. Cambridge University Press, New York, 1996.

microns and the velocities from meters per second to less than millimeters per second in the capillaries. As illustrated, the vessel constantly curves and branches and repeatedly changes dimensions [8]. The blood flow is pulsatile, and the velocity rapidly changes both magnitude and direction, as can be seen from the large difference between peak and mean velocities. In addition, the peak Reynolds number is often higher than 2000, which indicates disturbed or turbulent flow in parts of the cardiac cycle.

All of these factors should be considered when devising a system for measuring blood velocity in the human circulation. Foremost, it must be a fast measurement system with 1 to 20 ms temporal resolution to follow the changes in velocity due to the acceleration from pulsation. It should also have a submillimeter spatial resolution to see changes in flow over space and to visualize small vessels. A real-time system is also beneficial because the flow patterns change rapidly, and it must be possible to quickly find the sites of flow. The system should also ideally be capable of finding and visualizing flow in

Chapter 3

all directions because turbulent flow and vortices exist throughout the human circulation. Modern medical ultrasound systems can fulfill many of these demands, and the remaining part of this chapter describes the physics and signal processing needed to perform velocity estimation along the ultrasound beam and in other directions.

3.3   Ultrasound Scattering from Blood

The constituents of blood are shown in Table 3.2 [1,9–12]. The most prominent part is the erythrocytes (red blood cells), where a square millimeter of blood contains roughly 5 million cells. The resolution of ultrasound systems is at best on the order of the cube of the wavelength λ^3, which is given by $\lambda = c/f_0$, where c is the speed of sound and f_0 is the center frequency. For a 6 MHz system, $\lambda = 0.26$ mm, which is much larger than a single cell. An ultrasound system, thus, only observes a large, random collection of cells.

The scattering from a single point scatterer or small cell can be described by the differential scattering cross section σ_d. It is defined as the power scattered per unit solid angle at some angle Θ_s divided by the incident intensity [13,14]:

$$\sigma_d(\Theta_s) = \frac{V_e^2 \pi^2}{\lambda^4} \left[\frac{\kappa_e - \kappa_0}{\kappa_0} + \frac{\rho_e - \rho_0}{\rho_e} \cos\Theta_s \right]^2, \tag{3.1}$$

where V_e is the volume of the scatterer, and ρ_e is a small perturbation in density and κ_e in compressibility from their mean values ρ_0 and κ_0, respectively. This results in a scattered field, as shown in Figure 3.2, where the scattering is dependent on the angle. However, it can be seen that a signal is received in all directions.

The signal received by the ultrasound transducer is therefore independent of the orientation of the vessel, and it can be modeled as a random, Gaussian signal as it emanates from a large collection of independent random scatterers. This gives rise to the speckle pattern seen in ultrasound images. The scattering is very weak, as the density

Table 3.2   Properties of the Main Components of Blood

	Mass Density (g/cm³)	Adiabatic Compressibility (10⁻¹² cm/dyne)	Size (μm)	Particles (per mm³)
Erythrocytes	1.092	34.1	2 × 7	5 × 10⁶
Leukocytes	–	–	9–25	8 × 10³
Platelets	–	–	2–4	250–500 × 10³
Plasma	1.021	40.9	–	–
0.9% saline	1.005	44.3	–	–

Source: Data from E. L. Carstensen et al., *Journal of the Acoustical Society of America*, 25:286–289, 1953; F. Dunn et al., Absorption and dispersion of ultrasound in biological media. In *Biological Engineering*. McGraw-Hill, New York, 1969; R. J. Ulrick, *Journal of Applied Physics*, 18:983–987, 1947; and W. R. Platt, *Color Atlas and Textbook of Hematology*, J. B. Lippincott, Toronto, 1969. Table adapted from J. A. Jensen, *Estimation of Blood Velocities Using Ultrasound: A Signal Processing Approach*. Cambridge University Press, New York, 1996.

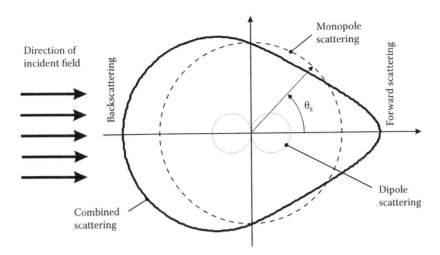

FIGURE 3.2 Monopole and dipole scattering from a perturbation in either compressibility or density. (From J. A. Jensen, *Estimation of Blood Velocities Using Ultrasound: A Signal Processing Approach,* Cambridge University Press, New York, 1996.)

and compressibility perturbations are small compared with the surrounding tissue, and often the signal from blood is 20 to 40 dB lower than from the surrounding tissue. Therefore, vessels in an ultrasound image appear black.

It should be noted that the signal received is random, but the same signal is received if the experiment is repeated for the same collection of scatterers. This is a very important feature of ultrasound blood velocity signals and is heavily used in these systems, as will be clear in the following sections. A small motion of the scatterers will therefore yield a second signal highly correlated with the first signal, when the motion $|v|T_{prf}$ is small compared with the beam width, and the velocity gradient is small across the scatterer collection observed. Here $|v|$ is the velocity magnitude and T_{prf} is the time between the measurements.

3.4 Ultrasound Signals from Flowing Blood

The derivation of velocity estimation methods is based on a model of ultrasound interaction between the moving blood and the ultrasound pulse. The model has to capture all the main features of the real received signal without unnecessary features and complications. This section will present a simple model for this interaction, which can readily be used for deriving new estimators and also give basic explanations for how to optimize the techniques. It can also be further expanded to the 2-D and 3-D velocity estimation methods presented in Sections 3.9 and 3.10.

The scattering from blood emanates from a large collection of independent, small point scatterers as described in Section 3.3. The 1-D received signal $y(t)$ can therefore be modeled as follows:

$$y(t) = p(t)^*s(t), \tag{3.2}$$

which is the convolution of the basic ultrasound pulse $p(t)$ with the random Gaussian signal scattering $s(t)$. The ultrasound pulse for flow estimation consists of several sinusoidal

Chapter 3

oscillations (M_p = 4–8) at the transducer's center frequency, f_0, convolved with the electromechanical impulse response of the transducer (from excitation voltage to pressure and from pressure to received voltage) [15]. Modern transducers are so broadband that a simple approximation is:

$$p(t) = g(t)\sin(2\pi f_0 t),\tag{3.3}$$

where the value of envelope $g(t)$ is 1 from $t = 0$ to M_p/f_0 and 0 elsewhere.

For a single point scatterer, the scattering can be modeled as a δ-function, and the received signal is:

$$y_1(t) = p(t) * a\delta\left(t - \frac{2d}{c}\right) = ap\left(t - \frac{2d}{c}\right),\tag{3.4}$$

where d is the distance to the point, c is the speed of sound (1540 m/s in tissue), and a is the scattering amplitude. Thus, there is a propagation delay $2d/c$ between transmitting the signal and receiving the response. For a moving scatterer, this response will change as the scatterer moves farther away from the transducer. For a scatterer velocity of $v_z = |v|\cos\theta$ in the axial direction, and a time between measurements of T_{prf}, the second received signal is:

$$y_2(t) = p(t) * a\delta\left(t - \frac{2d}{c} - \frac{2v_z T_{prf}}{c}\right) = ap\left(t - \frac{2d}{c} - \frac{2v_z T_{prf}}{c}\right).\tag{3.5}$$

Note that $v_z T_{prf} = T_{prf}|v|\cos\theta$ is the distance the scatterer moved along the ultrasound beam direction, where θ is the beam-to-flow angle. There is therefore an added delay of

$$t_s = \frac{2v_z}{c} T_{prf}\tag{3.6}$$

before the second signal is received. This is directly proportional to the velocity of the moving scatterer, and the axial velocity can directly be calculated by determining this delay. This can be achieved either through the phase shift estimator described in Section 3.7 or the time shift estimator in Section 3.8. Both estimators essentially just compare or correlate two received signals and find the shift in position between the received responses.

Combining Equations 3.3 and 3.5 gives a model for the received signal for several pulse emissions, as follows [16,17]:

$$y(t,i) = ap\left(t - \frac{2d}{c} - i\frac{2v_z}{c}T_{prf}\right) = ag\left(t - \frac{2d}{c} - it_s\right)\sin\left(2\pi f_0\left(t - \frac{2d}{c} - it_s\right)\right),\tag{3.7}$$

where i is the pulse emission number. For pulsed systems, the received signal is measured at a single depth corresponding to a fixed time t_x relative to the pulse emission.

This is performed by taking out one sample from each pulse emission to create the sampled signal. To simplify things, it can be assumed that the pulse is long enough and the motion slow enough, so that the pulse amplitude stays roughly constant during the observation time. The sampled signal can then be simply written as

$$y(t_x, i) = a \sin\left(2\pi f_0 \left(t_x - \frac{2d}{c} - it_s \right) \right)$$

$$= -a \sin\left(2\pi \frac{2v_z}{c} f_0 i T_{prf} - \phi_x \right), \tag{3.8}$$

where $\phi_x = 2\pi f_0 \left(\frac{2d}{c} - t_x \right)$ is a fixed phase shift depending on the measurement depth. The sampling interval is T_{prf}, and the frequency of the sampled signal is:

$$f_p = \frac{2v_z}{c} f_0, \tag{3.9}$$

which is directly proportional to the velocity. Thus, estimating the frequency of this sampled signal $y(t_x, i)$ can directly reveal the axial velocity as described in Section 3.6 on spectral velocity estimation. Often, it is advantageous to also sample the signal in the depth direction, and this gives

$$y_s(n, i) = ag\left(n\Delta T - \frac{2d}{c} - it_s \right) \sin\left(2\pi f_0 \left(n\Delta T - \frac{2d}{c} - it_s \right) \right), \tag{3.10}$$

which is often used for averaging along the depth direction. Here $\Delta T = 1/f_s$ is the sampling interval, and f_s is the sampling frequency.

An illustration of the sampling process for a single moving scatterer is shown in Figure 3.3, where the individually received signals are shown (Figure 3.3a). The scatterer is moving away from the transducer, and the time between emission and reception of the signal increases. A single sample is taken out for each received signal at the position of the solid gray line. The resulting sampled signal is shown in Figure 3.3b, where the basic emitted pulse can be recognized. A low velocity will yield a long pulse and hence a low frequency of the received signal, as indicated by Equation 3.9. A high velocity will compress the signal and thereby give a high frequency. The velocity can be found by three different methods. A Fourier transform can be applied on the signal in Figure 3.3b to find the frequency as described in Section 3.6. The phase shift can be determined between the received signals. This is described in Section 3.7. The time shift can be found by correlating consecutive received signals, that is, cross correlating received signals in Figure 3.3a as described in Section 3.8.

The signal model can also be expanded to include a collection of scatterers moving at the same velocity,

$$y(t, i) = p(t)^*s(t - it_s - iT_{prf}) = p(t - it_s - iT_{prf})^*s(t) = p(t - it_s)^*s(t), \tag{3.11}$$

Chapter 3

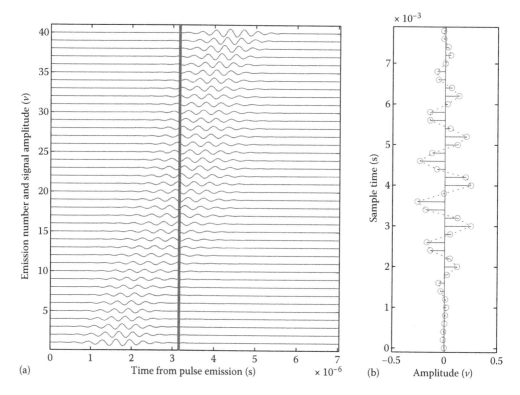

FIGURE 3.3 Received signals from a single scatterer moving away from the transducer. The solid gray line in (a) indicates the sampling instance for the samples shown in (b).

as the same pulse is emitted every time. Using superposition, this can just be seen as a summation of individual scatterers moving at their velocity, and the model and observations made previously can readily be applied.

3.4.1 Is It the Doppler Effect?

The ultrasound systems for measuring velocity are often called Doppler systems, implying that they rely on the Doppler shift to find the velocity. It is debatable whether this is really true. The Doppler effect can be described as a frequency shift on the received signal. The received signal can be written as

$$y_D(t) = a \sin\left(2\pi\left(1 + \frac{2v_z}{c}\right)f_0 t\right),$$ (3.12)

for a simple continuous wave source. Here the received frequency is scaled by $(1 + 2v_z/c)$. Thus, there is a frequency difference of $f_D = \frac{2v_z}{c} f_0$ between the emitted and the received signal. This is easy to determine for a continuous wave (CW) source and is used in CW systems, but it is extremely difficult for a pulsed source. First, the shift in frequency is very small, on the order of 1/1000 of the emitted frequency. The shift is therefore minute compared with the bandwidth of the signal, which is roughly $B = f_0/M_p$, M_p being the number of cycles emitted. Second, finding this frequency shift assumes that the signal is not affected by other physical effects that shift the mean frequency of the received signal. This is not true in medical ultrasound as the Rayleigh scattering from blood is frequency dependent and the dispersive attenuation in tissue is quite strong. Even for moderate tissue depths, it can span from ten to hundreds of kilohertz of shift in the mean frequency of the received signal, thus obscuring the Doppler shift [1]. This shift will be different and unknown from patient to patient and would therefore greatly influence the accuracy of the Doppler shift estimation. The Doppler effect is therefore not a reliable estimator of velocity, and no pulsed system relies on this effect. Consequently, it was not included in the model given in Equation 3.10. Using the time or phase shift between consecutively received signal is a more reliable method, which is not affected much by attenuation, scattering, beam modulation, or nonlinear propagation that all affect the received ultrasound signal [1]. The reason for this is that two signals are compared (correlated), and only the difference between two emissions is used in the velocity estimation. This results in a much more reliable system, which does not require an extremely precise frequency content to work. Therefore, I prefer not to call these systems Doppler systems, because it is confusing to what effect is actually used for the velocity estimation.

3.5 Simulation of Flow Signals

The model presented previously can also be used for simulating the signal from flowing blood. The Field II simulation system [18,19] uses a model based on spatial impulse responses. The received voltage signal is described by [20]:

$$v_r(t,\vec{r}_1) = v_{pe}(t) \underset{t}{\star} f_m(\vec{r}_2) \underset{r}{\star} h_{pe}(\vec{r}_1,\vec{r}_2,t), \tag{3.13}$$

where f_m accounts for the scattering by the medium, h_{pe} describes the spatial distribution of the ultrasound field, and v_{pe} is the 1-D pulse emanating from the pulse excitation and conversion from voltage to pressure and back again. $\vec{r}_1$ is the position of the transducer, and $\vec{r}_2$ is the position of the scatterers. $\underset{t}{\star}$ and $\underset{r}{\star}$ denote temporal and spatial convolution, respectively. This equation describes a summation of the responses from all the point scatterers properly weighted by the ultrasound field strength as described by h_{pe} and convolved with v_{pe}. This yields the signal for a collection of scatterers when the Doppler effect is neglected. For the next emission, the scatterer's positions should be propagated as follows:

$$\vec{r}_2(i+1) = \vec{r}_2(i) + T_{prf}\vec{v}(\vec{r}_2(i), iT_{prf}), \tag{3.14}$$

where $\vec{v}(\vec{r}_2(i),t)$ denotes the velocity vector for this point scatterer for emission i at time $t = iT_{prf}$. Propagating the scatterers and calculating the received signal will then yield a realistic flow signal, which is usable for developing, validating, and evaluating pulsed ultrasound flow systems.

Chapter 3

A typical parabolic velocity profile for stationary, laminar flow is:

$$v(r) = \left(1 - \frac{r^2}{R^2}\right)v_0,$$

(3.15)

where R is the vessel radius and v_0 is the peak velocity in the vessel. More realistic velocity profiles can be generated by using Womersley–Evans' description of pulsatile flow [21,22]. Here a few parameters can be used to describe the full temporal and spatial evolution of the pulsatile flow in, for example, the carotid or femoral arteries, and this can readily be included in the simulation. It is also possible to combine this method with computational fluid dynamics using finite element modeling for the flow [23] to capture the formation of turbulence and vortices.

3.6 Estimation of the Velocity Distribution

The frequency of the flow signals measured is directly proportional to the blood velocity, as shown in Equation 3.9. Finding the frequency content of the signal therefore reveals the velocity distribution in the vessel under investigation. This is used in spectral estimation systems, which often combine the measurement of velocity with an anatomic B-mode image, as shown in Figure 3.4. The top image shows the anatomy, and the measurement range gate for the flow is indicated by the broken line. The flow measurement is conducted within the vessel, and the "wings" indicate the assumed flow direction. The bottom display shows the velocity distribution as a function of time for five heart beats in the carotid artery.

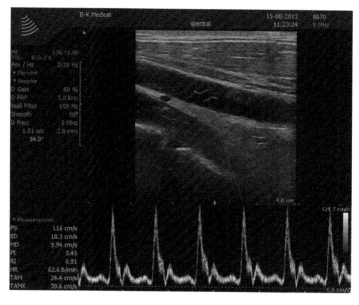

FIGURE 3.4 Duplex mode ultrasound imaging showing the anatomic B-mode image on the top and the spectral velocity distribution on the bottom. (Image courtesy of MD Peter M. Hansen.)

The velocity distribution changes over the cardiac cycle because of the pulsation of the flow, and thus the frequency content of the received signal is not constant. The direction of the flow can also be seen here. For the carotid artery, the spectrum is one sided (only positive frequency components) because the flow is unidirectional toward the brain. Thus, it is important to have signal processing that can differentiate between velocities toward or away from the transducer. This can be achieved by using complex signals with a one-sided spectrum. Making a Hilbert transform on the received radio frequency (RF) signal and forming the analytic signal [24] then gives

$$r_s(n_x, i) = a g_s(i) \exp\left(j\left(2\pi \frac{2v_z}{c} f_0 i T_{\text{prf}} - \phi_x \right) \right). \tag{3.16}$$

Here the emitted frequency f_0 is scaled by $\frac{2v_z}{c}$, which can be positive or negative depending on the sign of the velocity. Making a Fourier transform of $r_s(n_x, i)$ along the i direction will yield a one-sided spectrum, so that the velocity direction can be determined.

The signal from a vessel consists of a weighted superposition from all the primarily red blood cell scatterers in the vessel. They each flow at slightly different velocities, and calculating the power density spectrum of the signal will give the corresponding velocity distribution of the cells [1,25]. Displaying the spectrum therefore visualizes the velocity distribution. This has to be shown as a function of time to reveal the dynamic changes in the spectrogram.

Modern ultrasound scanners use a short time Fourier transform. The complex signal is divided into segments of typically 128 to 256 samples weighted by, for example, a von Hann window before Fourier transformation. The process is repeated every 1 to 5 ms as the spectra are displayed side by side as a function of time as shown in Figure 3.4. Compensating for the beam-to-flow angle then gives a quantitative and real-time display of the velocity distribution at one given position in the vessel.

The range gate can be selected to be small or large, depending on whether the peak velocity or the mean velocity is investigated. The averaging over the range gate can be made either by selecting the length of the emitted pulse or by averaging the spectra across the depth direction by calculating one spectrum for each depth sample n. The spectrogram acquisition method is used clinically when quantitative parameters such as peak velocity, mean velocity, or resistive index for the flow must be calculated.

3.7 Axial Velocity Estimation Using the Phase

Spectral systems only display the velocity distribution at one single position in the vessel. Often, it is preferred to visualize the velocity in a region using the so-called color flow mapping (CFM) method. Here data are acquired in several directions to construct a real time image of the velocity [26]. The method gives a single value for the velocity at each spatial position, but only a very limited amount of data are available to maintain a reasonable frame rate. Often, 8 to 16 emissions are made in the same direction, and the velocity is found in this direction as a function of depth. The acquisition is then repeated

Chapter 3

in other directions, and an image of velocity is made and superimposed on the normal B-mode image as shown in Figure 3.5.

These systems find the velocity from the phase shift between the acquired lines. The complex received signal is in continuous time written as

$$r_t(t) = ag_s(t)\exp\left(-j\left(2\pi\frac{2v_z}{c}f_0t - \phi_x\right)\right). \tag{3.17}$$

Taking the derivate of its phase gives

$$\phi' = \frac{d\phi}{dt} = \frac{d(-2\pi\frac{2v_z}{c}f_0t + \phi_x))}{dt} = -2\pi\frac{2v_z}{c}f_0, \tag{3.18}$$

so that the estimated velocity is

$$\hat{v}_z = -\frac{\phi'}{4\pi f_0}c. \tag{3.19}$$

The discrete version can be written as

$$r_s(n_x,i) = ag_s(i)\exp\left(-j\left(2\pi\frac{2v_z}{c}f_0iT_{prf} - \phi_x\right)\right) = x(i) + jy(i). \tag{3.20}$$

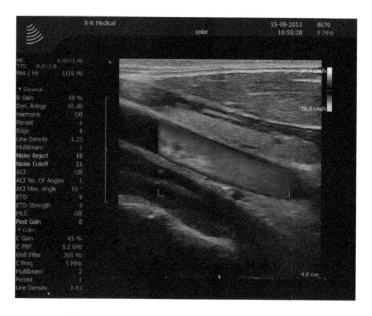

FIGURE 3.5   CFM image of the carotid artery. The red colors indicate velocity toward the transducer and blue away from the transducer. (Image courtesy of MD Peter M. Hansen.)

The phase difference can be found from

$$\Delta\phi = \phi(i+1) - \phi(i) = \arctan\frac{y(i+1)}{x(i+1)} - \arctan\frac{y(i)}{x(i)}$$

$$= -\left(2\pi\frac{2v_z}{c}f_0(i+1)T_{\text{prf}} - \phi_x\right) + \left(2\pi\frac{2v_z}{c}f_0 i T_{\text{prf}} - \phi_x\right) \tag{3.21}$$

$$= -2\pi\frac{2v_z}{c}f_0 T_{\text{prf}},$$

and the velocity can be estimated as

$$\hat{v}_z = -\frac{\Delta\phi}{4\pi T_{\text{prf}} f_0}c = \frac{\Delta\phi}{2\pi}\frac{f_{\text{prf}}}{2f_0}c = -\frac{\Delta\phi}{2\pi}\frac{f_{\text{prf}}}{2}\lambda = -\frac{\Delta\phi}{4\pi}\frac{\lambda}{T_{\text{prf}}}. \tag{3.22}$$

The maximum unique phase difference that can be estimated is $\Delta\phi = \pm\pi$, and the largest unique velocity is therefore $\hat{v}_{z,\text{max}} = \dfrac{\lambda}{4T_{\text{prf}}}$. This is determined by the wavelength used and the pulse repetition time or essentially by the sampling interval. The pulse repetition time of course has to be sufficiently large to cover the full depth and must be larger than $T_{\text{prf}} > 2d/c$. Combining the two limitations gives the following depth–velocity limitation:

$$\hat{v}_{z,\text{max}} < \frac{c}{4}\frac{\lambda}{2d} = \frac{c}{8df_0}c, \tag{3.23}$$

which limits the maximum detectable velocity for a given depth and is a limitation imposed by the use of a phase estimation system.

The velocity estimation is performed not by taking the phase difference between two measurements but rather by combining the two arctan operations from the following:

$$\tan(\Delta\phi) = \tan\left(\arctan\left(\frac{y(i+1)}{x(i+1)}\right) - \arctan\left(\frac{y(i)}{x(i)}\right)\right)$$

$$= \frac{\dfrac{y(i+1)}{x(i+1)} - \dfrac{y(i)}{x(i)}}{1 + \dfrac{y(i+1)}{x(i+1)}\dfrac{y(i)}{x(i)}} \tag{3.24}$$

$$= \frac{y(i+)x(i) - y(i)x(i+1)}{x(i+1)x(i) + y(i+1)y(i)},$$

using that

$$\tan(A-B) = \frac{\tan(A) - \tan(B)}{1 + \tan(A)\tan(B)}. \tag{3.25}$$

Chapter 3

Then

$$\arctan\left(\frac{y(i+1)x(i)-y(i)x(i+1)}{x(i+1)x(i)+y(i+1)y(i)}\right)=-2\pi f_0\frac{2v_z}{c}T_{prf} \tag{3.26}$$

or

$$\hat{v}_z=-c\frac{f_{prf}}{4\pi f_0}\arctan\left(\frac{y(i+1)x(i)-y(i)x(i+1)}{x(i+1)x(i)+y(i+1)y(i)}\right). \tag{3.27}$$

This simple algebraic equation directly yields the velocity from comparing two emissions. Often, the signal-to-noise ratio from blood signals is low because of the weak scattering from blood, and therefore it is advantageous to average over several emissions as follows:

$$\hat{v}_z=-c\frac{f_{prf}}{4\pi f_0}\arctan\left(\frac{\displaystyle\sum_{i=1}^{M}y(i+1)x(i)-y(i)x(i+1)}{\displaystyle\sum_{i=1}^{M}x(i+1)x(i)+y(i+1)y(i)}\right), \tag{3.28}$$

where M is the number of emissions. This can also be calculated as the phase of the lag one autocorrelation of the received signal [26]. Further averaging can be made along the depth direction as the data are highly correlated over a pulse length. This is calculated as [27]:

$$\hat{v}_z(N_x)=-c\frac{f_{prf}}{4\pi f_0}\arctan\left(\frac{\displaystyle\sum_{i=1}^{M}\sum_{n=-N_p/2}^{N_p/2}y(n+N_x,i+1)x(n+N_x,i)-y(n+N_x,i)x(n+N_x,i+1)}{\displaystyle\sum_{i=1}^{M}\sum_{n=-N_p/2}^{N_p/2}x(n+N_x,i+1)x(n+N_x,i)+y(n+N_x,i+1)y(n+N_x,i)}\right) \tag{3.29}$$

when the real part of the received data is given as $x(n,i)$ and $y(n,i)$ is the imaginary part. Here n is the time index (depth), and i is the pulse emission number. N_x is the starting sample for the depth to estimate the velocity. N_p is the number of RF samples to average over and typically corresponds to one pulse length. This is the approach suggested by Loupas et al. [27] and is the one used in nearly all modern scanners.

3.7.1   Stationary Echo Canceling

At vessel boundaries, the received signal consists of both reflections from the vessel boundary and the scattered signal from blood. Often, the reflection signal is 20 to 40 dB larger in amplitude compared with the signal from blood, and it therefore makes velocity estimation heavily biased or impossible. The reflection signal is often assumed to be stationary and thereby constant over the number of pulse emissions. Subtracting two consecutive signals will therefore remove the stationary component and leave a signal suitable for velocity estimation as follows:

$$r_{es}(n,i) = r_s(n,i) - r_s(n,i+1). \tag{3.30}$$

For a fully stationary signal, this will give zero, whereas the flow signal will be filtered depending on the correlation between the two emissions. This filtration on the flow part $r_f(n,i)$ of the signal can be calculated as follows:

$$r_e(n,i) = r_f(n,i) - r_f(n,i+1) = r_f(n,i) - r_f(n,i)\exp\left(j2\pi\frac{2v_z}{c} f_0 T_{prf} \right), \tag{3.31}$$

assuming that the signals are so highly correlated that the shift in position due to the flow can be described by a simple phase shift. In the Fourier domain this gives

$$R_e(f) = R_f(f)\left(1 - \exp\left(j2\pi\frac{2v_z}{c} f T_{prf} \right) \right), \tag{3.32}$$

where the Fourier transform is taken along the emissions (i). The transfer function of the filter is therefore

$$H(f) = \left(1 - \exp\left(j2\pi\frac{2v_z}{c} f T_{prf} \right) \right) = 2j\sin\left(\pi\frac{2v_z}{c} T_{prf} f \right). \tag{3.33}$$

The transfer function of this filter is shown in Figure 3.6 for $v_z = 0.385$ m/s and $f_{prf} = 3$ kHz. It can be seen that the filter reduces the energy of the RF signal significantly

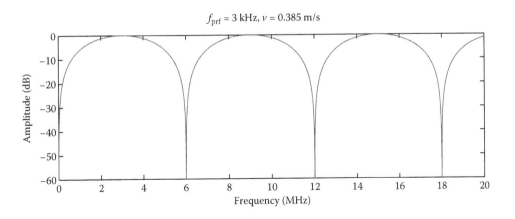

FIGURE 3.6   Transfer function of echo canceling filter.

Chapter 3

around a band of 6 MHz and at low frequencies. It is an unavoidable side effect of echo-canceling filtration that the energy of the flow signal is reduced and the signal-to-noise ratio is therefore reduced. The effect is especially noticeable at low flow velocities. Here the second signal is very similar to the first measurement because the time shift $t_s = 2v_z/cT_{prf}$ is small. The subtraction therefore removes most of the energy, and the noise power in the two measurements is added.

The reduction in signal-to-noise ratio due to the stationary echo canceling can be analytically calculated for a Gaussian pulse and is [1,28]:

$$R_{snr} = \sqrt{\frac{2\sqrt{2} + \exp\left(-\frac{2}{B_r^2}\right)}{2\sqrt{2} + \exp\left(-\frac{2}{B_r^2}\right)\xi_1 - 2\sqrt{2}\xi_2 \cos\left(2\pi\frac{f_0}{f_{sh}}\right)}}$$

$$\xi_1 = 1 - \exp\left(-\frac{1}{2}\left(\frac{\pi B_r f_0}{f_{sh}}\right)^2\right)$$

$$\xi_2 = \exp\left(-\left(\frac{\pi B_r f_0}{f_{sh}}\right)^2\right)$$

$$f_{sh} = \frac{c}{2v_z}f_{prf},$$

(3.34)

where the pulse is given as follows:

$$p(t) = \exp(-2(B_r f_0 \pi)^2 t^2)\cos(2\pi f_0 t).$$

(3.35)

Here B_r is the relative bandwidth and f_0 is the center frequency. The reduction for $f_0 = 3$ MHz, $f_{prf} = 3$ kHz, and $B_r = 0.08$ is shown in Figure 3.7. At zero velocity, the decrease is infinite as the two signals are identical and no velocity can be found. The reduction decreases progressively for increasing velocity, and a gain in SNR is found at the maximum detectable velocity. Here the two signals are inverted and compared with each other, and the subtraction then yields an addition of the two. The amplitude is therefore doubled, and the noise power is doubled, hence giving an improvement of 3 dB in SNR. The curve in Figure 3.7 is dependent on the echo-canceling filter used and on the pulse emitted $p(t)$, but there will also be an infinite loss at zero velocity.

Another method is to subtract the mean value of all the received lines as

$$r_e(n,i) = r_s(n,i) - \frac{1}{M}\sum_{k=1}^{M} r_s(n,k)$$

(3.36)

This gives a sharper cutoff in the transfer function of the filter and less noise added to the response.

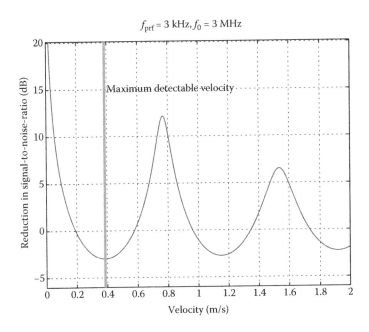

$f_{prf} = 3$ kHz, $f_0 = 3$ MHz

FIGURE 3.7   Loss in signal-to-noise ratio due to the echo canceling filter.

Many different echo-canceling filters have been suggested [29,30], but it remains a very challenging part of velocity estimation and probably the one factor most affecting the outcome of the estimation.

3.8 Axial Velocity Estimation Using the Time Shift

It is also possible to estimate the velocity directly from the time shift between the signals. Two consecutive signals are related by the following equation:

$$y_c(t,i+1) = y_c\left(t - \frac{2v_z}{c}T_{prf},i\right) = y_c(t-t_s,i) \tag{3.37}$$

Cross correlating two consecutive signals can then be used for finding the time shift and, hence, the velocity. This is calculated by [16,17]

$$R_{12}(\tau,i) = \int_T y_c(t,i)y_c(t+\tau,i+1)dt = \int_T y_c(t,i)y_c(t-t_s+\tau,i)dt = R_{11}(\tau-t_s) \tag{3.38}$$

Using Equation 3.2, the autocorrelation can be rewritten as

$$R_{11}(\tau-t_s,i) = \int_T p(t)*s(t,i)p(t)*s(t-t_s+\tau,i)dt = R_{pp}(\tau)*\int_T s(t,i)s(t-t_s+\tau,i)dt$$

$$= R_{pp}(\tau)P_s\delta(\tau-t_s) = P_sR_{pp}(\tau-t_s), \tag{3.39}$$

Chapter 3

where P_s is the scattering power, and the scattering signal is assumed to be random and white. $R_{pp}(\tau)$ is the autocorrelation of the emitted pulse, and this has a unique maximum value at $\tau = 0$. $R_{pp}(\tau - t_s)$ therefore has a unique maximum at $\tau = t_s$. The velocity can be derived from the following equation:

$$\hat{v}_z = c \frac{\hat{t}_s}{2T_{prf}}. \tag{3.40}$$

The cross correlation is calculated from the discrete sampled signals $y_s(n,i)$ as

$$R_{12d}(k,N_x) = \sum_{i=1}^{N_e-1} \sum_{n=-N_n/2}^{N_n/2} y_s(n+N_x,i) y_s(n+k+N_x,i+1), \tag{3.41}$$

where N_e is the number of emissions to average over, N_n is the number of samples to average over, and N_x is the sample number (depth) at which to find the velocity at. The position of the peak n_s in $R_{12}(k, N_x)$ is then found at the velocity calculated from

$$\hat{v}_z = c \frac{\hat{n}_s/f_s}{2T_{prf}}, \tag{3.42}$$

where f_s is the RF sampling frequency.

The cross-correlation function is shown in Figure 3.8. Figure 3.8a shows the two signals used where the time shift readily can be seen. Figure 3.8b shows the cross correlation of the signals along with an indication of the peak position. A 3 MHz Gaussian pulse was used in the simulation along with $f_{prf} = 5$ kHz and a 20 MHz sampling frequency. The velocity was 0.35 m/s, which gave a time shift of $t_s = 0.156$ μs.

The time shift is usually comparable to the sampling interval $1/f_s$, and the velocity estimates will be heavily quantized. This can be solved by fitting a second-order polynomial around the cross-correlation peak and then finding the peak value of the polynomial. The interpolation is calculated by [31]

$$n_{int} = n_s - \frac{\hat{R}_{12d}(n_s+1) - \hat{R}_{12d}(n_s-1)}{2(\hat{R}_{12d}(n_s+1) - 2\hat{R}_{12d}(n_s) + \hat{R}_{12d}(n_s-1))}, \tag{3.43}$$

and the interpolated estimate is given by

$$\hat{v}_{int} = \frac{c}{2} \frac{n_{int} f_{prf}}{f_s}. \tag{3.44}$$

This gives an increased resolution, if the cross-correlation estimate is sufficiently noise free.

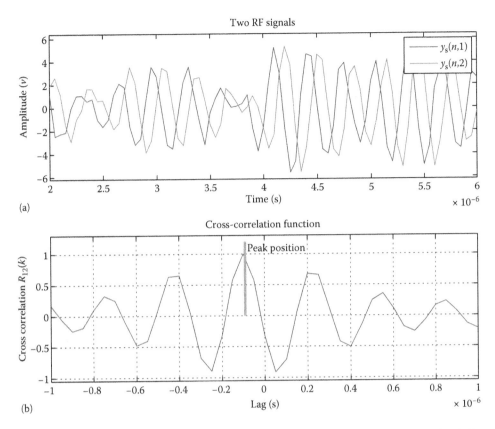

FIGURE 3.8 Illustration of the cross correlation. The graph shows the two signals to be correlated (a) and their cross correlation (b).

Several factors affect how well this estimator works. A short pulse should be used for the emission to provide the narrowest possible cross-correlation function. The average should be derived from a sufficient number of samples to give a good estimate, but should also be limited to a region where the velocity can be assumed to be constant in space. Also, the number of emissions averaged over time should be sufficient to ensure a good result without lowering the frame rate too much.

The cross-correlation method can find velocities higher than the autocorrelation approach, as it is not restricted to a phase shift of $\pm \pi$. In theory, any kind of velocity can be found, but this can lead to false peak detection [32]. Here a global maximum in the correlation function beyond the correct peak is found. As the cross correlation is determined by the autocorrelation of the pulse, these erroneous peaks will reside at $k/f_0 + n_s$, leading to a large error in the estimated velocity. This can be difficult to correct and gives spike artifacts in the image display. Often, the search range for finding the maximum is therefore limited to lie around zero velocity. No false peaks arise if the search range is limited to $-f_s/(2f_0) < k < f_s/(2f_0)$, which gives the same maximum velocity as the autocorrelation or phase shift approach.

It is difficult to decide which of the two methods are best. Often, the cross correlation gives better accuracies on the estimates, but this is offset by its lower sensitivity, which comes from using a shorter pulse. In general, the decision is dependent on the actual measurement situation and setup.

Chapter 3

3.9 Two-Dimensional Vector Velocity Estimation

The methods described so far only find the velocity along the ultrasound beam direction, and this is often perpendicular to the flow direction. The velocity component found is therefore often the smallest and least important. Many angle compensation schemes have been devised [33,34], but they all rely on the assumption that a single angle is applied for the whole cardiac cycle and region of interest, which in general is not correct. The velocity will often be in all directions and changes both magnitude and direction over the cardiac cycle because of the pulsating nature of the flow. There is thus a real need for vector velocity estimation methods.

The problem has been acknowledged for many years, and several authors have suggested schemes for finding the velocity vector. Fox [35] used two crossing beams to find the velocity for two directions and then combine it to yield the 2-D velocity vector. Newhouse et al. [36] suggested using the bandwidth of the received signal to determine the lateral component. Trahey et al. [37] used a speckle tracking approach for searching for the velocity vector.

Currently, the only method that has been introduced on commercial FDA-approved scanners is the transverse oscillation (TO) approach developed by Jensen and Munk [38,39]. A similar approach was also suggested by Anderson [40].

The traditional axial velocity estimation methods rely on the axial oscillation to find the velocity. The TO method introduces an oscillation transverse to the ultrasound propagation direction to make the received signals sensitive to a transverse motion. Such a transverse field is shown in Figure 3.9, which shows a contour plot of the linear point spread function (PSF), and oscillations can be seen both in the lateral and axial directions. Two fields are needed to make a complex field with an in-phase and quadrature component that can be used for finding the sign of the velocity in the lateral direction.

The lateral oscillation is generated by using a special apodization on the transducer during the receive processing. At the focus, there is a Fourier relation between the transducer's apodization function and the ultrasound field [41]. To generate a sinusoidal oscillation, the receive apodization derived for a continuous field should ideally consist of two sinc-shaped peaks with a distance of D. This will give a lateral wavelength of

$$\lambda_x = \lambda \frac{2D}{P_d}, \tag{3.45}$$

where P_d is the depth in tissue. Sending out a fairly broad beam and focusing the two fields in reception with this apodization function will yield the fields shown in Figure 3.9. The signals at the dashed line are shown in Figure 3.10a, where the solid curve is from the left field and the dashed curve is from the right field. These two signals should ideally be 90° phase shifted compared with each other to generate a one-side spectrum. For a pulsed field, Equation 3.45 is not accurate enough to ensure this, and a computer optimization has been made to adjust the focus to give the best possible result [42]. This has been performed for a convex array probe by minimizing the amplitude spectrum of the complex field for negative spatial frequencies, as shown in Figure 3.10b. This ensures a nearly one-sided spectrum and thereby the best possible estimate in terms of bias and standard deviation.

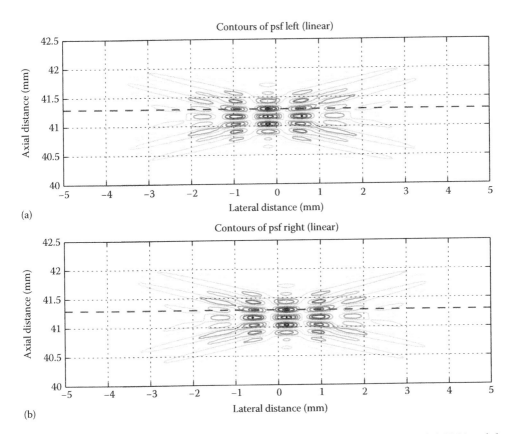

FIGURE 3.9 Ultrasound fields used in TO velocity estimation. The graph shows the left field (a) and the the right field (b).

The receive beamforming yields signals usable for the vector velocity estimation. They are combined into the complex signal $r_{sq}(i)$, and a temporal Hilbert transform of this gives the signal $r_{sqh}(i)$. The following equations are derived from these two new signals:

$$r_1(i) = r_{sq}(i) + jr_{sqh}(i),$$
$$r_2(i) = r_{sq}(i) - jr_{sqh}(i).$$

(3.46)

The velocity components are then estimated by the TO estimators derived by Jensen [39]. They are given by

$$v_x = \frac{\lambda_x}{2\pi 2 T_{prf}} \arctan\left(\frac{\Im\{R_1(1)\}\Re\{R_2(1)\} + \Im\{R_2(1)\}\Re\{R_1(1)\}}{\Re\{R_1(1)\}\Re\{R_2(1)\} - \Im\{R_1(1)\}\Im\{R_2(1)\}}\right),$$

(3.47)

and

$$v_z = \frac{c}{2\pi 4 T_{prf} f_0} \arctan\left(\frac{\Im\{R_1(1)\}\Re\{R_2(1)\} - \Im\{R_2(1)\}\Re\{R_1(1)\}}{\Re\{R_1(1)\}\Re\{R_2(1)\} + \Im\{R_1(1)\}\Im\{R_2(1)\}}\right).$$

(3.48)

Chapter 3

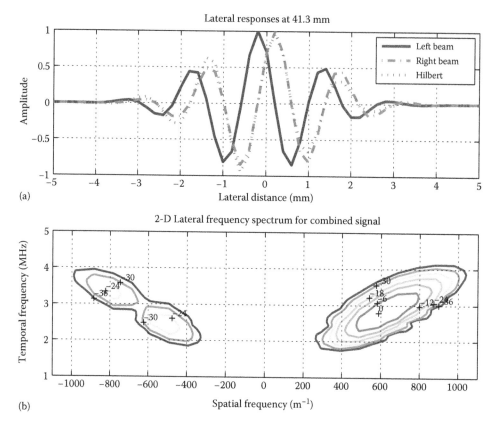

FIGURE 3.10 (a) Lateral left and right responses in the TO fields at the maximum compared with the Hilbert transform of the left field. (b) The 2-D Fourier transform of the complex TO PSF.

where $R_1(1)$ is the complex lag one autocorrelation value for $r_1(i)$, and $R_2(1)$ is the complex lag one autocorrelation value for $r_2(i)$. $\Im$ denotes the imaginary part and $\Re$ the real part, providing the velocity vector in the imaging plane.

Figure 3.11 [43] shows a vector flow image (VFI) of the carotid bifurcation measured by a linear array probe and the TO approach. The image is acquired right after the peak systole. The vectors show magnitude and direction of the flow, while the color intensities show velocity magnitude. A vortex can be seen in the carotid bulb. The vortex appears right after the peak systole and disappears in roughly 100 ms. This is a normal flow pattern in humans and shows the value of vector flow imaging. It is important to note that there is no single correct beam-to-flow angle in this image. Both magnitude and direction change rapidly as a function of both time and space, making it essential to have a vector flow estimation system to capture the full complexity of the hemodynamics.

3.10   Three-Dimensional Vector Velocity Estimation

The TO approach can also be extended to full 3-D imaging by using a 2-D matrix transducer. Here five lines are beamformed in parallel during the receive processing to obtain a set of lines for the transverse, elevation, and axial velocity components. The beamforming is visualized in Figure 3.12 [44], where the active elements used in the beamformation

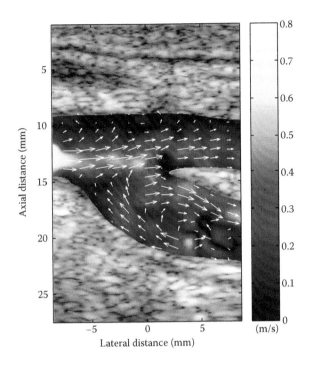

FIGURE 3.11 Vector flow image of the carotid bifurcation right after the peak systole, where a vortex is present in the carotid bulb. (From J. Udesen et al., *Ultrasound in Medicine and Biology*, 33:541–548, 2007.)

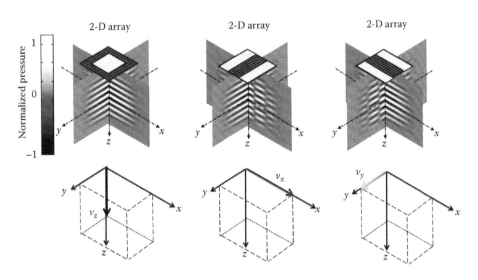

FIGURE 3.12 Receive apodization profiles applied to generate the TO fields for all three velocity components. The white shaded areas indicate the active elements in the 32 × 32 array. (From M. J. Pihl et al., *Proceedings of the IEEE Ultrasonics Symposium*, July 2013.)

Chapter 3

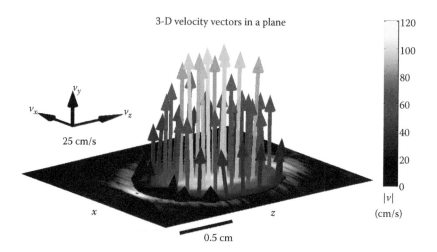

FIGURE 3.13 Three-dimensional vector velocity image for a parabolic, stationary flow. (From M. J. Pihl et al., *Proceedings of SPIE, Medical Imaging*, 8675:86750H-1–86750H-12, 2013.)

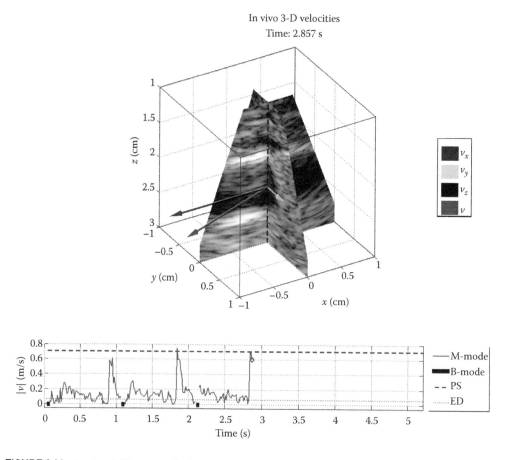

FIGURE 3.14 In vivo 3-D vector velocity image taken around the peak systole in the carotid artery of a healthy volunteer. (Image courtesy of Dr. Michael Johannes Pihl.)

are shaded white and the corresponding velocity direction is indicated below the array. Slices into the corresponding ultrasound fields are also shown below the transducer, and the oscillations in the axial, transverse, and elevation planes can be seen.

The approach has been implemented on the SARUS experimental ultrasound scanner [45] connected to a Vermon 32 × 32 3MHz matrix array transducer [46–48]. The parabolic flow in a recirculating flow rig was measured, and the result is shown in Figure 3.13. The flow is in the elevation direction (out of the imaging plane) of the image shown in the bottom, and both 1-D and 2-D velocity estimation systems would show no velocity. The arrows indicate the out-of-plane motion amplitude and direction and show the parabolic velocity profile.

The approach has also been used in vivo, as shown in Figure 3.14 for the carotid artery. Two intersecting B-mode images have been acquired, and the 3-D velocity vectors have been found at the intersection of the two planes. The estimated velocity magnitude as a function of time is shown in the lower graph, and the velocity vector is shown around the peak systole in the cardiac cycle. This method has the potential of showing the full dynamics of the complex flow in the human circulation in real time for a complete evaluation of the hemodynamics.

3.11   SA and Plane Wave Flow Estimation

The measurement systems described so far are all sequential in nature. They acquire the flow lines in one direction at a time, and this makes the measurement slow, especially when images consist of many directions or many emissions have to be used for flow estimation. Triplex imaging shows the B-mode, the CFM image, and the spectral information simultaneously and therefore needs to split the acquisition time between the three modes. This often makes the resulting frame rate unacceptably low for clinical use for large depths. This will also be a very limiting factor for 3-D flow imaging, which often has to resort to ECG gating when acquiring full volumes. Another drawback of traditional imaging is the use of transmit focusing. This cannot be made dynamic, and the images are only optimally focused at one single depth.

These problems can be solved by using new imaging schemes based on SA imaging [49–56] and plane wave imaging [57–59]. Both these techniques insonify the whole region to interrogate and reconstruct the images during receive beamforming. This process can potentially lead to very fast imaging and can also be used for flow imaging with very significant advantages.

The SA method is shown in Figure 3.15 [60]. The transducer on the top emits a spherical wave, and the scattered signal is then received on all elements of the transducer. This process is repeated for several emission sources N_i on the aperture, and the data are collected for all receiving elements. From the received data for a single emission, a full low-resolution (LR) image can be made. It is only focused in reception, but combining all the LR images yields a high-resolution (HR) image. This is also focused during transmission as all the emitted fields are summed in phase [61]. The approach gives better focused images than traditional beamforming [62] with at least a preserved penetration depth when coded excitation is used.

The imaging scheme can also be used for flow estimation, although the data are acquired over several emissions and therefore are shifted relative to one another. This is

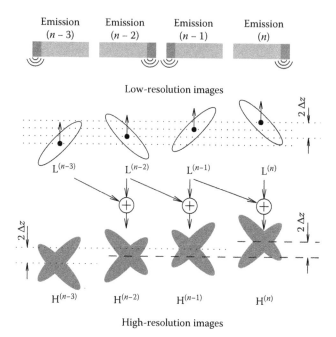

Emission (n − 3)   Emission (n − 2)   Emission (n − 1)   Emission (n)

Low-resolution images

$L^{(n-3)}$ $L^{(n-2)}$ $L^{(n-1)}$ $L^{(n)}$

$H^{(n-3)}$ $H^{(n-2)}$ $H^{(n-1)}$ $H^{(n)}$

High-resolution images

FIGURE 3.15 Acquisition of SA flow data and beamforming of high-resolution images. (From S. I. Nikolov and J. A. Jensen, *IEEE Transactions on Ultrasonics, Ferroelectrics, and Frequency Control*, 50(7):848–856, 2003.)

also illustrated in Figure 3.15 for a short sequence. The PSF for the low-resolution images is shown in Figure 3.15 for a point scatterer moving toward the transducer. The LR PSFs are different for the different emissions and therefore cannot be directly correlated to find the velocity. Adding the low-resolution images gives the PSF for the HR images, and it can be seen that these images have the same shape when the emission sequence combined is the same apart from the motion in position. The basic idea is therefore only to correlate the HR PSFs with the same emission sequence for finding the flow. This can also be performed recursively so that a new correlation function is made for every new LR image [56].

The approach is illustrated in Figure 3.16. The HR signals in one direction is shown on the top divided into segments. The length of the emission sequence is N_l, and therefore emission n and $n + N_l$ can be correlated. This can then be averaged with $n + 1$ correlated with $n + 1 + N_l$ as the time shift t_s is the same. It is therefore possible to continuously average the correlation function and therefore use all data to obtain a very precise estimate of the correlation and thereby the velocity. This can be performed for all directions in the HR image continuously.

The method has several advantages. The data can be acquired continuously. Therefore, data for flow imaging are continuously available everywhere in the image, which makes it possible to average over very large amounts of data and makes echo canceling much easier [60]. Initialization effects for the filter can be neglected as the data are continuous, and this makes a large difference for, e.g., low velocity flow. The cutoff frequency of the traditional echo-canceling filter is proportional to f_{prf}/M, where M can be made arbitrarily large. The correlation estimates can also be averaged over a larger time interval

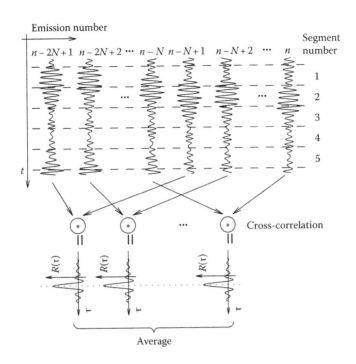

FIGURE 3.16 Averaging of cross-correlation functions in SA flow imaging. (From S. I. Nikolov and J. A. Jensen, *IEEE Transactions on Ultrasonics, Ferroelectrics, and Frequency Control*, 50(7):848–856, 2003.)

T_i. The length N_h is only limited by the acceleration a_f of the flow. For a cross-correlation system, there should be at most a $\frac{1}{2}$ sampling interval shift because of acceleration, that is,

$$a_f N_l T_{prf} N_h < \frac{1}{2 f_s T_{prf}} \frac{c}{2} \tag{3.49}$$

or

$$T_i = N_l T_{prf} N_h < \frac{f_{prf}}{2 f_s} \frac{c}{2 a_f}, \tag{3.50}$$

to avoid decorrelation in the estimate of the cross-correlation function.

The data can also be focused in any direction as complete data sets are acquired, and the position of both the emitting sources and the receivers are known. The signals for velocity estimation can therefore be focused along the flow lines, if the beam-to-flow angle is known. This focusing scheme is shown in Figure 3.17. For each depth, the data are focused along the flow and then used in a cross-correlation scheme to find the velocity [63,64].

The estimated profiles for such a scheme are shown in Figure 3.18 at a beam-to-flow angle of 60°. A linear array was used with an eight-emission SA sequence using a chirp

Chapter 3

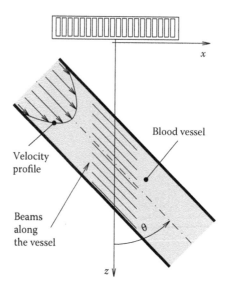

FIGURE 3.17   Directional beamforming along the flow lines. (From J. A. Jensen and S. I. Nikolov, *IEEE Transactions on Ultrasonics, Ferroelectrics, and Frequency Control*, 51:1107–1118, 2004.)

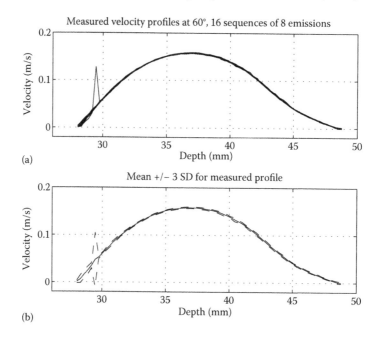

FIGURE 3.18   Estimated velocity profiles (a) at a beam-to-flow angle of 60° and (b) mean value ± 3 SD for SA vector flow imaging. (From J. A. Jensen and S. I. Nikolov, *IEEE Transactions on Ultrasonics, Ferroelectrics, and Frequency Control*, 51:1107–1118, 2004.)

pulse. Data from 64 elements were acquired for each emission, and 16 sequences, for a total of 128 emissions, were averaged. All the 20 estimated velocity profiles are shown in Figure 3.18a, and the mean ± 3 SD values are shown in Figure 3.18b. The mean relative standard deviation was 0.36% [63]. The approach also works for fully transverse flow and can yield a fast and quantitative display of the vector velocity.

It is also possible to determine the angle from the data. Here the directional lines are beamformed in all directions, and the one with the highest relative correlation indicates the angle [65]. An example of an in vivo SA VFI from the carotid artery is shown in Figure 3.19, where both velocities and angles have been estimated.

The data can also be used for visualizing the location of flow. This is performed by finding the energy of the signals after echo canceling in a B-flow system [66] or power Doppler mode. The intensity of the signal is then roughly proportional to the velocity. An example of an SA B-flow image is shown in Figure 3.20 [67] at two different time instances in the cardiac cycle.

Another method for making fast and continuous imaging is to use plane wave emission. Here the full transducer is used to transmit a plane wave, and then data are acquired for all the receiving elements [57]. The full image can then be reconstructed as for SA imaging. The image is only focused during reception and will have a lower resolution and higher side lobes than conventional images. This can be compensated by using several plane waves at different angles as illustrated in Figure 3.21. Combining these methods with a proper apodization can then lead to a full HR image [68]. This imaging scheme has the same advantages as SA imaging with a continuous data stream that can be used for increasing the sensitivity of flow estimation.

These imaging schemes can also be made very fast, and this is beneficial for looking at transitory and very fast flow phenomena, which are abundant in the human circulation [69,70]. A plane wave VFI is shown in Figure 3.22. A single plane wave was continuously emitted, and the full image was beamformed for each emission. This was used in a speckle tracking scheme to find the velocity vectors and resulted in 100 independent vector velocity images per second [71]. A valve in the jugular vein and the carotid artery were imaged. Figure 3.22a shows the open valve on the top, where a clockwise vortex is

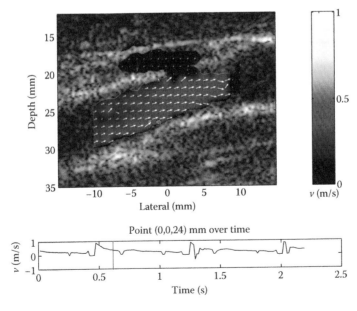

FIGURE 3.19 In vivo SA vector flow imaging from the carotid artery. (From J. A. Jensen and N. Oddershede, *IEEE Transactions on Ultrasonics, Ferroelectrics, and Frequency Control*, 25:1637–1644, 2006.)

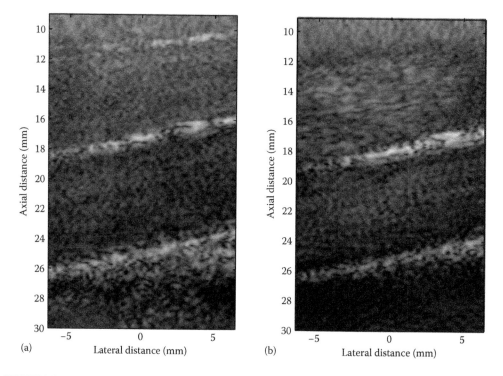

FIGURE 3.20   SA B-flow image of jugular vein and carotid artery. (a) Frame 166 of 375. (b) Frame 356 of 375. (From J. A. Jensen, *Proceedings of SPIE, Progress in Biomedical Optics and Imaging*, 5373:44–51, 2004.)

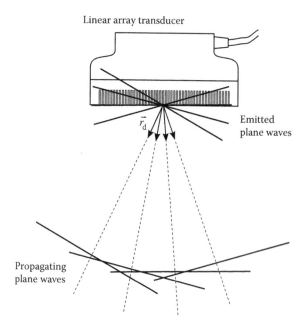

FIGURE 3.21   Plane wave imaging.

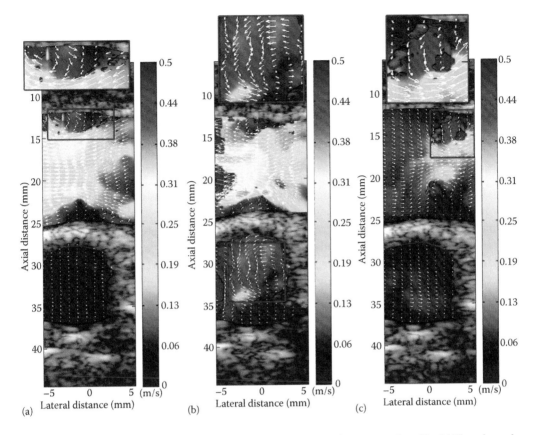

FIGURE 3.22 In vivo plane wave vector flow images acquired at a frame rate of 100 Hz. (a) Flow through a valve in the jugular vein at the peak systole. (c) Reverse flow during diastole. Note how the vortices behind the valve leaflet change rotation direction. Secondary flow is also seen in (b) in the carotid artery below the jugular vein. (From K. L. Hansen et al., *Ultraschall in der Medizin*, 30:471–476, 2009.)

found behind the valve leaflets. The valve is incompetent and does not close correctly as shown in Figure 3.22c, where a noticeable reverse flow is seen. The vortex behind the leaflet has also changed direction. Figure 3.22b shows secondary rotational flow in the carotid artery during the peak systole, indicating the importance of having a full 3-D flow system.

SA imaging and plane wave flow imaging are excellent for observing slow moving flow because of the long observation time possible. This has been demonstrated in Refs. [68] and [72], which used plane wave imaging for mapping the brain function of a rat. The new acquisition methods can, thus, obtain data suitable for both fast vector velocity imaging and slow flow estimation for functional ultrasound imaging.

3.12 Motion Estimation and Other Uses

The methods described can also be used for motion estimation in general. Tissue motion can be found by leaving out the echo-canceling filter, and then all the methods can be applied for strain imaging [73], radiation force imaging [74,75], shear wave imaging [57], tissue Doppler [76], and other methods relying on the detection of motion or velocity. In general, the methods have an improved performance for tissue because of the increased signal-to-noise ratio and the lack of the echo-canceling filter. It is therefore possible to calculate the derivatives necessary for some of these methods.

The velocity estimates are also used in deriving quantitative numbers useful for diagnostic purposes. In particular, the new vector velocity estimates can be used for making the diagnosis more quantitative by calculating, for example, the volume flow [77], by deriving quantities for indicating turbulence [78], and by finding mean or peak velocities. It is also possible to use the vector velocity data for calculating flow gradients by solving the Navier–Stokes equations [79].

The development within velocity estimation is by no means complete. The combination of SA and plane wave imaging with 2-D and 3-D vector velocity and functional imaging is still a very active research area, and more complete information about the complex flow in the human body can be obtained. It will in real time reveal the many places for transient turbulences, vortices, and other multidirectional flow, and it will become possible to derive many more quantitative parameters for characterizing the patient's circulation.

References

1. J. A. Jensen. *Estimation of Blood Velocities Using Ultrasound: A Signal Processing Approach*. Cambridge University Press, New York, 1996.
2. D. H. Evans, W. N. McDicken, R. Skidmore, and J. P. Woodcock. *Doppler Ultrasound, Physics, Instrumentation, and Clinical Applications*. John Wiley & Sons, New York, 1989.
3. D. H. Evans and W. N. McDicken. *Doppler Ultrasound, Physics, Instrumentation, and Signal Processing*. John Wiley & Sons, New York, 2000.
4. T. L. Szabo. *Diagnostic Ultrasound Imaging Inside Out*. Elsevier, 2004.
5. R. S. C. Cobbold. *Foundations of Biomedical Ultrasound*. Oxford University Press, 2006.
6. C. G. Caro, T. J. Pedley, R. C. Schroter, and W. A. Seed. Mechanics of the circulation. In *MTP International Review of Science Physiology. Series 1, Volume 1: Cardiovascular Physiology*, 15–48. Butterworth, London, 1974.
7. D. W. Holdsworth, C. J. D. Norley, R. Fraynek, D. A. Steinman, and B. K. Rutt. Characterization of common carotid artery blood-flow waveforms in normal human subjects. *Physiological Measurement*, 20:219–240, 1999.
8. W. W. Nichols and M. F. O'Rourke. *McDonald's Blood Flow in Arteries, Theoretical, Experimental and Clinical Principles*. Lea & Febiger, Philadelphia, 1990.
9. E. L. Carstensen, K. Li, and H. P. Schwan. Determination of the acoustic properties of blood and its components. *Journal of the Acoustical Society of America*, 25:286–289, 1953.
10. F. Dunn, P. D. Edmonds, and W. J. Fry. Absorption and dispersion of ultrasound in biological media. In *Biological Engineering*. McGraw-Hill, New York, 1969.
11. R. J. Ulrick. A sound velocity method for determining the compressibility of finely divided substances. *Journal of Applied Physics*, 18:983–987, 1947.
12. W. R. Platt. *Color Atlas and Textbook of Hematology*. J. B. Lippincott, Toronto, 1969.
13. P. M. Morse and K. U. Ingard. *Theoretical Acoustics*. McGraw-Hill, New York, 1968.
14. L. Y. L. Mo and R. S. C. Cobbold. Theoretical models of ultrasonic scattering in blood. In *Ultrasonic Scattering in Biological Tissues*. CRC Press, Boca Raton, FL, 1993.
15. G. S. Kino. *Acoustic Waves, Devices, Imaging, and Analog Signal Processing*. Prentice-Hall, Englewood Cliffs, NJ, 1987.
16. O. Bonnefous and P. Pesqué. Time domain formulation of pulse-Doppler ultrasound and blood velocity estimation by cross correlation. *Ultrasonic Imaging*, 8:73–85, 1986.
17. O. Bonnefous, P. Pesqué, and X. Bernard. A new velocity estimator for color flow mapping. *Proceedings of the IEEE Ultrasonics Symposium*, 855–860, 1986.

18. J. A. Jensen and N. B. Svendsen. Calculation of pressure fields from arbitrarily shaped, apodized, and excited ultrasound transducers. *IEEE Transactions on Ultrasonics, Ferroelectrics, and Frequency Control*, 39:262–267, 1992.

19. J. A. Jensen. Field: A program for simulating ultrasound systems. *Medical and Biological Engineering and Computing*, 10th Nordic-Baltic Conference on Biomedical Imaging, vol. 4, suppl. 1, Part 1:351–353, 1996.

20. J. A. Jensen. A model for the propagation and scattering of ultrasound in tissue. *Journal of the Acoustical Society of America*, 89:182–191, 1991.

21. J. R. Womersley. Oscillatory motion of a viscous liquid in a thin-walled elastic tube. I: The linear approximation for long waves. *Philosophical Magazine*, 46:199–221, 1955.

22. D. H. Evans. Some aspects of the relationship between instantaneous volumetric blood flow and continuous wave Doppler ultrasound recordings III. *Ultrasound in Medicine and Biology*, 8:617–623, 1982.

23. A. Swillens, L. Løvstakken, J. Kips, H. Torp, and P. Segers. Ultrasound simulation of complex flow velocity fields based on computational fluid dynamics. *IEEE Transactions on Ultrasonics, Ferroelectrics, and Frequency Control*, 56(3):546–556, 2009.

24. A. V. Oppenheim and R. W. Schafer. *Discrete-Time Signal Processing*. Prentice-Hall, Englewood Cliffs, NJ, 3rd ed., 2010.

25. D. W. Baker. Pulsed ultrasonic Doppler blood-flow sensing. *IEEE Transactions on Sonics and Ultrasonics*, SU-17:170–185, 1970.

26. C. Kasai, K. Namekawa, A. Koyano, and R. Omoto. Real-time two-dimensional blood flow imaging using an autocorrelation technique. *IEEE Transactions on Sonics and Ultrasonics*, 32:458–463, 1985.

27. T. Loupas, J. T. Powers, and R. W. Gill. An axial velocity estimator for ultrasound blood flow imaging, based on a full evaluation of the Doppler equation by means of a two-dimensional autocorrelation approach. *IEEE Transactions on Ultrasonics, Ferroelectrics, and Frequency Control*, 42:672–688, 1995.

28. J. A. Jensen. Stationary echo canceling in velocity estimation by time-domain cross-correlation. *IEEE Transactions on Medical Imaging*, 12:471–477, 1993.

29. S. Bjærum, H. Torp, and K. Kristoffersen. Clutter filter design for ultrasound colour flow imaging. *IEEE Transactions on Ultrasonics, Ferroelectrics, and Frequency Control*, 49:204–209, 2002.

30. A. C. H. Yu and L. Løvstakken. Eigen-based clutter filter design for ultrasound color flow imaging: A review. *IEEE Transactions on Ultrasonics, Ferroelectrics, and Frequency Control*, 57(5):1096–1111, 2010.

31. S. G. Foster. *A pulsed ultrasonic flowmeter employing time domain methods*. PhD thesis, Department of Electrical Engineering, University of Illinois, Urbana, IL, 1985.

32. J. A. Jensen. Artifacts in velocity estimation using ultrasound and cross-correlation. *Medical and Biological Engineering and Computing*, 32/4 suppl.:s165–s170, 1994.

33. B. Dunmire, K. W. Beach, K-H Labs., M. Plett, and D. E. Strandness. Cross-beam vector Doppler ultrasound for angle independent velocity measurements. *Ultrasound in Medicine and Biology*, 26:1213–1235, 2000.

34. P. Tortoli, G. Bambi, and S. Ricci. Accurate Doppler angle estimation for vector flow measurements. *IEEE Transactions on Ultrasonics, Ferroelectrics, and Frequency Control*, 53(8):1425–1431, Aug. 2006.

35. M. D. Fox. Multiple crossed-beam ultrasound Doppler velocimetry. *IEEE Transactions on Sonics and Ultrasonics*, SU-25:281–286, 1978.

36. V. L. Newhouse, D. Censor, T. Vontz, J. A. Cisneros, and B. B. Goldberg. Ultrasound Doppler probing of flows transverse with respect to beam axis. *IEEE Transactions on Biomedical Engineering*, BME-34:779–788, 1987.

37. G. E. Trahey, J. W. Allison, and O. T. von Ramm. Angle independent ultrasonic detection of blood flow. *IEEE Transactions on Biomedical Engineering*, BME-34:965–967, 1987.

38. J. A. Jensen and P. Munk. A new method for estimation of velocity vectors. *IEEE Transactions on Ultrasonics, Ferroelectrics, and Frequency Control*, 45:837–851, 1998.

39. J. A. Jensen. A new estimator for vector velocity estimation. *IEEE Transactions on Ultrasonics, Ferroelectrics, and Frequency Control*, 48(4):886–894, 2001.

40. M. E. Anderson. Multi-dimensional velocity estimation with ultrasound using spatial quadrature. *IEEE Transactions on Ultrasonics, Ferroelectrics, and Frequency Control*, 45:852–861, 1998.

41. R. Bracewell. *The Fourier Transform and Its Applications*. McGraw-Hill, New York, 3rd ed., 1999.

42. J. A. Jensen. Optimization of transverse oscillating fields for vector velocity estimation with convex arrays. *Proceedings of the IEEE Ultrasonics Symposium*, pp. 1753–1756, July 2013.

43. J. Udesen, M. B. Nielsen, K. R. Nielsen, and J. A. Jensen. Examples of in-vivo blood vector velocity estimation. *Ultrasound in Medicine and Biology*, 33:541–548, 2007.

Chapter 3

44. M. J. Pihl, M. B. Stuart, B. G. Tomov, P. M. Hansen, M. B. Nielsen, and J. A. Jensen. In vivo three-dimensional velocity vector imaging and volumetric flow rate measurements. *Proceedings of the IEEE Ultrasonics Symposium*, pp. 72–75, July 2013.

45. J. A. Jensen, H. Holten-Lund, R. T. Nilsson, M. Hansen, U. D. Larsen, R. P. Domsten, B. G. Tomov, M. B. Stuart, S. I. Nikolov, M. J. Pihl, Y. Du, J. H. Rasmussen, and M. F. Rasmussen. SARUS: A synthetic aperture real-time ultrasound system. *IEEE Transactions on Ultrasonics, Ferroelectrics, and Frequency Control*, 60(9):1838–1852, 2013.

46. M. J. Pihl, M. B. Stuart, B. G. Tomov, J. M. Hansen, M. F. Rasmussen, and J. A. Jensen. Preliminary examples of 3D vector flow imaging. *Proceedings of SPIE, Medical Imaging*, 8675:86750H-1–86750H-12, 2013.

47. M. J. Pihl and J. A. Jensen. A transverse oscillation approach for estimation of three-dimensional velocity tensors. Part I: Concept and simulation study. *IEEE Transactions on Ultrasonics, Ferroelectrics, and Frequency Control*, 61(10):1599–1607, 2014.

48. M. J. Pihl and J. A. Jensen. A transverse oscillation approach for estimation of three-dimensional velocity tensors. Part II: Experimental investigation. *IEEE Transactions on Ultrasonics, Ferroelectrics, and Frequency Control*, 61(10):1608–1618, 2014.

49. J. T. Ylitalo and H. Ermert. Ultrasound synthetic aperture imaging: Monostatic approach. *IEEE Transactions on Ultrasonics, Ferroelectrics, and Frequency Control*, 41:333–339, 1994.

50. M. O'Donnell and L. J. Thomas. Efficient synthetic aperture imaging from a circular aperture with possible application to catheter-based imaging. *IEEE Transactions on Ultrasonics, Ferroelectrics, and Frequency Control*, 39:366–380, 1992.

51. J. A. Jensen, S. Nikolov, K. L. Gammelmark, and M. H. Pedersen. Synthetic aperture ultrasound imaging. *Ultrasonics*, 44:e5–e15, 2006.

52. M. Karaman, P. C. Li, and M. O'Donnell. Synthetic aperture imaging for small scale systems. *IEEE Transactions on Ultrasonics, Ferroelectrics, and Frequency Control*, 42:429–442, 1995.

53. M. Karaman and M. O'Donnell. Subaperture processing for ultrasonic imaging. *IEEE Transactions on Ultrasonics, Ferroelectrics, and Frequency Control*, 45:126–135, 1998.

54. G. R. Lockwood and F. S. Foster. Design of sparse array imaging systems. *Proceedings of the IEEE Ultrasonics Symposium*, 1237–1243, 1995.

55. G. R. Lockwood, J. R. Talman, and S. S. Brunke. Real-time 3-D ultrasound imaging using sparse synthetic aperture beamforming. *IEEE Transactions on Ultrasonics, Ferroelectrics, and Frequency Control*, 45:980–988, 1998.

56. S. I. Nikolov, K. Gammelmark, and J. A. Jensen. Recursive ultrasound imaging. *Proceedings of the IEEE Ultrasonics Symposium*, 2:1621–1625, 1999.

57. M. Tanter, J. Bercoff, L. Sandrin, and M. Fink. Ultrafast compound imaging for 2-D motion vector estimation: Application to transient elastography. *IEEE Transactions on Ultrasonics, Ferroelectrics, and Frequency Control*, 49:1363–1374, 2002.

58. J. Udesen, F. Gran, K. L. Hansen, J. A. Jensen, C. Thomsen, and M. B. Nielsen. High frame-rate blood vector velocity imaging using plane waves: Simulations and preliminary experiments. *IEEE Transactions on Ultrasonics, Ferroelectrics, and Frequency Control*, 55(8):1729–1743, 2008.

59. J. Udesen, F. Gran, and J. A. Jensen. Fast color flow mode imaging using plane wave excitation and temporal encoding. *Proceedings of SPIE, Progress in Biomedical Optics and Imaging*, 5750:427–436, February 2005.

60. S. I. Nikolov and J. A. Jensen. In-vivo synthetic aperture flow imaging in medical ultrasound. *IEEE Transactions on Ultrasonics, Ferroelectrics, and Frequency Control*, 50(7):848–856, 2003.

61. K. L. Gammelmark and J. A. Jensen. Multielement synthetic transmit aperture imaging using temporal encoding. *IEEE Transactions on Medical Imaging*, 22(4):552–563, April 2003.

62. M. H. Pedersen, K. L. Gammelmark, and J. A. Jensen. In-vivo evaluation of convex array synthetic aperture imaging. *Ultrasound in Medicine and Biology*, 33:37–47, 2007.

63. J. A. Jensen and S. I. Nikolov. Directional synthetic aperture flow imaging. *IEEE Transactions on Ultrasonics, Ferroelectrics, and Frequency Control*, 51:1107–1118, 2004.

64. O. Bonnefous. Measurement of the complete (3D) velocity vector of blood flows. *Proceedings of the IEEE Ultrasonics Symposium*, 795–799, 1988.

65. J. A. Jensen and N. Oddershede. Estimation of velocity vectors in synthetic aperture ultrasound imaging. *IEEE Transactions on Ultrasonics, Ferroelectrics, and Frequency Control*, 25:1637–1644, 2006.

66. R. Y. Chiao, L. Y. Mo, A. L. Hall, S. C. Miller, and K. E. Thomenius. B-mode blood flow (B-flow) imaging. *Proceedings of the IEEE Ultrasonics Symposium*, 1677–1680, 2000.

67. J. A. Jensen. Method for in-vivo synthetic aperture B-flow imaging. *Proceedings of SPIE, Progress in Biomedical Optics and Imaging*, 5373:44–51, 2004.

68. E. Mace, G. Montaldo, B. Osmanski, I. Cohen, M. Fink, and M. Tanter. Functional ultrasound imaging of the brain: Theory and basic principles. *IEEE Transactions on Ultrasonics, Ferroelectrics, and Frequency Control*, 60(3):492–506, 2013.

69. J. Bercoff, G. Montaldo, T. Loupas, D. Savery, F. Meziere, M. Fink, and M. Tanter. Ultrafast compound Doppler imaging: Providing full blood flow characterization. *IEEE Transactions on Ultrasonics, Ferroelectrics, and Frequency Control*, 58(1):134–147, January 2011.

70. B-F. Osmanski, M. Pernot, G. Montaldo, A. Bel, E. Messas, and M. Tanter. Ultrafast Doppler imaging of blood flow dynamics in the myocardium. *IEEE Transactions on Medical Imaging*, 31(8):1661–1668, 2012.

71. K. L. Hansen, J. Udesen, F. Gran, J. A. Jensen, and M. B. Nielsen. In-vivo examples of complex flow patterns with a fast vector velocity method. *Ultraschall in der Medizin*, 30:471–476, 2009.

72. E. Mace, G. Montaldo, I. Cohen, M. Baulac, M. Fink, and M. Tanter. Functional ultrasound imaging of the brain. *Nature Methods*, 8(8):662–664, 2011.

73. E. Konofagou and J. Ophir. A new elastographic method for estimation and imaging of lateral displacements, lateral strains, corrected axial strains and poisson's ratios in tissues. *Ultrasound in Medicine and Biology*, 24(8):1183–1199, 1998.

74. K. Nightingale, M. S. Soo, R. Nightingale, and G. Trahey. Acoustic radiation force impulse imaging: In-vivo demonstration of clinical feasibility. *Ultrasound in Medicine and Biology*, 28:227–235, 2002.

75. K. R. Nightingale, M. L. Palmeri, R. W. Nightingale, and G. E. Trahey. On the feasibility of remote palpation using acoustic radiation force. *Journal of the Acoustical Society of America*, 110(1):625–634, 2001.

76. W. N. McDicken, G. R. D. Sutherland, C. M. Moran, and L. N. Gordon. Color Doppler velocity imaging of the myocardium. *Ultrasound in Medicine and Biology*, 18(6–7):651–654, 1992.

77. K. L. Hansen, J. Udesen, C. Thomsen, J. A. Jensen, and M. B. Nielsen. In vivo validation of a blood vector velocity estimator with MR angiography. *IEEE Transactions on Ultrasonics, Ferroelectrics, and Frequency Control*, 56(1):91–100, 2009.

78. M. M. Pedersen, M. J. Pihl, J. M. Hansen, P. M. Hansen, P. Haugaard, M. B. Nielsen, and J. A. Jensen. Secondary arterial blood flow patterns visualised with vector flow ultrasound. *Proceedings of the IEEE Ultrasonics Symposium*, 1242–1245, 2011.

79. J. B. Olesen, M. S. Traberg, M. J. Pihl, and J. A. Jensen. Non-invasive measurement of pressure gradients using ultrasound. *Proceedings of SPIE, Medical Imaging*, 1–7, March 2013, 86750G.

Chapter 3

Diagnostic Ultrasound Imaging

4. Ultrasound Elastography

Timothy J. Hall, Assad A. Oberai, Paul E. Barbone, and Matthew Bayer

Chapter 4

Ultrasound Imaging and Therapy. Edited by Aaron Fenster and James C. Lacefield © 2015 CRC Press/Taylor & Francis Group, LLC. ISBN: 978-1-4398-6628-3

4.1   Introduction

Ultrasound is a commonly used imaging modality that is still under active development with great potential for future breakthroughs although it has been used for decades. One such breakthrough is the recent commercialization of methods to estimate and image the (relative and absolute) elastic properties of tissues. Most leading clinical ultrasound system manufacturers offer some form of elasticity imaging software on at least one of their ultrasound systems. The most common elasticity imaging method is based on a surrogate of manual palpation.

There is a growing emphasis in medical imaging toward quantification. Ultrasound imaging systems are well suited toward those goals because a great deal of information about tissues and their microstructure can be extracted from ultrasound wave propagation and motion tracking phenomena. Estimating tissue viscoelasticity using ultrasound as a quantitative surrogate for palpation is one of those methods.

This chapter will review the methods used in *quasi-static* (palpation-type elastography). The primary considerations in data acquisition and analysis in commercial implementations will be discussed. Moreover, methods to extend palpation-type elastography from images of relative deformation (mechanical strain) to quantitative images of elastic modulus and even the elastic nonlinearity of tissue will also be presented. Section 4.4 will review promising clinical results obtained to date.

4.1.1   Brief Clinical Motivation

Manual palpation has been a common component of medical diagnosis for millennia. It is well understood that physiological and pathological changes alter the stiffness of tissues. Common examples are the breast self-examination (or clinical breast examination) and the digital rectal examination. Although palpation is commonly used, it is known to lose sensitivity for smaller and deeper isolated abnormalities. Palpation is also limited in its ability to estimate the size, depth, and relative stiffness of an inclusion or to monitor changes over time.

Given the long history of successful use of palpation, even with its limits, there was a strong motivation to develop a surrogate that could remove a great deal of the subjectivity, provide better spatial localization, provide spatial context of surrounding tissues, and improve estimates of tumor size and relative stiffness. The spatial and temporal sampling provided by clinical ultrasound systems, as well as their temporal stability, make them very well suited to this task. The first clinically viable real-time elasticity imaging system was reported in 2001 [1], and significant improvements have been made since then.

The typical commercial elasticity imaging system provides real-time elasticity imaging with either a side-by-side display of standard B-mode and strain images or a color overlay of elasticity image information registered on the B-mode image (or both options). Some metric of feedback to the user is also often provided so the user knows if the scanning methods are appropriate and/or if the data acquired are high quality.

4.1.2 Theory Supporting Quasi-Static Elastography

A basic assumption commonly used in palpation-type elastography is that the loading applied to deform the tissue is quasi-static, meaning that motion is slow enough that inertial effects—time required and inertial mass—are irrelevant. Following the presentation of Fung [2], a basic description of the underlying principles can be used to understand the assumptions used in elasticity imaging.

There are a variety of descriptors of the motion associated with solid mechanics, but a useful one for our purposes is the Cauchy–Almansi strain tensor:

$$e_{ij} = \frac{1}{2}\left[\frac{\partial u_j}{\partial x_i} + \frac{\partial u_i}{\partial x_j} - \frac{\partial u_k}{\partial x_i}\frac{\partial u_k}{\partial x_j}\right], \tag{4.1}$$

$$\approx \frac{1}{2}\left[\frac{\partial u_j}{\partial x_i} + \frac{\partial u_i}{\partial x_j}\right] = \varepsilon_{ij}, \tag{4.2}$$

where $u(x_1, x_2, x_3, t)$ is the displacement of a particle instantaneously located at x_1, x_2, x_3, and time t, $i = 1, 2, 3$ (three-dimensional [3-D] space), and repeat indices imply summation over that index (Einstein's summation notation). The particle velocity, v_i, is given by the *material* derivative of the displacement as follows:

$$v_i = \frac{\partial u_i}{\partial t} + v_j\frac{\partial u_i}{\partial x_j}, \tag{4.3}$$

and the particle acceleration, α_i, is similarly defined as follows (replacing particle displacements with particle velocities in Equation 4.3):

$$\alpha_i = \frac{\partial v_i}{\partial t} + \alpha_j\frac{\partial v_i}{\partial x_j}. \tag{4.4}$$

The conservation of mass is expressed by

$$\frac{\partial \rho}{\partial t} + \frac{\partial(\rho v_i)}{\partial x_i} = 0, \tag{4.5}$$

where ρ is the mass density in the neighborhood of point u_i, and the conservation of momentum is expressed by

$$\rho\alpha_i = \frac{\partial \sigma_{ij}}{\partial x_j} + X_i, \tag{4.6}$$

where σ_{ij} are the stresses and X_i represents body forces.

Often, the medium of interest may be accurately modeled as a linear elastic and iso-tropic material. In that case, Hooke's law relating stress to linearized strain is written as follows:

$$\sigma_{ij} = \lambda \varepsilon_{kk} \delta_{ij} + 2\mu \varepsilon_{ij}, \tag{4.7}$$

where λ and μ are the first and the second Lamé constants, respectively.

Equations 4.1 through 4.7 represent 22 equations with 22 unknowns (ρ, u_i, v_i, α_i, ε_{ij}, σ_{ij}, $i, j = 1, 2, 3$). The tensors ε_{ij} and σ_{ij} are both symmetric, so they contain only six inde-pendent values each.

In many elasticity imaging contexts, displacements and particle velocities are suffi-ciently small in that their products may be neglected. With that assumption, Equations 4.1 through 4.6 may be linearized (disregarding products and cross terms of small quan-tities). Dropping cross terms simplifies the Cauchy–Almansi strain tensor in Equation 4.1 into the infinitesimal strain tensor Equation 4.2 and the material derivatives in Equations 4.3 and 4.4 into simple derivatives. Making these substitutions and then fur-ther substituting Equation 4.7 into Equation 4.6 yield the well-known Navier equation in solid mechanics. In the indicial notation used so far, this equation is presented as follows:

$$\mu \frac{\partial^2 u_i}{\partial x_j \partial x_j} + (\lambda + \mu) \frac{\partial^2 u_j}{\partial x_i \partial x_j} + X_i - \rho \frac{\partial^2 u_i}{\partial t^2} = -\frac{\partial \lambda}{\partial x_i} \frac{\partial u_j}{\partial x_j} - \frac{\partial \mu}{\partial x_j} \left(\frac{\partial u_i}{\partial x_j} + \frac{\partial u_j}{\partial x_i} \right). \tag{4.8}$$

In a homogeneous material, the right-hand side of Equation 4.8 vanishes, which gives the following equation, with the further assumption $X_i = 0$:

$$\mu \frac{\partial^2 u_i}{\partial x_j \partial x_j} + (\lambda + \mu) \frac{\partial}{\partial x_i} \frac{\partial u_j}{\partial x_j} = \rho \frac{\partial^2 u_i}{\partial t^2}. \tag{4.9}$$

This equation can also be expressed in a more streamlined vector notation, in which we again drop the body forces for simplicity:

$$\mu \nabla^2 \mathbf{u} + (\lambda + \mu) \nabla (\nabla \cdot \mathbf{u}) = \rho \frac{\partial^2 \mathbf{u}}{\partial t^2}. \tag{4.10}$$

Two wave equations may be derived from the Navier equation (Equation 4.10), one for compressional waves and one for shear waves. The fundamental theorem of vector calculus (i.e., Helmholtz theorem) states that any smooth nonsingular vector field can be broken into a sum of a divergence-free vector field and a curl-free vector field. We call these two components $\mathbf{u}_s$ and $\mathbf{u}_c$, respectively, with the subscripts standing for "shear" and "compressional"; thus, the equation is presented as follows:

$$\mathbf{u} = \mathbf{u}_s + \mathbf{u}_c. \tag{4.11}$$

Equation 4.10 can then be split into two equations, one for each component of the vector field. For $\mathbf{u}_s$, the divergence vanishes, and we obtain

$$\mu\nabla^2\mathbf{u}_s = \rho\frac{\partial^2\mathbf{u}_s}{\partial t^2}. \tag{4.12}$$

For $\mathbf{u}_c$, we can use the identity $\nabla(\nabla \cdot \mathbf{u}) = \nabla^2\mathbf{u} + \nabla \cdot (\nabla \cdot \mathbf{u})$, noting that the curls vanish, to obtain

$$(\lambda + 2\mu)\nabla^2\mathbf{u}_c = \rho\frac{\partial^2\mathbf{u}_c}{\partial t^2}. \tag{4.13}$$

In a heterogeneous medium, the two types of wave fields are coupled.

Equations 4.12 and 4.13 are wave equations. The divergence-free vector field $\mathbf{u}_s$ represents a shear wave (sometimes called an *S-wave*), with the wave speed dependent on density and on the Lamé constant μ, also known as the *shear modulus*. The curl-free vector field $\mathbf{u}_c$ represents a compressional wave (sometimes called a *P-wave*), which is the type of wave emitted and collected by ultrasound imaging devices. The compressional wave speed is determined by the density and the sum $(\lambda + 2\mu)$, called the *P-wave modulus*. In soft tissues, $\lambda \gg \mu$, and hence $\lambda + 2\mu \approx \lambda \approx K$, the bulk modulus.

4.1.3   Clinical Implementation of Strain Imaging

Quasi-static elastography is usually performed with freehand scanning, analogous to other forms of clinical ultrasound imaging. Software packages to perform elasticity imaging are now implemented on clinical ultrasound imaging systems from most manufacturers. These systems provide images of relative deformation (mechanical strain), which are mapped either to grayscale images (black showing effectively no strain, white showing highest strain in the field), as shown in Figure 4.1, or some other color map, where the latter option can be displayed separately or as an overlay on standard B-mode images.

A limiting factor for current implementations of quasi-static elastography is the dominance of 1-D array transducers used in clinical ultrasound imaging systems. With these transducers, generally only 2-D radio frequency (RF) data fields are available, and from these, only 2-D displacement and strain fields can be estimated. This restriction prevents tracking motion perpendicular to the image plane and limits the ability to track a particular volume element in tissue over relatively large (<5% strain) deformation. This places a practical limit on the ability to track large single-step deformations. Another practical limit to tracking large deformations with a 1-D array is the difficulty in keeping the tissue in the image plane during deformation. Even with a frame-to-frame deformation of 1% strain, a sequence of only a few images results in enough strain to require some training and skill in obtaining high-quality sequences of strain images [3]. As a result, most commercial implementations of quasi-static elastography are optimized for relatively small (~0.3% strain) deformations.

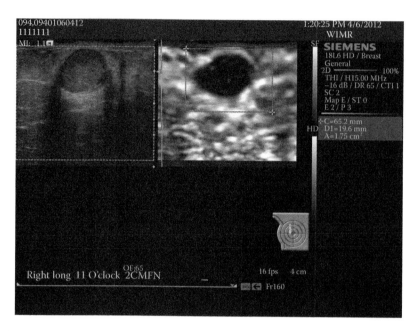

FIGURE 4.1 Typical B-mode and strain images of a fibroadenoma in a breast. The strain image of the tumor is comparable in size with that seen in the B-mode image (consistent with benign disease).

It is also well established that the appearance of a strain image, at least in breast tissues, is highly dependent on the amount of preload [3], and this is supported by recent studies demonstrating that shear wave speeds in the breast depend on preload [4]. The implications of elastic nonlinearity in quasi-static elastography are discussed in Section 4.3.

4.2 Motion Tracking and Strain Imaging

4.2.1 Basics

The core goal in ultrasound elastography is to deduce the elastic properties of tissue by observing how it moves. The motion being observed may be intrinsic to the body part being observed, as in imaging of the heart, or it may be induced. There are several ways of inducing motion in tissue. Acoustic radiation force impulse (ARFI) imaging, for example, uses a strong ultrasound pulse to create a force on tissue below the skin surface. Force can also be applied at the skin surface, either by the ultrasound transducer itself or by some external device.

The types of induced motion can also be categorized according to time scale. Dynamic elastography tracks transient or sinusoidal motion at higher frequencies, whereas quasi-static elastography tracks motion that is slow compared with relaxation processes in the tissue. For the quasi-static setup, the motion may be as simple as holding the transducer in place while a patient breathes. Viscous effects are important in dynamic elastography but are generally not considered in quasi-static methods, although there are exceptions.

Whatever the character of the motion, a series of ultrasound image frames is then acquired while the tissue deforms. The image frames may be 2-D planes or, less commonly, 3-D volumes, depending on the imaging hardware. Signal and imaging processing techniques are used to estimate a map of the tissue displacements that have occurred between any pair of image frames.

Finally, the spatial or temporal patterns in these displacements may be analyzed to reveal the elastic properties of tissue, such as stiffness or viscosity. Stiffness is the most common material property of interest and is often displayed in images of relative strain. However, as described in Section 4.3, quantitative images of fundamental material properties (e.g., shear modulus) are possible based on these data.

4.2.2   Ultrasound Image Formation

Many features of the ultrasonic motion tracking problem depend on the ultrasound image formation process. Briefly, an ultrasound system creates an image by transmitting a series of short, focused pulses of sound. Tissue contains dense, semirandom variations in acoustic impedance, which causes echo signals to return to the system. The position, shape, organization, number, and relative impedance difference of those scattering sources affect the individual echo signals from those scatterers, averaged over the volume of the pulse. This variation is the source of ultrasound (brightness mode [B-mode]) image contrast. The transmitted acoustic pulse therefore forms the basis for the system's point spread function (PSF), which is convolved over the tissue's scatterers to form an image [5,6]. In an ultrasound imaging system, the PSF may vary significantly with position because it depends on the focal properties of the ultrasound transducer, but over moderately sized regions, the PSF can be approximated as invariant.

Because the ultrasound pulse is oscillatory, it does not simply blur the response from these small scatterers. Instead, the scattered waves interfere, creating randomly distributed regions of constructive and destructive interference corresponding to high and low wave amplitudes. This random interference results in the distinctive patterns of ultrasound speckle [7]. The patterns are random to the extent that the underlying scatterer distribution is random, but they are deterministic in the sense that over multiple imaging experiments, the same piece of tissue imaged under the same conditions will produce identical speckle patterns. If the underlying piece of tissue translates, the speckle will move with it.

Speckle detracts from conventional ultrasound imaging because its variation in brightness does not correspond to real tissue structures, and it decreases the visibility of any real structures present. It is ideal for the motion tracking problem, however, because it contains a high level of detail and is stable over multiple images within limits. These properties make it possible to effectively track a small region of interest in tissue by tracking the speckle pattern it produces.

4.2.3   Motion Tracking Algorithms

Many algorithms exist to perform this motion tracking. Most algorithms operate on two image frames, a predeformation frame and a postdeformation frame. The data used are different from the usual ultrasound images. Standard ultrasound (B-mode) images are log-compressed versions of the envelope of the returned echo signal. The enveloping procedure discards the high-frequency oscillations (RF carrier) in the returning echo, only retaining the wave envelope amplitude. Motion tracking algorithms more often use the raw RF echo signal because rapid phase changes in these oscillations allow more precise displacement estimation.

Given this RF data, a data window is selected for each site in the predeformation frame at which it will estimate displacement. The size and dimensionality of this window, also known as the *correlation window* or *correlation kernel*, can vary. In general, smaller kernels obtain better resolution, whereas larger kernels incorporate more data and thus decrease tracking error.

From this point, two general methods can be used. In the first, called the *block matching* or *correlation-based method*, this data kernel is compared with a set of similarly sized kernels within a search region in the postdeformation image [8]. The best match among these kernels is evaluated by a matching function such as sum of absolute differences (SAD), sum of squared differences (SSD), or normalized cross correlation (NCC) [9]. The difference between the position of the best matching kernel and the first kernel's original position is taken to be the displacement at that point in the predeformation image.

The block matching search can only move data kernels and compute matching functions one sample at a time, so the process so far has only produced a displacement value rounded to the nearest sample. To obtain more precise measurements, some means of displacement estimation at subsample scales is required. Again, several methods exist. One simple method is to upsample the original ultrasound data so that the ultimate sample-level estimates are finer. Another approach is to fit the matching function values to a function model, such as a parabola or a cosine, and compute the location of the peak [10]. More complex methods have also been developed. One example, devised by Viola et al. [11,12], uses the spline representation of one of the signals to construct the SSD as a polynomial function of displacement then solves for the location of the polynomial's minimum.

In the second general tracking method, known as the *phase-based method*, the displacement is not found by finding the peak of a matching function but rather by the measurement of a phase difference between the predeformation and the postdeformation kernels. Tissue Doppler [13], which applies traditional Doppler signal processing to tissue rather than blood, could be classified as a phase-based motion tracking. Other common variants include the phase-zero estimation [14] and the Loupas algorithm [15,16] used for ARFI elastography. These estimation methods can be more accurate or faster than correlation-based equivalents for small displacements, but aliasing prevents their use in tracking displacements that exceed half a wavelength.

For either of these methods, data kernels and search regions can have varying dimensionality. The initial experiments in ultrasound elastography used 1-D kernels and 1-D search regions, measuring only axial displacements [17]. These initial methods were analogous to previous work on time-delay estimation for sonar signals and similar 1-D data [18,19]. Extensions to two dimensions were soon made, however, finding that 2-D kernels reduced tracking error [20,21] and that 2-D search regions account for lateral motion [22] and enable the measurement of lateral tissue displacements [23].

4.2.4   Strain Imaging

Raw displacement maps are difficult to interpret. One option for producing an image of clinical value is to compute and display strain instead of displacement. Strain is the gradient of displacement and measures how much a tissue has been compressed, stretched, or sheared. In quasi-static elastography, the most revealing quantity is usually axial normal strain, the amount of compression or expansion in the direction of the ultrasound beams, which is also the direction of applied force. This type of strain is a useful surrogate for tissue stiffness, and images of axial normal strain have been repeatedly shown to have clinical value [24,25].

The diagnostic potential of lateral strain [26,27] and shear strain [28,29] are also under investigation. Lateral strain can be used to measure tissue relaxation under a constant axial displacement, for example, and shear strain patterns can indicate the degree of slip at a tumor boundary. Malignant tumors tend to be more tightly bound to the background tissue than benign ones and can be distinguished by larger image areas undergoing shear strain [29].

The technical challenge of computing strain from displacement is equivalent to the challenge of estimating derivatives from noisy data. Common methods compute simple finite differences or make piecewise least squares linear fits to the displacement data [30]. Both methods can be analyzed as forms of linear filters on the displacement data, approximating the ideal differentiator while providing some level of noise immunity [31].

Strain imaging has been incorporated into many clinical systems, so engineers must consider topics such as image display and user feedback as well as algorithm design. Because strain images are most likely to be used in conjunction with traditional B-mode images [25], it is important to display both at once. Strain images are thus usually displayed as either a color overlay on the B-mode or as a separate grayscale image next to the B-mode.

Collecting good-quality data for tracking can be challenging in tissue, so methods have also been developed to provide automated feedback on strain image quality. Jiang et al. [32], for example, developed a metric that is the product of the correlation between the predeformation RF frame and a motion-compensated version of the postdeformation RF frame (a measure of tracking accuracy) and the correlation between consecutive strain images (a measure of strain image consistency). Displayed in real time, such a metric may help clinicians know when to rely on a strain image and when to adjust their imaging technique.

The major drawback of strain imaging is its relative nature. Strain values are not intrinsic to tissue; they depend on the force applied and a tissue's surroundings. A simple inclusion in a homogeneous background, for example, naturally gives rise to strain patterns radiating outward from the inclusion. This false contrast in the background is known as a *stress concentration artifact*. Moreover, although sometimes strain itself is the desired diagnostic quantity—as in cardiac elastography, where low contractility can indicate tissue damage—strain is more often a surrogate for an intrinsic parameter like shear modulus. For these reasons, it is desirable to estimate these intrinsic parameters directly. Modulus reconstruction methods are addressed in detail in Section 4.3.

4.2.5   Motion Tracking Performance and Error

There are three main classes of error for motion tracking algorithms. The first is known as *peak-hopping error* and occurs when the predeformation data window is matched

Chapter 4

with the wrong area in the postdeformation image. The algorithm "hops" from the true peak in the matching function to a false peak. This error often appears in a displacement map as an isolated point bearing no relation to its neighbors. Figure 4.2 illustrates this type of error as well as others to be described. The magnitude of a peak-hopping error can be as large as the search region.

Peak hops are most directly linked to the sample-level displacement estimation and can be greatly reduced by the use of multilevel or guided-search strategies, which will be described later. The other two classes of error are more connected with the subsample estimation method. One of these classes is sometimes called *jitter error*, which is simply the variance of measured displacement values around the true value. Assuming the correct peak in the matching function has been selected, the exact location of that peak is still a noisy measurement. Several theoretical analyses exist for this type of error [33–36].

The last class of error is the bias inherent in most subsample estimation methods [10] and is thus systematic rather than random. These methods begin with a sample-level displacement estimate and then calculate a refinement to it. For most methods, the refinement is biased toward the original sample-level estimate. For example, an image point with a true displacement of 2.3 samples will have a sample-level estimate of 2, and the expected value of its subsample estimate may be 2.2. This bias is most significant when there is low overall displacement and causes a characteristic banding pattern in the resulting strain images. The magnitude of this bias depends on the subsample estimation method used; simpler methods tend to have a greater bias than more sophisticated methods.

Peak-hopping and jitter errors have the same ultimate causes, although it is helpful to distinguish between them. The first and most expected source of error is electronic noise in the ultrasound equipment. This noise will cause differences between two images of

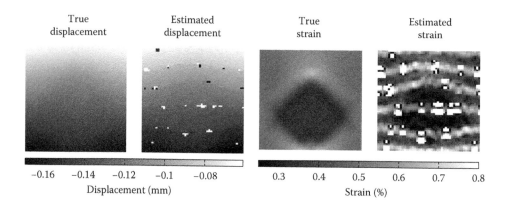

FIGURE 4.2 Simulated example of displacement maps and strain images, illustrating the various types of motion tracking error. The simulated tissue is a hard spherical inclusion in uniform background material. Peak-hopping errors appear as isolated, obvious errors in the estimated displacement map (second from left). In the same image, jitter errors are evident in its generally noisier appearance compared with the true displacement (far left). There is also a subtle banding pattern visible in the estimated displacement map. This pattern would be more obvious in the estimated strain image (far right) if it were not so dominated by the results of peak-hopping errors.

the same region of tissue and thus degrade the accuracy of a matching function. Other noise or artifacts from the ultrasound imaging process would also affect tracking performance. Reverberation artifacts, for example, which can arise from multiple specular reflections between layers of tissue, do not move in the same way as the underlying tissue and therefore will mislead a tracking algorithm. Another source of error for 2-D images is motion perpendicular to the image plane, in the elevational direction (see Figure 4 of Hall et al. [3]). Motion that cannot be seen in the image cannot be tracked. With small deformations and steady hands, out-of-plane motion can be minimized, but it is guaranteed to be significant for large deformations.

The last and most complex source of error arises from the very process being measured: tissue deformation. In particular, any kind of motion that deviates from the implicit assumptions in the block matching algorithm will cause problems. The implicit assumption is that each small speckle patch does not change or deform but only translates. For small tissue deformations and small speckle patches, this assumption approximately holds. If compression, rotation, or shear becomes large enough to change the relative position of the acoustic scattering sources within an ultrasound pulse, the block matching assumption breaks down, and speckle patches will no longer remain stable between images, an effect called *strain decorrelation*.

To a first approximation, the primary effect of tissue deformation is a corresponding distortion in the image mirroring that deformation [37]. If the strain is high or the correlation windows are large, the data inside a correlation window in the postdeformation image will deform, enough to degrade the matching function. The importance of this part of strain decorrelation depends on the size of the correlation windows used because larger windows will span more deformation and be poorer matches for their undeformed counterparts. A second effect results from the tissue deformation within the volume of the ultrasound pulse. If deformation is large enough, pulses in the postdeformation image will have scattered from a different collection of scatterers, or the same collection at slightly different locations, than they did in the predeformation image. This will cause the shape of the speckle to change in a way that does not correspond to the underlying tissue motion [5].

For elastography methods with very small deformations, such as ARFI, errors due to electronic noise may be most important, and peak-hopping errors can be entirely avoided. For the larger deformations of quasi-static elastography, generally all types and sources of error are relevant.

4.2.6   Refinements

The basic algorithm of motion tracking by kernel comparison has been subject to countless modifications and improvements. Often, an advanced algorithm will target a particular type or source of motion tracking error.

One strategy for decreasing peak-hopping errors is guided search, where certain displacement estimates are used as a first guess for the displacements at their neighbors [8,38,39]. A multilevel approach serves a similar purpose, where displacements are first estimated on a coarse grid that is interpolated and used as guidance for a denser grid [40]. In both approaches, the search region of the guided displacement estimation sites can be reduced far enough to eliminate the threat of a peak hop, provided the guess displacement is not in error. These strategies have the further benefit of reducing computation time because only a subset of the total number of estimation sites has to use a full-size search region.

Another strategy for reducing peak hops is regularization, which selects displacement estimates subject to some kind of informative constraint, such as continuity between neighboring estimation sites. Various types of regularization have been attempted [41–43]. It is also possible to use regularized displacement estimates to initialize a guided search, thus combining the two strategies [32].

One example of these types of strategies is the quality-guided algorithm of Chen et al. [39,44] and its extension using regularization by Jiang and Hall [45]. The quality-guided algorithm begins by estimating displacements in a coarse grid over the RF echo signal frames, using a full search region. Those initial estimates, or "seeds," are then used to guide neighboring estimation sites. When a new site's displacement is estimated, that displacement provides guidance for its neighbors. Estimation sites with higher-quality guidance, measured by the correlation value of the guiding estimate, are processed first. In this way, estimates with higher correlations propagate to guide large regions of the image, and estimates with low correlations do not spread and are replaced. Jiang's addition to this method was to "validate" seed estimates by regularization with respect to the spatial continuity of displacements in their immediate neighborhood.

Another type of refinement to the motion tracking procedure is to compensate for some of the deformation-induced errors by adapting the RF echo data to the deformation it is experiencing. This process is known as *companding* (for compressing/expanding) or *temporal stretching*. Methods include global stretching of the postdeformation image by the average strain, local redeformation according to initial strain estimates from the uncorrected images [40], and an adaptive search of possible strain values, taking the best fit as a direct measure of tissue strain [46,47].

Three-dimensional motion tracking, making use of 3-D ultrasound imaging, is also an active area of research [48,49]. This development directly addresses errors due to elevational motion. It also has the potential to make elastography easier for users, with less need for very precise motions to avoid out-of-plane motion. The basic principles of motion tracking are directly carried over from the 2-D case, but the abundance of data and the lower frame rates introduce practical challenges.

4.2.7   Displacement Accumulation

A final modification to the standard motion tracking algorithm is that motion may be tracked in multiple steps, known as an *accumulation* or *multicompression* strategy [50–55]. This is required for large total deformations—exceeding approximately 5% strain—because the change in the speckle pattern is too great to simply track between image frames at the beginning and end of the deformation. It may also be used for elastography methods that need a record of displacement through time, such as ARFI or shear wave elastography. In either case, ultrasound image frames are acquired at intervals over the course of a deformation. Displacements are estimated between these intermediate images and then accumulated if necessary for the application.

Because each estimation step carries its own error, accumulated displacement estimates also tend to accumulate error. Reducing the number of steps decreases the accumulated error but increases the strain and decorrelation in each step. Intuitively, there would exist some optimum strain step size for displacement accumulation [52]. With smaller steps, there is an unnecessary accumulation of estimation error; with larger steps, strain decorrelation degrades the results.

Recent work has discovered that significant covariances exist between steps [54,55], which has important effects on the accumulated displacement error. Errors induced by strain are correlated between accumulation steps, so that they tend to build up quickly. Errors induced by electronic noise, in contrast, are anticorrelated and tend to cancel one another. Together, these properties reduce the expected dependence of accumulated error on strain step size, resulting in a broader optimum for multicompression techniques.

4.3   Modulus Reconstruction

In this section, we describe the process of inferring the spatial distribution of elastic properties of tissue from the knowledge of its displacement field. This is accomplished by making use of mechanical balance laws, assumed stress–strain models (called *constitutive models*), and measured displacement fields. Evaluating material properties instead of relying on strain as an inverse measure of stiffness has several benefits, including the following:

1. Material properties are largely independent of operating conditions and therefore offer an objective assessment of the tissue.
2. Certain behavior, such as changes in stiffness with increasing strain, is difficult to comprehend by examining strain images alone. By contrast, mechanical images provide a clearer picture.
3. Quantitative material property images have applications beyond detection and diagnosis of disease. They can be used for treatment monitoring and for generating patient-specific models for surgical planning.

In Section 4.3.1, we first consider the mathematical models that are derived from the mechanical balance laws and constitutive models. These models describe a relationship between tissue displacement and the spatial distribution of its material properties. Because this relationship determines the amount of displacement data required to obtain a unique distribution of the material parameters, we also discuss the uniqueness of the underlying inverse problem in this section.

Thereafter, we describe two alternate computational strategies for determining the spatial distribution of material properties from displacement estimates in Section 4.3.2. The first strategy relies on directly using the measured displacements in the equations of equilibrium to determine the material properties. We refer to this as the direct approach. The second strategy uses the displacement data in an objective function that measures the difference between a predicted and a measured displacement field. The material property distribution is then obtained by minimizing this difference. We refer to this approach as the minimization method. Generally speaking, the direct method is computationally less demanding; however, it is more sensitive to noise, and it cannot handle incomplete data as easily as the minimization method.

Chapter 4

Finally, we present some of our recent work in the area of elasticity imaging in Section 4.3.3. This includes new constitutive models that are based on the underlying tissue microstructure and using force data to create quantitative elasticity images using quasi-static ultrasound elastography.

4.3.1 Mathematical Models and Uniqueness

4.3.1.1 Linear Elasticity

During quasi-static ultrasound elastography, the speed of compression is much slower than any of the mechanical wave speeds within the medium. The inertial term may be neglected as a result in the equation for the balance of linear momentum. Given this and the fact that tissue is primarily composed of water, it may also be assumed to be incompressible. Further, when dealing with glandular tissue, one may assume that its response is isotropic. Consequently, the equations for the quasi-static infinitesimal deformation of an incompressible isotropic linear elastic material are a good starting point for a mathematical model for this application. The only material parameter that appears in these equations is the shear modulus, denoted by μ.

In three dimensions, the equation of the balance of linear momentum enforces a differential relationship between the shear modulus μ, the displacement u, and the pressure p. This may be written as

$$N(u, p; \mu) = 0. \tag{4.14}$$

We note that although the pressure field is not the primary quantity of interest, it still needs to be treated as an unknown because it is not measured. Thus, the problem we wish to solve is as follows: given u, find μ and p, such that together they satisfy Equation 4.14. This leads to a system of three partial differential equations (PDEs) for the two unknowns, μ and p. Despite having more equations than unknowns, this system is not overdetermined. In fact, it is underdetermined. In particular, we have shown that given a single displacement field, there is an infinite number of compatible shear modulus distributions and pressure fields that satisfy Equation 4.14. The reason for this overwhelming nonuniqueness is the lack of boundary data for μ or p. This nonuniqueness of the inverse problem can be corrected easily by including an additional measured displacement field whose principal strain directions are different from the first field. Then the dimension of the space of shear modulus distributions that are compatible with both measured displacement fields is five. The shear modulus can be written as a linear combination of at most five independent fields. Thus, an additional measured displacement helps tremendously in correcting the ill-posedness of the underlying problem and in removing ambiguity in reconstructed images [56].

In quasi-static ultrasound elastography, displacement data are typically measured in a plane. As a result, the 3-D elasticity problem has to be simplified to a 2-D setting. There are two options available: plane strain or plane stress. Both assume that the property distribution does not vary in the out-of-plane direction. Further, one assumes that out-of-plane strains vanish in plane strain and out-of-plane stresses vanish in plane stress. Consequently, plane strain is more appropriate for thick specimens that are confined in the out-of-plane direction, whereas plane stress assumption is appropriate for thin, unconfined specimens. Although it is not clear as to which assumption is more appropriate during ultrasound elastography, perhaps the fact that the breast is typically not confined as it is compressed makes plane stress a better assumption.

In the plane strain assumption, the form of the equations of equilibrium is unchanged from three dimensions (Equation 4.14). The 2-D version of this equation implies that a single displacement field is compatible with an infinite number of shear modulus fields, and so the problem is (very) nonunique [57,58]. By contrast, by requiring the modulus to be compatible with two independent displacement fields, the dimension of the set of possible shear modulus distributions reduces to four [58].

In the plane stress hypothesis, one can determine the pressure field completely in terms of the measured strains and the shear modulus by equating the out-of-plane normal stress to zero. Once this expression is inserted in the equations of equilibrium written for the in-plane directions, we arrive at the following equation:

$$N(u; \mu) = 0. \tag{4.15}$$

Note that the previously mentioned equations do not contain pressure. Using a single measured displacement field in these equations, one can determine the shear modulus everywhere up to a multiplicative constant [59]. Thus, the assumption of plane stress yields a nearly unique solution for μ with a single displacement field.

In small-deformation quasi-static elastography, the equations of equilibrium for the quasi-static infinitesimal deformation of an incompressible isotropic linear elastic material provide the relationship between the material parameters and the displacement field. These equations must be solved to determine the material parameters. In three dimensions and in two dimensions under the plane strain hypothesis, a single deformation field yields an infinite number of shear modulus distributions that are compatible with these equations. Hence, the problem of recovering the shear modulus is very nonunique. By measuring another independent displacement field, this nonuniqueness can be addressed to a large extent. By contrast, the 2-D plane stress problem provides a unique solution (up to a multiplicative constant) with a single displacement field.

4.3.1.2 Nonlinear Elasticity

Over the last couple of decades, several ex vivo studies on the mechanical response of different breast tissues have revealed that the nonlinear elastic response of benign and malignant tumors is significantly different [60,61]. In particular, malignant tumors appear to start stiffening with strain at a smaller value of applied strain when compared with benign tumors. This observation is consistent with the microstructural arrangement of collagen bundles observed in these tumors [62,63]. The collagen fiber bundles (which are the primary structural element of glandular tissue) are straight and less tortuous in malignant tumors and tend to be more tortuous and wavy in benign tumors. Thus, one would expect that fiber bundles in the former would uncoil to their arc length at a smaller applied strain than the latter. Further, once any fiber has reached this state, it would offer greater resistance to a deformation because of its large tensile stiffness. On the stress–strain curve, this would correspond to an earlier (at smaller strain) onset of nonlinear behavior.

Chapter 4

This has led researchers to consider a nonlinear stress–strain relationship of the type, that is,

$$\sigma = -p1 + \mu\left(e^{\gamma(I_1-3)}G + E\right),$$

(4.16)

in elasticity imaging. In Equation 4.16, p and μ are the pressure and the shear modulus, respectively; γ is a nonlinear parameter that determines the nonlinear response of the tissue; G and E are finite-deformation measures of strain; 1 is the identity tensor; and I_1 is the trace of the Cauchy–Green strain. A model of this type was used by Goenezen et al. [64] to determine the value of the nonlinear parameter in five fibroadenomas and five invasive ductal carcinomas (IDCs). It was found that the value of the nonlinear parameter was elevated for the malignant tumors, and one could correctly diagnose malignancy based on this value in 9 of 10 cases. The typical images of the shear modulus and nonlinear parameter for a fibroadenoma and an IDC are shown in Figures 4.3 and 4.4.

Clearly, the addition of a new nonlinear parameter implies that we need to measure displacement fields at small and finite strains. Using the study of Ferreira et al. [65], we determined just how much displacement data are necessary for a unique reconstruction

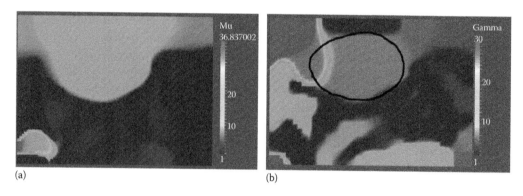

(a)   (b)

FIGURE 4.3   (a) Shear modulus image for a typical fibroadenoma. (b) Corresponding image of the nonlinear parameter. Note that within the tumor boundary, the value of the nonlinear parameter is small.

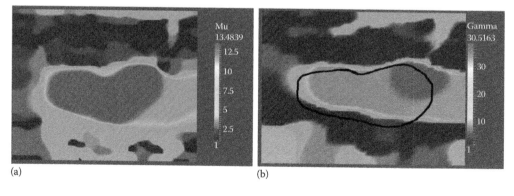

(a)   (b)

FIGURE 4.4   (a) Shear modulus image for a typical invasive ductal carcinoma (IDC). (b) Corresponding image of the nonlinear parameter. Note that within the tumor boundary the value of the nonlinear parameter is elevated.

of the shear modulus and the nonlinear parameter in two dimensions. Not surprisingly, our conclusions depended on whether we considered plane stress or plane strain and were analogous to the linear case. In particular, we derived the following conclusions:

1. For plane stress, we concluded that one displacement field at a small value of strain (say less than 1%) was sufficient to determine the shear modulus everywhere in the tissue. Thereafter, a single displacement field at a finite strain (say 15%) was sufficient to determine the nonlinear parameter everywhere.
2. For plane strain, one displacement field at small strain still allowed an infinite number of independent shear modulus distributions. However, the addition of one more independent field reduced this number to just four and made the problem almost unique. This result was the same as for the linear case. Thereafter, assuming that the shear modulus was already determined, one displacement field at finite strain allowed an infinite number of independent nonlinear parameter distributions. The addition of one more independent field reduced this number to just four and made the problem almost unique.

4.3.2   Direct and Minimization–Based Solution Methods

Broadly speaking, there are two types of methods available for solving the inverse elasticity problem: the direct method and the minimization method. Each has its benefits and drawbacks and therefore a class of problems to which it is best suited.

4.3.2.1   Direct Method

When the displacement field is measured everywhere in the tissue, we can view the equations of motion (Equations 4.14 and 4.15) as PDEs for the shear modulus (and pressure, for Equation 4.14), where the displacements and the resulting strains appear as spatially varying known parameters. Thus, one may attempt to solve these PDEs directly to determine the shear modulus. There are however several aspects of this problem that make its direct solution challenging:

1. There is no boundary data available for the material parameters. Thus, we need to develop methods that recognize this and solve the PDEs without the need for any boundary conditions.
2. As described in Section 4.3.1, the lack of boundary data makes this problem severely nonunique, and this nonuniqueness can be alleviated through the use of multiple measurements. Thus, any numerical method must be able to use data from multiple displacement measurements.
3. In all cases, the PDE system for the material parameters is a hyperbolic system, and such systems require special numerical methods with enhanced stability for their solution.
4. Finally, any noise in the displacement measurements implies noise in the measured strain, which in turn means rough (with large spatial gradients) parameters in the PDEs. The proposed numerical method should be designed to handle this situation.

Over the last several years, our group has developed a class of variation formulations, which we refer to as the adjoint-weighted equations (AWEs), that take into account the

Chapter 4

difficulties described earlier [66–68]. The discretization of these variational equations through standard finite element methods has led to efficient and robust numerical techniques for solving the inverse elasticity problem.

In the context for the plane stress problem, the AWE formulation is given by Equation 4.17, that is, find μ such that

$$\int_{\Omega} \sum_{i=1}^{M} N^* \left(u_m^{(i)}; w \right) \cdot N \left(u_m^{(i)}; \mu \right) dx = 0, \tag{4.17}$$

for all weighting function w. In Equation 4.17, N^* denotes the adjoint of the operator N, $u_m^{(i)}$ is the ith measured displacement field, Ω is the spatial domain over which the shear modulus distribution is sought, and M is the number of measured displacement fields. Under certain restrictions on the measured data, we have proven that this formulation will lead to numerical methods that are stable and convergent. We note that for linear elasticity, the discrete form of AWE leads to a simple, linear algebraic system, which needs to be solved to determine the nodal values of the shear modulus. This makes this method very fast and efficient.

One slight drawback of this method is its sensitivity to noise in the measured displacements. This can be overcome by adding a regularization term or by smoothing the displacements prior to their use. In either case, this adds somewhat to the complexity of the algorithm. The major shortcoming of the AWE method is its inability to handle missing data and displacement components with varying accuracy. The latter is particularly important in quasi-static elastography because the measured displacement component in the axial direction (along the transducer axis) is much more accurate than the component in the lateral direction. These shortcomings are overcome by the minimization method, at the expense of increased computational effort, which is described in the next section.

4.3.2.2 Minimization Method

In the minimization method, the inverse elasticity problem is solved as a minimization problem [30,69,70]. In particular, we seek a material parameter distribution that minimizes

$$\pi = \sum_{I=1}^{M} \frac{1}{2} \left\| T \left(u^{(i)} - u_m^{(i)} \right) \right\|^2 + \alpha \mathcal{R}[\mu] \tag{4.18}$$

under the constraint that each of the predicted displacement fields, $u^{(i)}$, satisfy the equations of equilibrium (Equation 4.14 or 4.15) with the given estimate of the shear modulus (or any other material parameter). In the previous equations, $\|\cdot\|$ denotes the L_2 norm, the matrix T is selected to weigh the more accurate displacement measurement directions more strongly, $\mathcal{R}$ is a regularization term, and α is the regularization parameter. The minimization formulation offers flexibility in that the matrix T can be selected to de-emphasize noisy data, and it can be set to zero matrix in regions where data are not available. Further, even in the presence of noisy data, the smoothness of the solution can be ensured by increasing the regularization parameter. A popular choice of the regularization term is the total variation regularization. This term penalizes fluctuations in the solution without regard to their slope. This makes it particularly useful for solving problems with abrupt changes in material properties (such as those observed in tumors).

The minimization problem is typically solved by computing the derivative of the objective function with respect to the optimization parameters (the nodal values of the shear modulus). Repeated evaluations of this vector (which is called the *gradient vector*) at different values of the shear modulus are then used to construct second derivative information embodied in an approximate Hessian matrix. The Hessian is used in a Newton-like algorithm to solve the minimization problem. The most expensive component of the approach described previously is the evaluation of the gradient vector. These costs can be reduced significantly by evaluating the solution of an adjoint problem [70,71]. In addition, when solving the nonlinear inverse elasticity problem, a continuation strategy in material parameters can be used to further bring down these costs [64,72,73]. Yet another approach to solving the minimization problem involves writing it as a constrained minimization problem, computing the nonlinear equations corresponding to the saddle-point solution, and solving these equations [74,75].

4.3.3 Recent Advances in Modulus Reconstruction

We end this section with descriptions of two recent advances in modulus reconstruction.

4.3.3.1 Quantitative Reconstruction

With recent advances in experimental capabilities and instrumentation, it is now possible to measure forces on several patches on the tissue as it is compressed. This additional data can be used to generate maps of the absolute value of elastic parameters, as opposed to maps that are relative to an unknown value.

We have developed two methods for accomplishing this. One is a postprocessing method, where we reconstruct relative modulus images using the displacement data. Thereafter, using this modulus distribution, we evaluate the force on the patch where the measured force data is available and rescale the shear modulus by the ratio of the measured to the predicted force so that they are rendered to be the same. This relatively simple and quick method makes quasi-static elasticity reconstructions quantitative. However, it does not make effective use of forces measured on several patches. Further, it is not easily extended to nonlinear elasticity models. Our second approach overcomes these limitations.

In this approach, we modify the displacement matching term when solving the minimization problem by appending to it a force-matching term. The force-matching term is equal to the sum of the square of the difference between measured and predicted forces. It depends on the material parameters directly through their appearance in the definition of the traction vector and indirectly through the predicted displacement field. Both these dependencies are accounted for while calculating the gradient vector when solving this problem. In tests of this method, we have found that the addition of the force-matching term to the objective function in Equation 4.18 is not a good strategy. Moreover, in any practical case, even a little (as small as 0.5%) force data yield unsatisfactory results. The

Chapter 4

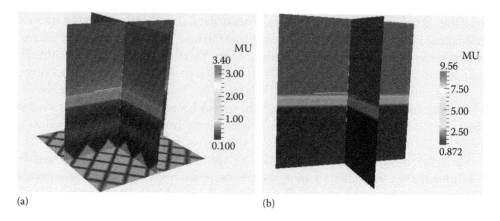

FIGURE 4.5 Reconstructed modulus distribution for a synthetic two-layer phantom with 1% noise in the displacement field. The actual value of the shear moduli in the bottom and top layers is 1 and 10 units, respectively, and the force is measured on the bottom face. When the force matching term is added to the objective function and it is solved for μ, the reconstruction on (a) is obtained. Notice the appearance of the oscillations in the bottom layer and an artificial boundary layer. Using the same data when the objective function is written in terms of log μ, the reconstruction on (b) is obtained. The absolute value of the shear modulus is close to the right answer.

reason for this is the opposing tendencies of the force matching and the regularization terms. The latter forces the shear modulus to be as small as possible (a value determined by the lower bound set in the minimization algorithm), whereas the former selects it to best match the force measurement. This leads to artificial boundary layers in the shear modulus as it tries to satisfy both of these requirements (see Figure 4.5).

A simple solution to this problem is to define a new material parameter $\psi = \log \mu$ and reformulate the minimization problem in terms of ψ. Now, to match the force data, the minimization algorithm alters ψ by an additive constant, and the regularization term is unchanged by the inclusion of this constant. As a result, the conflict between these two terms is resolved (see Figure 4.5).

4.3.3.2 Microstructure-Based Constitutive Models

Soft glandular tissue is well modeled as a composite material comprising stiff fiber bundles with varying tortuosity embedded in a soft matrix, where the fiber bundles can be used to represent the collagen content of the tissue. This describes the microstructure of tissue at the scale of approximately 50 μm. However, the displacement measurements made in elasticity imaging are at a resolution of approximately 1–3 mm. Consequently, the microstructure must be averaged, or homogenized, to effectively represent the displacement that is measured.

A simple but useful homogenized model for a fibrous microstructure was developed by Cacho et al. [76]. The authors assume that every point in the tissue contains a certain concentration of fiber bundles, which is defined in terms of two number density functions: one that determines the orientation of the fibers and another that determines the tortuosity of the fibers. The tortuosity is defined as the arc-length of a fiber bundle divided by the distance between its end points. Thereafter, several simplifying assumptions are made. These include assuming that the fibers offer no resistance until they are stretched beyond the value of their tortuosity, all the fiber bundles are stretched by the same macroscopic stretch, and the response of one fiber family is independent of the others. The result is a macroscopic stress–strain law that contains the microscopic variables as material parameters. For example, if it is assumed that the distribution of the fibers is isotropic, the fibers are very stiff compared with the matrix, and the tortuosity distribution is centered about τ, then this model contains τ as a material parameter and produces a stress–stretch behavior of the type shown in Figure 4.6.

Remarkably, the simple result in Figure 4.6 explains and makes the connection between observations made regarding the microstructure and mechanical behavior of breast tumors. On the one hand, it has been observed that malignant tumors present collagen fiber bundles that are less tortuous than those observed in benign tumors in SHG images of breast tumors [62,63]. On the other hand, it has been observed that malignant tumors start to stiffen with strain at a smaller value of the overall applied strain when compared with benign tumors in ex vivo mechanical tests of breast tumors [60,61]. We note that the stress–stretch curve shown in Figure 4.6 is consistent with both these observations and can be used to understand the macroscopic mechanical behavior based on microstructural differences.

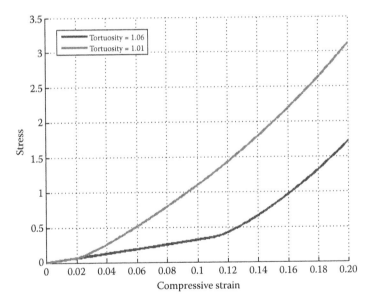

FIGURE 4.6 Stress versus (compressive) strain curve for two uniform synthetic tissue samples described by the microstructural constitutive law. Each sample contains an isotropic distribution of around 4000 collagen fiber bundles with a tortuosity distribution sharply centered on τ. We note that the sample with smaller tortuosity (τ = 1.01) stiffens with increasing strain at a lower compressive strain. To that extent, it is representative of a malignant tumor, whereas the sample with larger tortuosity (τ = 1.06) is representative of a benign tumor. We also observe that the knee of both curves is located at a compressive strain that is approximately two times their tortuosity. This is explained by the fact that the fibers offer resistance only by stretching and that they begin to stretch only when the lateral strain (which is extensional) reaches the value of their tortuosity. The lateral strain is in turn roughly half of the applied compressive axial strain.

Chapter 4

Our effort in the area of microstructural modeling is focused on two tracks:

1. Using the constitutive model described previously and the large strain displacement data acquired from quasi-static imaging to solve the inverse problem to create images of the average microstructural parameters of tissue in vivo.
2. Developing more accurate homogenized constitutive models that make fewer assumptions on the tissue response.

4.4   Clinical Applications Literature

There is a large and growing body of literature describing clinical trials of palpation-type elastography. Enough studies have been performed to lead to consensus documents on the use of elastography methods from both the European Federation of Ultrasound in Medicine and Biology [77,78] and the World Federation of Ultrasound in Medicine and Biology [79,80].

With little practice, freehand elasticity imaging is relatively easy. The methods for data acquisition vary depending on the optimization of the motion tracking algorithm of a specific commercial implementation. Some of these systems are optimized for negligible transducer motion (muscle quiver for the person holding the transducer or patient motion from beating heart or respiration is sufficient). In other cases, mild or moderate (~1% frame-average strain) is desired. In any case, in acknowledging the nonlinear elastic response of tissue, it is widely recognized that minimal preloading ("precompression") is desired (indeed, it is required in some cases).

Interpretation methods vary depending on the clinical application, but a starting point for all approaches is based on the initial findings of Garra et al. [24], later confirmed and extended [3], in which the tumor size seen in strain images tends to be larger for cancerous lesion than that seen in B-mode images whereas the size of benign tumors tends to be the same size or smaller in strain images compared with B-mode. That concept was extended to include the heterogeneity of the strain distribution in the five-point classification scheme suggested by Itoh et al. [81]. This latter scheme is now broadly applied to strain imaging in several organ systems with mixed results, as described in the next section.

4.4.1   Breast

Inarguably, the most successful application of palpation-type elasticity imaging, to date, is in breast ultrasound imaging, in which it has clearly demonstrated improved differentiation of benign from malignant disease over B-mode imaging alone in numerous clinical trials. For example, in a prospective study including 188 lesions (127 benign and 61 malignant) in 175 women, using the elasticity image scoring method proposed by Itoh et al. [81], Raza et al. [82] reported a sensitivity of 92.7% and a specificity of 85.5% in differentiating between benign and malignant lesions. Importantly, of the 76 benign lesions assigned an ultrasound BI-RADS 4a, 82.9% had an elasticity score of 1 or 2 (suggesting normal tissue). There were four false-negative findings in their study demonstrating that elasticity imaging alone is not sufficient for breast lesion diagnosis at this stage of development.

In a large multicenter, unblinded study evaluating 635 breast masses in 578 women, Barr et al. [83] used the ratio of lesion size in the strain image to the lesion size in the corresponding B-mode image to classify lesions as benign or malignant. They found that 361 of the 413 benign lesions had a lesion size ratio less than 1.0 and 219 of the 222 malignant lesions had a lesion size ratio of at least 1.0, resulting in a sensitivity of 99% and a specificity of 87%. They report that sensitivity at individual sites ranged from 96.7% to 100% and specificities ranged from 66.7% to 95.4%. A more recent study by some of the same authors and using the same criteria for differentiation [84] involving 230 lesions reported 99% sensitivity, 91.5% specificity, 90% positive predictive value, and 99.2% negative predictive value.

4.4.2   Other Clinical Applications

Numerous other attempts to use strain imaging for differentiation of benign and malignant masses have had mixed results. Thyroid imaging is a compelling problem, but no consensus has been reached regarding performance [78]. Prostate is another organ where palpation is common, and the use of elastography seems reasonable. However, strain imaging is difficult, in part, because the normal prostate is stiffer than its surrounding tissues, and it is difficult to track deformation with a 1-D array and 2-D tracking. Perhaps when 2-D arrays and 3-D tracking become available, prostate strain imaging will become more practical. Endoscopic ultrasound elastography has also been attempted, but again, mixed results are found [78].

4.4.3   Conclusions

A great deal of progress has been made since the first real-time elasticity imaging systems were introduced. Elasticity image quality has improved significantly, tools to help select the high-quality strain images have been developed, and numerous studies have demonstrated the benefit of this and similar techniques. Modulus reconstruction, especially for large deformation data, provides an exciting extension of that work with the potential for extracting quantitative information about tissue properties and the underlying collagen structure. There is great potential supporting continued research and development of this modality.

References

1. T. J. Hall, Y. Zhu, C. S. Spalding, and L. T. Cook. In vivo results of real-time freehand elasticity imaging, *Proceedings of the IEEE Ultrasonics Symposium*, 1653–1657, 2001.
2. Y.-C. Fung. *A First Course in Continuum Mechanics* (Chap. 12). Prentice Hall, 1994. ISBN 0-13-61524-2.
3. T. J. Hall, Y. Zhu, and C. S. Spalding. In vivo real-time freehand palpation imaging. *Ultrasound in Medicine and Biology*, 29:427–435, March 2003.
4. R. G. Barr and Z. Zhang. Effects of precompression on elasticity imaging of the breast. *Journal of Ultrasound in Medicine*, 31:895–902, 2012.
5. J. Meunier and M. Bertrand. Ultrasonic texture motion analysis: Theory and simulation. *IEEE Transactions on Medical Imaging*, 14:293–300, January 1995.
6. J. Ng, R. Prager, N. Kingsbury, G. Treece, and A. Gee. Modeling ultrasound imaging as a linear, shift-variant system. *IEEE Transactions on Ultrasonics, Ferroelectrics and Frequency Control*, 53:549–563, March 2006.

Chapter 4

7. R. F. Wagner, S. W. Smith, J. M. Sandrik, and H. Lopez. Statistics of speckle in ultrasound B-scans. *IEEE Transactions on Sonics and Ultrasonics*, 30(3):156–163, 1983.
8. Y. Zhu and T. Hall. A modified block matching method for real-time freehand strain imaging. *Ultrasonic Imaging*, 24(3):161–176, 2002.
9. F. Viola and W. Walker. A comparison of the performance of time-delay estimators in medical ultrasound. *IEEE Transactions on Ultrasonics, Ferroelectrics and Frequency Control*, 50(4):392–401, 2003.
10. I. Céspedes, Y. Huang, J. Ophir, and S. Spratt. Methods for estimation of subsample time delays of digitized echo signals. *Ultrasonic Imaging*, 17(2):142–171, 1995.
11. F. Viola and W. F. Walker. A spline-based algorithm for continuous time-delay estimation using sampled data. *IEEE Transactions on Ultrasonics, Ferroelectrics and Frequency Control*, 52:80–93, January 2005.
12. F. Viola, R. L. Coe, K. Owen, D. A. Guenther, and W. F. Walker. Multi-Dimensional Spline-Based Estimator (MUSE) for motion estimation: Algorithm development and initial results. *Annals of Biomedical Engineering*, 36:1942–1960, December 2008.
13. K. Miyatake, M. Yamagishi, N. Tanaka, M. Uematsu, N. Yamazaki, Y. Mine, A. Sano, and M. Hirama. New method for evaluating left ventricular wall motion by color-coded tissue Doppler imaging: In vitro and in vivo studies. *Journal of the American College of Cardiology*, 25:717–724, March 1995.
14. A. Pesavento, A. Lorenz, and H. Ermert. Phase root seeking and the Cramer-Rao-Lower bound for strain estimation. *1999 IEEE Ultrasonics Symposium Proceedings. International Symposium (Cat. No. 99CH37027)*, 2:1669–1672, 1999.
15. T. Loupas, J. Powers, and R. Gill. An axial velocity estimator for ultrasound blood flow imaging, based on a full evaluation of the Doppler equation by means of a two-dimensional autocorrelation approach. *IEEE Transactions on Ultrasonics, Ferroelectrics and Frequency Control*, 42:672–688, July 1995.
16. G. F. Pinton, J. J. Dahl, and G. E. Trahey. Rapid tracking of small displacements with ultrasound. *IEEE Transactions on Ultrasonics, Ferroelectrics and Frequency Control*, 53:1103–1117, June 2006.
17. J. Ophir, I. Céspedes, H. Ponnekanti, Y. Yazdi, and X. Li. Elastography: A quantitative method for imaging the elasticity of biological tissues. *Ultrasonic Imaging*, 1991.
18. W. Remley. Correlation of signals having a linear delay. *Journal of the Acoustical Society of America*, 35:65, 1963.
19. G. Carter. Coherence and time delay estimation. *Proceedings of the IEEE*, 75(2):236–255, 1987.
20. B. H. Friemel, L. N. Bohs, and G. E. Trahey. Relative performance of two-dimensional speckle-tracking techniques: Normalized correlation, nonnormalized correlation and sum-absolute-difference. *Proceedings of the IEEE Ultrasonics Symposium*, 1481–1484, 1995.
21. R. G. P. Lopata, M. M. Nillesen, H. H. G. Hansen, I. H. Gerrits, J. M. Thijssen, and C. L. de Korte. Performance evaluation of methods for two-dimensional displacement and strain estimation using ultrasound radio frequency data. *Ultrasound in Medicine and Biology*, 35:796–812, May 2009.
22. T. J. Hall. AAPM/RSNA physics tutorial for residents: Topics in US: Beyond the basics: Elasticity imaging with US. *Radiographics*, 23(6):1657–1671, 2003.
23. L. Bohs and G. Trahey. A novel method for angle independent ultrasonic imaging of blood flow and tissue motion. *IEEE Transactions on Biomedical Engineering*, 38(3):280–286, 1991.
24. B. S. Garra, E. I. Céspedes, J. Ophir, S. R. Spratt, A. Zuurbier, M. Magnant, and M. F. Pennanen. Elastography of breast lesions: Initial clinical results. *Radiology*, 79–86, 1997.
25. E. Burnside, T. Hall, A. Sommer, G. Hesley, G. Sisney, W. Svensson, J. Fine, J. Jiang, and N. Hangiandreou. Differentiating benign from malignant solid breast masses with US strain imaging. *Radiology*, 245(2):401, 2007.
26. E. Konofagou. A new elastographic method for estimation and imaging of lateral displacements, lateral strains, corrected axial strains and Poisson's ratios in tissues. *Ultrasound in Medicine and Biology*, 24(8):1183–1199, 1998.
27. R. Righetti, J. Ophir, S. Srinivasan, and T. A. Krouskop. The feasibility of using elastography for imaging the Poisson's ratio in porous media. *Ultrasound in Medicine and Biology*, 30:215–228, February 2004.
28. E. E. Konofagou, T. Harrigan, and J. Ophir. Shear strain estimation and lesion mobility assessment in elastography. *Ultrasonics*, 38:400–404, 2000.
29. H. Xu, M. Rao, T. Varghese, A. Sommer, S. Baker, T. J. Hall, G. A. Sisney, and E. S. Burnside. Axial-shear strain imaging for differentiating benign and malignant breast masses. *Ultrasound in Medicine and Biology*, 36:1813–1824, November 2010.
30. F. Kallel and J. Ophir. A least-squares strain estimator for elastography. *Ultrasonic Imaging*, 1997.

31. J. E. Lindop, G. M. Treece, A. H. Gee, and R. W. Prager. The general properties including accuracy and resolution of linear filtering methods for strain estimation. *IEEE Transactions on Ultrasonics, Ferroelectrics and Frequency Control*, 55:2363–2368, November 2008.

32. J. Jiang, T. J. Hall, and A. M. Sommer. A novel performance descriptor for ultrasonic strain imaging: A preliminary study. *IEEE Transactions on Ultrasonics, Ferroelectrics and Frequency Control*, 53:1088–1102, June 2006.

33. W. Walker and G. Trahey. A fundamental limit on delay estimation using partially correlated speckle signals. *IEEE Transactions on Ultrasonics, Ferroelectrics and Frequency Control*, 42:301–308, March 1995.

34. M. Bilgen and M. F. Insana, Error analysis in acoustic elastography. I. Displacement estimation. *Journal of the Acoustical Society of America*, 101:1139–1146, Feb. 1997.

35. T. Varghese and J. Ophir. A theoretical framework for performance characterization of elastography: The strain filter. *IEEE Transactions on Ultrasonics, Ferroelectrics and Frequency Control*, 44:164–172, January 1997.

36. I. Céspedes, J. Ophir, and S. K. Alam. The combined effect of signal decorrelation and random noise on the variance of time delay estimation. *IEEE Transactions on Ultrasonics, Ferroelectrics and Frequency Control*, 44(1):220–225, 2002.

37. I. Céspedes and J. Ophir. Reduction of image noise in elastography. *Ultrasonic Imaging*, 15, no. 2:89–102, 1993.

38. J. Jiang, T. J. Hall, and A. M. Sommer. A novel image formation method for ultrasonic strain imaging. *Ultrasound in Medicine and Biology*, 33:643–652, April 2007.

39. L. Chen, G. M. Treece, J. E. Lindop, A. H. Gee, and R. W. Prager. A quality-guided displacement tracking algorithm for ultrasonic elasticity imaging. *Medical Image Analysis*, 13(2):286–296, 2009.

40. P. Chaturvedi, M. F. Insana, and T. J. Hall. 2-D companding for noise reduction in strain imaging. *IEEE Transactions on Ultrasonics, Ferroelectrics and Frequency Control*, 45:179–191, January 1998.

41. J. Jiang and T. J. Hall, A generalized speckle tracking algorithm for ultrasonic strain imaging using dynamic programming. *Ultrasound in Medicine and Biology*, 35:1863–1879, November 2009.

42. H. Rivaz, E. Boctor, P. Foroughi, R. Zellars, G. Fichtinger, and G. Hager. Ultrasound elastography: A dynamic programming approach. *IEEE Transactions on Medical Imaging*, 27:1373–1377, October 2008.

43. Y. Petrank, L. Huang, and M. O'Donnell. Reduced peak-hopping artifacts in ultrasonic strain estimation using the Viterbi algorithm. *IEEE Transactions on Ultrasonics, Ferroelectrics and Frequency Control*, 56:1359–1367, July 2009.

44. L. Chen, R. Housden, G. M. Treece, A. H. Gee, and R. W. Prager. A hybrid displacement estimation method for ultrasonic elasticity imaging. *IEEE Transactions on Ultrasonics, Ferroelectrics and Frequency Control*, 57(4):866–882, 2010.

45. J. Jiang and T. J. Hall. A fast hybrid algorithm combining regularized motion tracking and predictive search for reducing the occurrence of large displacement errors. *IEEE Transactions on Ultrasonics, Ferroelectrics and Frequency Control*, 58(4):730–736, 2011.

46. S. K. Alam, J. Ophir, and E. E. Konofagou. An adaptive strain estimator for elastography. *IEEE Transactions on Ultrasonics, Ferroelectrics and Frequency Control*, 45:461–472, January 1998.

47. S. Srinivasan, J. Ophir, and S. Alam. Elastographic imaging using staggered strain estimates. *Ultrasonic Imaging*, 24(2002):229–245, 2002.

48. G. M. Treece, J. E. Lindop, A. H. Gee, and R. W. Prager. Freehand ultrasound elastography with a 3-D probe. *Ultrasound in Medicine and Biology*, 34:463–474, March 2008.

49. T. G. Fisher, T. J. Hall, S. Panda, M. S. Richards, P. E. Barbone, J. Jiang, J. Resnick, and S. Barnes. Volumetric elasticity imaging with a 2-D CMUT array. *Ultrasound in Medicine and Biology*, 36:978–990, June 2010.

50. S. Y. Yemelyanov, A. R. Skovoroda, M. A. Lubinski, B. M. Shapo, and M. O'Donnell. Ultrasound elasticity imaging using Fourier based speckle tracking algorithm. *Proceedings of the IEEE Ultrasonics Symposium*, 1065–1068, 1992.

51. M. O'Donnell, A. Skovoroda, B. Shapo, and S. Emelianov. Internal displacement and strain imaging using ultrasonic speckle tracking. *IEEE Transactions on Ultrasonics, Ferroelectrics and Frequency Control*, 41(3):314–325, 1994.

52. T. Varghese and J. Ophir. Performance optimization in elastography: Multicompression with temporal stretching. *Ultrasonic Imaging*, 18(3):193–214, 1996.

53. H. Du, J. Liu, C. Pellot-Barakat, and M. F. Insana. Optimizing multicompression approaches to elasticity imaging. *IEEE Transactions on Ultrasonics, Ferroelectrics and Frequency Control*, 53(1):90–99, 2006.

Chapter 4

54. M. Bayer and T. J. Hall. Variance and covariance of accumulated displacement estimates. *Ultrasonic Imaging*, 35(2):90–108, 2013.

55. M. Bayer, T. J. Hall, L. P. Neves, and A. A. O. Carneiro. Two-dimensional simulations of displacement accumulation incorporating shear strain. *Ultrasonic Imaging*, 36(1):55–73, 2014.

56. U. Albocher, P. E. Barbone, A. A. Oberai, and I. Harari. Uniqueness of inverse problems of isotropic incompressible three-dimensional elasticity. *Journal of the Mechanics and Physics of Solids*, 73:55–68, 2014.

57. P. E. Barbone and J. C. Bamber. Quantitative elasticity imaging: What can and cannot be inferred from strain images. *Physics in Medicine and Biology*, 47:2147–2164, June 2002.

58. P. E. Barbone and N. H. Gokhale. Elastic modulus imaging: On the uniqueness and nonuniqueness of the elastography inverse problem in two dimensions. *Inverse Problems*, 20(1):283, 2004.

59. P. E. Barbone and A. A. Oberai. Elastic modulus imaging: Some exact solutions of the compressible elastography inverse problem. *Physics in Medicine and Biology*, 52(6):1577, 2007.

60. T. Krouskop, T. Wheeler, F. Kallel, B. Garra, and T. Hall. Elastic moduli of breast and prostate tissues under compression. *Ultrasonic Imaging*, 20(4):260–274, 1998.

61. P. Wellman, R. D. Howe, E. Dalton, and K. A. Kern. Breast tissue stiffness in compression is correlated to histological diagnosis. *Harvard BioRobotics Laboratory Technical Report*, 1999.

62. G. Falzon, S. Pearson, and R. Murison. Analysis of collagen fibre shape changes in breast cancer. *Physics in Medicine and Biology*, 53(23):6641, 2008.

63. M. W. Conklin, J. C. Eickhoff, K. M. Riching, C. A. Pehlke, K. W. Eliceiri, P. P. Provenzano, A. Friedl, and P. J. Keely. Aligned collagen is a prognostic signature for survival in human breast carcinoma. *American Journal of Pathology*, 178(3):1221–1232, 2011.

64. S. Goenezen, J.-F. Dord, Z. Sink, P. E. Barbone, J. Jiang, T. J. Hall, and A. A. Oberai. Linear and nonlinear elastic modulus imaging: An application to breast cancer diagnosis. *IEEE Transactions on Medical Imaging*, 31:1628–1637, August 2012.

65. E. R. Ferreira, A. A. Oberai, and P. E. Barbone. Uniqueness of the elastography inverse problem for incompressible nonlinear planar hyperelasticity. *Inverse Problems*, 28(6):065008, 2012.

66. P. E. Barbone, A. A. Oberai, and I. Harari. Adjoint-weighted variational formulation for a direct computational solution of an inverse heat conduction problem. *Inverse Problems*, 23(6):2325, 2007.

67. U. Albocher, A. A. Oberai, P. E. Barbone, and I. Harari. Adjoint-weighted equation for inverse problems of incompressible plane-stress elasticity. *Computer Methods in Applied Mechanics and Engineering*, 198(30):2412–2420, 2009.

68. P. E. Barbone, C. E. Rivas, I. Harari, U. Albocher, A. A. Oberai, and Y. Zhang. Adjoint-weighted variational formulation for the direct solution of inverse problems of general linear elasticity with full interior data. *International Journal for Numerical Methods in Engineering*, 81(13):1713–1736, 2010.

69. M. Doyley, P. Meaney, and J. Bamber. Evaluation of an iterative reconstruction method for quantitative elastography. *Physics in Medicine and Biology*, 45(6):1521, 2000.

70. A. A. Oberai, N. H. Gokhale, and G. R. Feijoo. Solution of inverse problems in elasticity imaging using the adjoint method. *Inverse Problems*, 19(2):297, 2003.

71. A. A. Oberai, N. H. Gokhale, M. M. Doyley, and J. C. Bamber. Evaluation of the adjoint equation based algorithm for elasticity imaging. *Physics in Medicine and Biology*, 49(13):2955, 2004.

72. N. H. Gokhale, P. E. Barbone, and A. A. Oberai. Solution of the nonlinear elasticity imaging inverse problem: The compressible case. *Inverse Problems*, 24:045010, August 2008.

73. S. Goenezen, P. Barbone, and A. A. Oberai. Solution of the nonlinear elasticity imaging inverse problem: The incompressible case. *Computer Methods in Applied Mechanics and Engineering*, 200(13–16):1406–1420, 2011.

74. G. Biros and O. Ghattas. Parallel Lagrange–Newton–Krylov–Schur methods for PDE-constrained optimization. Part I: The Krylov–Schur solver. *SIAM Journal on Scientific Computing*, 27(2):687–713, 2005.

75. M. A. Aguilo Valentin, D. Ridzal, and J. G. Young. Solving large-scale inverse problems in elasticity using sequential quadratic programming (SQP). Tech. Rep., Sandia National Laboratories (SNL-NM), Albuquerque, NM, 2012.

76. F. Cacho, P. Elbischger, J. Rodriguez, M. Doblare, and G. A. Holzapfel. A constitutive model for fibrous tissues considering collagen fiber crimp. *International Journal of Non-Linear Mechanics*, 42(2):391–402, 2007.

77. J. Bamber, D. Cosgrove, C. F. Dietrich, J. Fromageau, J. Bojunga, F. Calliada, V. Cantisani, J.-M. Correas, M. D'Onofrio, E. E. Drakonaki, M. Fink, M. Friedrich-Rust, O. H. Gilja, R. F. Havre, C. Jenssen, A. S. Klauser, R. Ohlinger, A. Săftoiu, F. Schaefer, I. Sporea, and F. Piscaglia. EFSUMB guidelines and recommendations on the clinical use of ultrasound elastography. Part 1: Basic principles and technology. *Ultraschall in der Medizin*, 34:169–184, 2013.

78. D. Cosgrove, F. Piscaglia, J. Bamber, J. Bojunga, J.-M. Correas, O. H. Gilja, A. S. Klauser, I. Sporea, F. Calliada, V. Cantisani, M. D'Onofrio, E. E. Drakonaki, M. Fink, M. Friedrich-Rust, J. Fromageau, R. F. Havre, C. Jenssen, R. Ohlinger, A. Săftoiu, F. Schaefer, and C. F. Dietrich. EFSUMB guidelines and recommendations on the clinical use of ultrasound elastography. Part 2: Clinical applications. *Ultraschall in der Medizin*, 34:238–253, 2013.

79. T. Shiina, K. R. Nightingale, M. L. Palmeri, T. J. Hall, J. C. Bamber, M. B. Nielsen, R. Barr, J. Bojunga, L. Castera, B. Choi, D. Cosgrove, A. Dominique, A. Farrokh, G. Ferraioli, C. Filice, M. Friedrich-Rust, K. Nakashima, F. Schafer, S. Suzuki, S. Wilson, M. Kudo, C. D. K. H. Seitz, W. K. Moon, and I. Sporea. WFUMB Guidelines and recommendations on the clinical use of ultrasound elastography: Part 1: Basic principles and terminology. *Ultrasound in Medicine and Biology*, in press, November 2014.

80. R. G. Barr, K. Nakashima, D. Amy, D. Cosgrove, A. Farrokh, F. Schaefer, J. C. Bamber, L. Castera, B. I. Choi, Y.-H. Chou, C. F. Dietrich, H. Ding, G. Ferraioli, C. Filice, M. Friedrich-Rust, T. J. Hall, K. R. Nightingale, M. L. Palmeri, T. Shina, S. Suzuki, I. Sporea, S. Wilson, and M. Kudo. WFUMB Guidelines and recommendations on the clinical use of ultrasound elastography: Part 2: Breast. *Ultrasound in Medicine and Biology*, in press, November 2014.

81. A. Itoh, E. Ueno, E. Tohno, and H. Kamma. Breast disease: Clinical application of US elastography for diagnosis. *Radiology*, 239(2):341–350, 2006.

82. S. Raza, A. Odulate, E. M. W. Ong, S. Chikarmane, and C. W. Harston. Using real-time tissue elastography for breast lesion evaluation our initial experience. *Journal of Ultrasound in Medicine*, 29:551–563, 2010.

83. R. G. Barr, S. Destounis, L. B. L. II, W. E. Svensson, C. Balleyguier, and C. Smith. Evaluation of breast lesions using sonographic elasticity imaging: A multicenter trial. *Journal of Ultrasound in Medicine*, 31:281–287, 2011.

84. S. Destounis, A. Arieno, R. Morgan, P. Murphy, P. Seifert, P. Somerville, and W. Young. Clinical experience with elasticity imaging in a community-based breast center. *Journal of Ultrasound in Medicine*, 32:297–302, 2013.

Chapter 4

5. Quantitative Ultrasound Techniques for Diagnostic Imaging and Monitoring of Therapy

Michael L. Oelze

Ultrasound Imaging and Therapy. Edited by Aaron Fenster and James C. Lacefield © 2015 CRC Press/Taylor & Francis Group, LLC. ISBN: 978-1-4398-6628-3

5.1 Introduction

Conventional ultrasound B-mode imaging is extensively used clinically for applications ranging from obstetrics, cardiology, abdominal imaging, and cancer detection. Conventional B-mode images are constructed from log-compressing the envelope of ultrasound signals reflected from different tissue structures. As such, these images can depict anatomical features with high quality and high resolution. The size of structures in the order of millimeters can be estimated using conventional B-mode imaging. However, aside from estimating the size and shape of anatomical features, conventional, B-mode imaging is qualitative.

As a result of the qualitative nature of B-mode ultrasound imaging, the diagnostic potential of ultrasound is limited. For example, in breast cancer detection, B-mode ultrasound is good at detecting lesions or abnormalities, that is, it has high sensitivity. However, B-mode ultrasound is not as good at classifying a lesion as malignant or benign, that is, it has low specificity. Therefore, developing techniques that can improve the specificity of ultrasonic imaging is highly medically significant.

The development of quantitative ultrasound (QUS) techniques provides the potential to improve the diagnostic content of ultrasonic imaging by increasing the sensitivity and specificity of ultrasound. Specifically, QUS techniques can provide numbers related to tissue microstructure that may enable the unique identification of a tissue or disease state, that is, increasing the specificity. These QUS estimates can be correlated to specific tissue regions through parametric maps or images. Because QUS parameters are related to tissue-specific properties, QUS parameters can be operator and system independent. QUS parameters may improve the sensitivity of ultrasonic imaging because these parameters can provide a new source of image contrast enabling the improved detection of lesions or abnormalities that may not be detectable with conventional B-mode imaging.

QUS techniques can encompass many different operational modes: spectral-based imaging, envelope statistics, time-domain parameter estimation, and elastography. In this chapter, we will focus on spectral-based estimation and estimation of envelope statistics parameters for QUS. The latest QUS techniques will be explored, and successful applications of QUS techniques for diagnostic ultrasound and for monitoring of therapy will be discussed.

5.2 Spectral-Based Techniques

5.2.1 Backscatter Coefficient

Spectral imaging techniques in ultrasound are based on obtaining a good representation of the power spectrum of the backscattered signal through the periodogram. The estimate of the power spectrum is normalized by taking into account the scattering volume and the characteristics of the transducer and excitation pulse. The backscatter coefficient (BSC) is related to the normalized power spectrum and is a fundamental property of the scattering medium. The BSC is the basis for spectral-based QUS estimates and is defined as follows (Insana et al. 1990):

$$\sigma_{BSC}(f) = \frac{R^2}{V} \frac{\langle I_{sc}(f) \rangle}{I_{inc}(f)} \propto W(f), \tag{5.1}$$

where R is the distance to the scattering volume of interest, V is the scattering volume defined by the beamwidth and range gate length, $I_{sc}(f)$ and $I_{inc}(f)$ are the scattered and incident fields, respectively, and $W(f)$ is the normalized power spectrum. Therefore, the BSC is both operator and system independent. In regions where the scattering is uniform, the BSC can be parameterized to yield estimates of the scatterer properties, which can then provide a geometrical interpretation of the underlying tissue microstructure.

5.2.2   Parameterization of the BSC

To parameterize the BSC, models of scattering from the underlying tissue must be adopted. In the simplest case, the BSC can be parameterized as a straight line, and estimates of the spectral slope (SS), spectral intercept (SI), and midband fit (MBF) can be obtained (Lizzi et al. 1997). Only two of these parameters combined yield independent information. In more complex models, the scatterers can be modeled as spheres, cylinders, or other more complex structures (Insana 1995). From more complex models, estimates of the effective scatterer diameter (ESD) and effective acoustic concentration (EAC) have been adopted (Lizzi et al. 1987; Insana et al. 1990; Oelze et al. 2002).

The model of scattering from tissues has been described by both a discrete scattering model (scatterers have definitive boundaries and definable locations) and a continuous scattering model (the medium is modeled as heterogeneous with acoustic impedance as a function of spatial location) (Insana and Brown 1991). The discrete scattering models are important for understanding how coherent scattering affects the estimated BSC (Wear et al. 1993) and how compounding and windowing techniques can smooth the spectral estimates of BSC. The continuous scattering models are used with weak scattering conditions to produce models of scatterers that can provide a geometrical interpretation of the underlying tissue microstructure (e.g., the ESD). For example, starting with the inhomogeneous wave equation, when plane wave incidence is assumed and multiple scattering is assumed to be negligible, the Born approximation results in a BSC that is proportional to the Fourier transform of the spatial autocorrelation of the impedance distribution of the medium. If the scattering comes from an isotropic medium, then the BSC can be described by the following equation (Insana 1995; Insana and Hall 1990; Insana and Brown 1991):

$$\sigma_{BSC} = \frac{k^3 V_s \gamma_{EAC}}{8\pi} \int_0^\infty b_\gamma(\Delta r) \sin(2k\Delta r) \Delta r d\Delta r, \tag{5.2}$$

Chapter 5

where V_s is the volume of a scatterer, γ_{EAC} is the EAC and is defined as the number density of scatterers times the square of the relative impedance mismatch between scatterer and background, Δr is a lag coefficient, and $b_\gamma(\Delta r)$ is the three-dimensional correlation coefficient of the acoustic impedance. From Equation 5.2, scattering from tissue with structures that can be described by simple geometrical shapes results in analytical expressions for the BSC, allowing estimates of the effective size of the scatterers. In the case where analytical solutions can be obtained, the BSC can be described by an intensity form factor defined as the ratio of the BSC for a scattering medium with scatterers of finite size to that of a similar medium containing point scatterers,

$$F(2k) = \frac{\sigma_{BSC}}{\sigma_0},$$

(5.3)

where σ_0 is the BSC from a medium containing only point scatterers. Some common form factors used in tissue characterization are as follows:

$$F_1(2k) = \left[\left(\frac{3}{2ka_{eff}}\right)j_1(ka_{eff})\right]^2 \text{ (fluid sphere)},$$

(5.4)

$$F_2(2k) = e^{-0.827k^2a_{eff}^2} \text{ (Gaussian)},$$

(5.5)

where a_{eff} is the effective radius of a scatterer. The effective scatterer radius represents a volume-weighted mean of the sizes of scatterers contributing to the scattered signal. In the case of a Gaussian scatterer, the radius corresponds to the characteristic dimension of a scatterer defined by the distance to where the impedance falls by one half its impedance maximum at the center.

Estimates of the scatterer properties (i.e., ESD and EAC) can be obtained by using an estimator that compares the BSC calculated from measurements to a theoretical BSC and minimizes some cost function versus trial values of ESD and EAC. In most instances, an estimator will provide a single value for the ESD. However, this value may represent a distribution of scatterer sizes, and the width and shape of this scatterer size distribution will influence the final ESD estimate (Lavarello and Oelze 2011). Different estimators can be used to reduce the effects of scatterer size distribution on the final ESD estimate (Lavarello and Oelze, 2012). The most common estimator in use for ESD and EAC calculation is the minimum average squared deviation (MASD), which minimizes the average squared difference between the measured BSC and a theoretical BSC (Insana et al. 1990),

$$\widehat{ESD} = \min_{ESD} \left\{ \frac{1}{M} \sum_{n=1}^{M} (X_n - \bar{X})^2 \right\},$$

(5.6)

where M is the number of data points in the analysis bandwidth, $X = 10\log_{10}\left[\hat{\sigma}_{BSC}/\sigma_{BSC}\right]$, where "^" designates the estimated value and

$$\bar{X} = \frac{1}{M} \sum_{n=1}^{M} X_n.$$

(5.7)

Different trial values of the ESD are selected over a range of values, and the estimate that minimizes the error between the estimated BSC and the theoretical BSC is selected. $\bar{X}$ represents a gain factor between the estimated and the theoretical BSC. Once the estimate of ESD is determined, $\bar{X}$ can be used to calculate the estimate of the EAC.

5.2.3   Calibration and the BSC

For the BSC to be system and operator independent, it is important to understand how the ultrasound signal interacts with scatterers in the field. From this understanding, techniques can be derived to account for system-dependent effects. Consider a sample medium filled with a large number of identical scatterers spaced at random locations in the medium as illustrated in Figure 5.1. Each scatterer will scatter the transmitted ultrasound pulse in all directions. Some of the scattered energy will propagate back to the transducer (backscatter), and the backscattered wavelets will be summed together coherently at the source to yield a backscattered time signal. This gives rise to a complex interference pattern, that is, speckle.

Assume in this scenario that each of the scatterers is identical. Therefore, each scatterer will modify the impulse response of the ultrasonic source by some frequency-dependent scattering function, $s(t)$, which we represent by the following equation:

$$r(t) = p(t)*s'(t)*a(t), \tag{5.8}$$

where $r(t)$ is the received signal, $p(t)$ is the pulse-echo impulse response but also incorporates the system-dependent effects (i.e., the transmit voltage level, diffraction effects, etc.), $a(t)$ is a function that accounts for the frequency-dependent attenuation of the signal, and

$$s'(t) = s(t + t_1) + s(t + t_2) + \ldots + s(t + t_N), \tag{5.9}$$

which relates the scatter function in time domain for N scatterers located at different temporal locations (depths in the tissue). Essentially, pulses are arriving at different times because scatterers are located at different distances from the transducer, and their arrival pulses add together coherently. The scattered pulses will be summed to make the backscattered time train. Equivalently,

$$r(t) = p(t)*[s(t - t_1) + s(t - t_2) + \ldots + s(t - t_N)]*a(t). \tag{5.10}$$

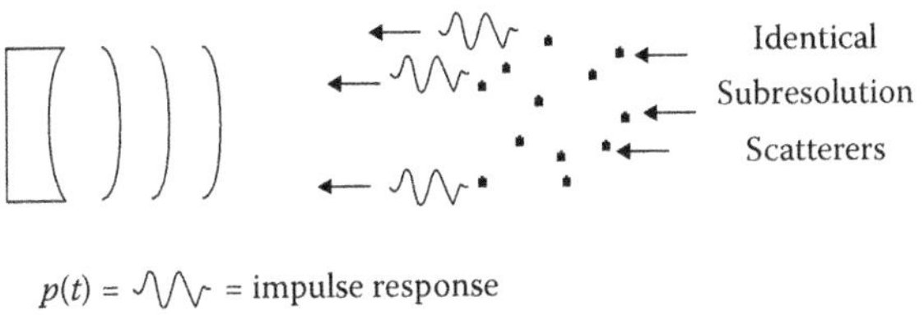

$p(t) = $ ⌇⋁⋁⋏ = impulse response

Identical
Subresolution
Scatterers

FIGURE 5.1   Graphic depiction of subresolvable scatterers and the scattered ultrasound pulse.

Taking the Fourier transform of the previous equation gives

$$R(f) = P(f)S'(f)A(f),$$

(5.11)

or

$$R(f) = P(f)A(f)S(f)\left[e^{-j2\pi ft_1} + e^{-j2\pi ft_2} + \ldots + e^{-j2\pi ft_N}\right].$$

(5.12)

An estimate of the power spectrum can be obtained by taking the magnitude squared of the previous equation,

$$|R(f)|^2 = |P(f)|^2 |A(f)|^2 |S'(f)|^2.$$

(5.13)

By considering each scatterer contribution individually, this can be further reduced to

$$|R(f)|^2 = |P(f)|^2 |A(f)|^2 |S(f)|^2 \left[N + 2 \sum_{n>m=1}^{N} \cos(2\pi f[t_n - t_m]) \right].$$

(5.14)

The first term on the right is called the incoherent scattering term and depends only on the scattering function and the number of scatterers contributing to the power spectrum estimate. The first term does not depend on the spatial locations of the scatterers. The second term is called the coherent scattering term and depends on the scattering function and the spatial locations of the scatterers. For many randomly spaced scatterers, the second term is assumed to be small and acts only as noise relative to the incoherent spectrum when trying to parameterize the BSC. If regularly spaced scatterers exist in the field, peaks in the coherent spectrum will occur corresponding to the scatterer spacings (Wear et al. 1993).

The scattering function depends on the size, shape, and mechanical properties of the underlying scatterers, which are assumed to be related to actual underlying tissue microstructure. To extract the scattering function from the received signal, the power spectrum is approximated as the incoherent spectrum, the frequency-dependent attenuation of the signal is accounted for, and the system transfer function is divided as follows:

$$|R(f)|^2 \approx |P(f)|^2 |A(f)|^2 N |S(f)|^2,$$

(5.15)

yielding

$$N|S(f)|^2 = \frac{|R(f)|^2}{|P(f)|^2 |A(f)|^2}.$$

(5.16)

From the previously mentioned equation, we can estimate properties of the underlying microstructure or scatterers using appropriate models. However, we need to determine the transducer contributions to the received signal.

To estimate $|P(f)|^2$, two techniques have been developed:

1. The planar reflector technique (Sigelmann and Reid 1973; Lizzi et al. 1983; Madsen et al. 1984; Insana et al. 1990): In the planar reflector technique, the same ultrasound equipment and settings used to interrogate the sample are used to transmit a pulse and receive the reflection from a planar surface of known reflectivity (see Figure 5.2). The pulse incident on the planar surface is at normal incidence. The technique works for weakly focused sources within the depth of field of the source. All frequencies are assumed to be reflected from the planar surface with a known reflection coefficient, γ. The received signal can be modeled as

$$r(t) = p(t)^* \gamma \delta(t - t_0),\tag{5.17}$$

or

$$|R_{planar}(f)|^2 = \gamma^2 |P(f)|^2.\tag{5.18}$$

This yields

$$|P(f)|^2 = \frac{|R_{planar}(f)|^2}{\gamma^2}.\tag{5.19}$$

The planar reference technique is used for weakly focused single-element sources but is not used for electronically steered arrays and strongly focused sources.

2. The reference phantom technique (Yao et al. 1990): In the reference phantom technique, the source used to interrogate a sample is used to interrogate a well-characterized reference phantom (see Figure 5.3). The same equipment and settings used to interrogate the sample are used to interrogate the reference phantom. The scattering function or the BSC for the reference phantom is known based on extensive measurements from the phantom using a technique such as the planar reflector method, that is, $N|S_{phantom}(f)|^2 \approx |S_{known}(f)|^2$. The signal received from the reference phantom can be modeled as

$$r_{phantom}(t) = p(t)^* s'_{phantom}(t),\tag{5.20}$$

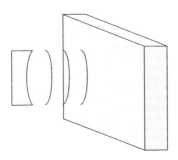

FIGURE 5.2   Graphical representation of the planar reference technique.

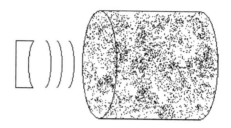

FIGURE 5.3 Graphical representation of the reference phantom technique.

or

$$\left|R_{\mathrm{phantom}}(f)\right|^2 \approx \left|P(f)\right|^2 N \left|S_{\mathrm{phantom}}(f)\right|^2. \tag{5.21}$$

Therefore,

$$\frac{\left|R_{\mathrm{phantom}}(f)\right|^2}{\left|S_{\mathrm{known}}(f)\right|^2} = \frac{\left|P(f)\right|^2 N \left|S_{\mathrm{phantom}}(f)\right|^2}{\left|S_{\mathrm{known}}(f)\right|^2}, \tag{5.22}$$

or

$$\left|P(f)\right|^2 = \frac{\left|R_{\mathrm{phantom}}(f)\right|^2}{\left|S_{\mathrm{known}}(f)\right|^2}. \tag{5.23}$$

The reference phantom technique can be used with strongly focused sources and array systems. However, the technique requires a well-characterized phantom and can have larger variance in estimates due to coherent noise from phantom reference.

5.2.3.1 Attenuation Compensation

Accurate estimation of the BSC requires a calibration spectrum and also compensation for frequency-dependent attenuation losses (O'Donnell and Miller 1981; Oelze and O'Brien 2002). When estimating the ESD, undercompensating for the frequency-dependent attenuation will result in an overestimate of the ESD, whereas overcompensating for the frequency-dependent attenuation will result in an underestimation of ESD. Frequency-dependent attenuation should be compensated for losses in the signal in the medium between the transducer and the location in the sample where the signal is being windowed. If the attenuation is large, compensation for frequency-dependent attenuation should also occur over the length of the windowed signal. In many cases, the actual frequency-dependent attenuation of the sample is not known, and either a priori knowledge of the sample is known to provide an approximate value for attenuation or additional processing must be performed to estimate attenuation.

5.2.3.2 Estimating the BSC with Finite Data Samples

To obtain the BSC, the normalized power spectrum is estimated from the backscattered RF time signal using one of the calibration techniques discussed earlier. However, the normalized power spectrum estimate is based on a periodogram. The periodogram is used to estimate the true power spectrum when only a finite sample size is available. Averaging the periodogram over a longer signal and from several independent samples will provide a better estimate of the true power spectrum related to the sample. In a practical sense, this is accomplished with the ultrasound backscattered data by windowing a segment of the signal and averaging the estimated power spectra from several independent windowed scan lines. Scan lines are considered to provide independent samples when they are at least a beamwidth apart.

A normalized power spectrum estimate will be associated with a windowed signal segment and the number of scan lines used in the estimate. Assuming that the scan lines are parallel and separated by a beam width, the normalized power spectrum comes from a data block. Figure 5.4 shows an image of a data block used for the estimate of the normalized power spectrum. The normalized power spectrum estimate can then be associated with a BSC and parameterized. The parameter estimates can be correlated to a specific sample region corresponding to the window location and the scan line locations. A map or image of the parameter estimates can then be constructed based on spatially correlating parameter estimates with the data blocks. Figure 5.5 shows a parametric image using the ESD estimate overlaid on a B-mode image of a tumor to give context. From the parametric image, colored pixels correspond to data blocks used to provide the spectral-based estimate of ESD.

The larger the time window and the larger the number of independent scan lines used in the periodogram estimate, the better the normalized power spectrum representation (Huisman and Thijssen 1996; Lizzi et al. 1997; Oelze and O'Brien 2004a). However, the larger the sample size used in the estimate, that is, the larger the data block size, the worse the spatial resolution of the corresponding parameter image. Therefore, a trade-off exists between obtaining a good estimate of the BSC from the data block and obtaining a good estimate of the spatial resolution of subsequent parameter images. By choosing a smaller window size, primarily the bias of BSC estimates is increased. By averaging the power spectra from fewer independent scan lines, primarily the variance of BSC estimates will increase.

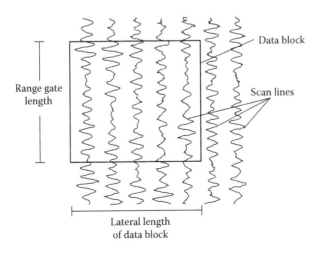

FIGURE 5.4   Diagram of the data block used to provide a localized estimate of the BSC.

Chapter 5

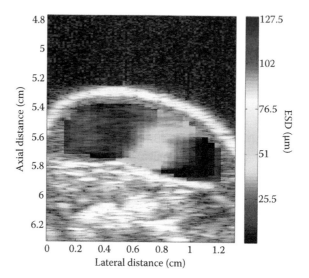

FIGURE 5.5 Parametric image of a mouse sarcoma tumor enhanced with estimates of the ESD using a Gaussian form factor.

5.2.4 Quality of Spectral Estimates

The quality of spectral estimates can be quantified in terms of the bias (accuracy) and variance (precision) of the estimates. As stated before, the main trade-off between providing good bias and providing good variance of estimates is the reduction in the spatial resolution of the estimator. Increasing the amount of data available to the periodogram estimate will result in a better estimate in terms of bias and variance. However, the estimation of parameters from models assumes that scattering is uniform within a data block. The larger the data block size, the lower the bias and variance of estimates if the data block contains uniform scattering. However, if the data block becomes too large, the likelihood of having nonuniform scattering within the data block increases. In other words, as the size of the data block increases, the greater the chances that two or more different kinds of tissue or sample will be included in the data block. This will lead to increases in variance and bias of estimates because the estimates are assuming that only one kind of scatterer is present. Smaller data blocks will result in a better capability to resolve regions containing different kinds of scattering properties.

Aside from the size of the data block, the bias of the spectral-based estimates is dependent on several additional factors. The choice of the window length and the choice of the windowing function will lead to differences in estimate bias. For example, tapered window functions (e.g., Hanning) provide better bias than nontapered windowing functions (e.g., rectangular window) (Oelze and O'Brien 2004b). Increasing the analysis bandwidth can improve the bias of estimates, and improving the signal-to-noise ratio (SNR) can result in improved bias of estimates (Chaturvedi and Insana 1996; Sanchez et al. 2009). Poor calibration procedures and poor models of scattering can result in increased bias of estimates.

In spectral estimation, it is typically more important to reduce the variance of spectral-based estimates. Reducing the variance of estimates improves the ability for

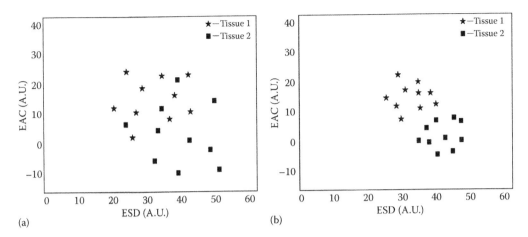

FIGURE 5.6 (a) Scatter plot of ESD versus EAC for a sample with larger variance and significant overlap of parameters and (b) scatter plot where the variance of estimates is smaller and improved separation exists between the parameters for the two tissue types.

estimates to uniquely classify tissues. For example, Figure 5.6 shows scatter plots of two distributions of scatterer property estimates for two kinds of tissues. Although the mean value of the estimates results in the separation of the two kinds of tissues, the large variance means it is likely that misclassification can occur. By reducing the variance of the estimates (Figure 5.6b), the misclassification error is reduced because the estimate overlap is reduced.

Similar to the bias, the variance of estimates can be improved by increasing the data block size, increasing the SNR, and increasing the analysis bandwidth. In addition, the variance of estimates is reduced when using a nontapered windowing function versus a tapered window function (Oelze and O'Brien 2004b). Other techniques have been adopted to reduce the variance of spectral estimates, including angular compounding techniques (Gerig et al. 2004; Lavarello et al. 2009), Welch's approach to windowing (Ghoshal and Oelze 2010; Liu and Zabsebski 2010), and multitaper methods (Liu and Zabsebski 2010). Compounding approaches reduce variance by reducing the amplitude of coherent spectrum terms. The reduction in variance from compounding N independent samples is calculated as follows (Oppenheim and Shafer 1975):

$$\mathrm{var}\overline{\left(\hat{x}\right)} = \frac{\mathrm{var}\left(\hat{x}\right)}{N}. \tag{5.24}$$

On the other hand, Welch's technique and multitaper methods reduce the variance of spectral estimates by reducing the number of coherent terms in the BSC calculation (Ghoshal and Oelze 2010).

5.2.5 Attenuation Estimation

As stated previously, compensating for frequency-dependent attenuation is important for providing accurate estimates from the BSC. In many cases, the exact frequency-dependent attenuation is not known for a sample. Therefore, either a best-guess value

Chapter 5

for the attenuation must be used to compensate for frequency-dependent losses or additional estimation procedures can be performed to provide attenuation. Several techniques have been explored based on the backscattered power spectrum to provide estimates of the frequency-dependent attenuation. These techniques are important for providing estimates of attenuation for compensating spectral-based estimates; however, estimates of attenuation can also be used to classify tissues. Therefore, the estimation of attenuation falls into two categories: total attenuation estimation and local attenuation estimation. The total attenuation estimation refers to estimating the total attenuation up to a point in the sample at a certain distance from the transducer. Total attenuation is used to provide compensation for BSC estimates. The local attenuation estimation is used to provide estimates of attenuation in some region of interest within the sample. In local attenuation estimation, maps or images of attenuation can be constructed. Local attenuation estimation is used to classify tissues.

Different techniques have been proposed to estimate both total and local attenuation. Total attenuation techniques include a reference phantom technique (Lu et al. 1995), a technique based on filtering the backscattered signal with a Gaussian function (Bigelow 2010), and a technique that assumes a model for scattering and simultaneously estimates the attenuation and BSC using a least squares method (Nam et al. 2011). In each of these techniques, the attenuation is assumed to be linearly increasing with frequency. Furthermore, in the last two techniques, the BSC is assumed to be power law. Accurate estimates of total attenuation have been observed using these techniques in phantom studies.

Local attenuation algorithms fall under three main categories: spectral shift algorithms (Fink et al. 1983; Baldeweck et al. 1995; Kuc and Li 1985), spectral difference algorithms (Parker and Waag 1983; Parker et al. 1988; Yao et al. 1990), and hybrid algorithms (Kim and Varghese 2008). The spectral shift algorithms use the shift in the center frequency of the backscattered power spectrum with depth to estimate the frequency-dependent attenuation. Practically, power spectra from several different windowed segments versus depth are calculated, and the shift versus depth is estimated to provide the attenuation slope. If the scatterer size for the different windowed segments changes over depth, then significant bias and variance is introduced into the attenuation estimates. The spectral difference method examines the change in the backscattered power over depth and relates this change to attenuation. Similar to the spectral shift method, different windowed segments over depth are used to provide the attenuation estimate. If the acoustic concentration of scatterers changes over the depth, then significant bias and variance of the estimates can occur. In hybrid approaches, both spectral shift and spectral difference methods are used in an attempt to make the estimate more robust to changes in scatterer properties, which might occur when comparing windowed segments over depth.

5.2.6 Generalized Spectrum Approaches

Techniques to parameterize the BSC assume that the scattering statistics in a data block are uniform, arise from many subresolution scatterers per resolution cell, and arise from scatterers with random spatial location. However, in many situations, one or more of these criteria are not met. Therefore, it is important to be able to detect when one or more of these criteria are not met to adjust strategies for processing data blocks and to extract potential additional information from the signal. One technique that has been used to categorize the scattering from a signal is the generalized spectrum (GS).

The GS enables the separation of the signal into diffuse and coherent components. The diffuse component is characterized by randomly spaced, subresolvable scatterers with sufficient concentrations to produce a signal with circular Gaussian statistics (Abeyratne et al. 1996). The coherent component consists of nonrandomly spaced scatterers and reflectors that introduce a constant phasor term that shifts the mean value of the Gaussian statistics. The coherent component can be further divided into long-range and short-range order if desired (Insana et al. 1986; Luchies 2011). A long-range order consists of regularly (nonrandomly) spaced scatterers, where the regular spacing exists throughout the tissue region. A short-range order consists of nonrandomly spaced reflectors or scatterers such as blood vessels or organ boundaries. Figure 5.7 shows graphs of different kinds of signals that may be encountered when there are dominate coherent scattering components. When there are periodic scatterers, the spectrum results in peak associated with the scatterer separation. For a large specular scatterer, the power spectrum is dominated by the spectral content of the specular scatterer.

Several studies have demonstrated that the presence of a coherent scattering component can be useful for tissue characterization purposes. For example, the mean scatterer spacing (MSS) parameter is an estimate that arises from the presence of regularly spaced scatterers in the medium. Fellingham and Sommer (1984) used the MSS parameter to distinguish between normal and cirrhotic liver tissue. Other studies have explored the GS for improving diagnostics of breast cancer and quantifying the duct structure of the breast (Donohue et al. 2001).

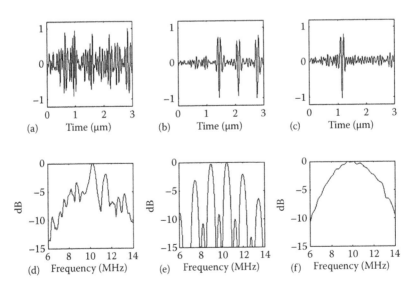

FIGURE 5.7 Examples of scan line segments from different types of scatterer collection scenarios, including (a) diffuse scatters, (b) diffuse scatterers with regularly spaced scatterers, and (c) diffuse scatterers with an isolated dominant scatterer. (d–f) Power spectra associated with the scatterer collection scenarios. (Adapted from Luchies A, *Quantitative ultrasonic characterization of diffuse scatterers in the presence of structures that produce coherent echoes.* MS thesis, University of Illinois at Urbana-Champaign, 2011.)

Chapter 5

The GS for a signal is defined over the bifrequency plane and calculated as follows:

$$G(f_1, f_2) = E[Y(f_1)Y^*(f_2)], \tag{5.25}$$

where $E[\cdot]$ represents the statistical expectation operator, $Y(f)$ is the Fourier transform of a signal, $y(t)$, and asterisk (*) denotes the complex conjugate. The GS is an expanded autocorrelation function that is formed by taking the outer product of the Fourier transform of the signal with itself. After estimating the GS, the collapsed average over the GS provides a means to extract parameters that are related to the presence or absence of the coherent component. The collapsed average is calculated as follows (Luchies et al. 2010):

$$C(f') = \frac{1}{M(f')} \left| \sum_{f_2 - f_1 = f'} G(f_1, f_2) \right|, \tag{5.26}$$

where f' is a frequency difference associated with the off-diagonal components of the bifrequency plane, and $M(f')$ is the total number of discrete GS points associated with the off-diagonal component $f_2 - f_1 = f'$. Figure 5.8 shows the GS and collapsed average

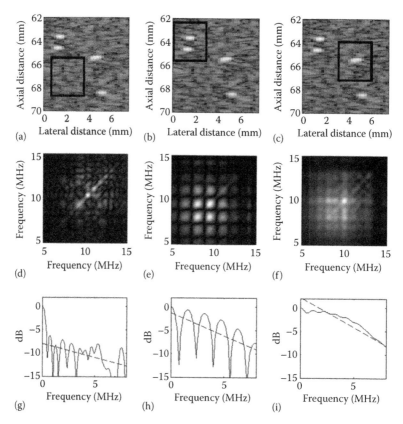

FIGURE 5.8 Example of data blocks (outlined using black boxes) with (a) diffuse, (b) periodic with diffuse scatterers, and (c) specular with diffuse scatterers. Associated GS are shown in (d–f), and the associated collapsed averages of the GS are shown in (g–i). (Adapted from Luchies A, *Quantitative ultrasonic characterization of diffuse scatterers in the presence of structures that produce coherent echoes*. MS thesis, University of Illinois at Urbana-Champaign, 2011.)

of the GS associated with diffuse, periodic, and a specular scatterer. From the collapsed average of the GS, estimates of the signal properties can be identified and used for characterization. For example, the spacing of the peaks in Figure 5.8h can be used to provide an estimate of the MSS. Similarly, if peaks are detected in a data block or a strong specular component is detected in a data block (because the collapsed average does not fall off rapidly versus frequency), then the data block may be sorted into a group that would not be analyzed for diffuse scattering properties, that is, ESD or EAC (Luchies et al. 2010).

5.3 Envelope Statistics

Several models for the statistics of the envelope of acoustic and optical signals have been proposed over the past few decades with applications to sea echo, medical ultrasound, and laser (Hruska and Oelze 2009). Some of these distributions include the Rayleigh distribution, the Rician distribution, the K distribution, the homodyned K distribution, and the Nakagami distribution. Because the derivations of these distributions have been covered extensively in the literature (e.g., Destrempes and Cloutier 2010), only a brief review is given.

5.3.1 Rayleigh Distribution

The Rayleigh distribution arises when a large number of nearly identical and randomly located scatterers contribute to the echo signal. The probability density function (pdf) is calculated as follows:

$$p_A(A) = \frac{A}{\sigma^2} \exp\left(-\frac{A^2}{2\sigma^2}\right),$$ (5.27)

where A (which is assumed to be positive) represents the envelope amplitude and σ^2 is the echo signal variance. If the signal follows a Rayleigh distribution, the only information about a sample that can be extracted from the envelope of the backscattered ultrasound is the mean backscattered power.

5.3.2 Rician Distribution

The Rician distribution extends the Rayleigh distribution to include a coherent signal, which could be the result of periodically located scatterers or specular scatterers in the signal (Dutt and Greenleaf 1994, 1995). The Rician pdf is calculated as follows:

$$p_A(A) = \frac{A}{\sigma^2} e^{-\frac{[A^2+s^2]}{2\sigma^2}} I_0\left(\frac{As}{\sigma^2}\right)$$ (5.28)

where I_0 is the modified Bessel function of the first kind and s is the coherent signal component. When $s \to 0$ (no coherent component), the pdf reduces to the Rayleigh pdf. When $s \to \infty$, the pdf reduces to a Gaussian. The Rician pdf enables the estimation of strength of the coherent signal.

Chapter 5

5.3.3 K Distribution

Jakeman and Pusey (1976) introduced the use of the K distribution, a generalization of the Rayleigh distribution, in the context of microwave sea echo to model situations where the number of scatterers per resolution cell is assumed to be small. In the case of few scatterers per resolution cell, the pdf is calculated as follows:

$$p_A(A) = \frac{2b}{\Gamma(\mu)} \left(\frac{bA}{2} \right)^{\mu} K_{\mu-1}(bA) \tag{5.29}$$

where $\Gamma(\bullet)$ is the Gamma function, $K_n(\bullet)$ is the modified Bessel function of the second kind, nth order, and μ is the effective number of scatterers per resolution cell. The b parameter can be expressed as follows:

$$b = 2\sqrt{\frac{\mu}{E[A^2]}}, \tag{5.30}$$

where $E[\bullet]$ is the expectation operator. The K distribution approaches the Rayleigh distribution in the limit $\mu \rightarrow \infty$. The K distribution enables the estimation of the number of scatterers per resolution cell. If the resolution cell of the ultrasound source can be calculated, the actual number density of scatterers can be estimated from the estimate of μ.

5.3.4 Nakagami Distribution

The Nakagami distribution, or m-distribution, was developed by Nakagami after extensive experiments on the long-distance multipath propagation of radio waves via the ionosphere and/or troposphere (Nakagami 1960; Suzuki 1977). The Nakagami distribution is calculated as follows:

$$p_A(A) = \frac{2m^m A^{2m-1}}{\Gamma(m)\Omega^m} \exp\left(\frac{-mA^2}{\Omega} \right), \tag{5.31}$$

where m is the Nakagami parameter and Ω is the scaling factor. The two parameters of the Nakagami pdf can be estimated as follows:

$$\Omega = \langle A^2 \rangle, \tag{5.32}$$

and

$$m = \frac{\Omega^2}{\langle (A^2 - \Omega)^2 \rangle}. \tag{5.33}$$

The value of the estimated Nakagami parameter enables the classification of the signal, and therefore the sample or tissue, into one of three categories. If the Nakagami parameter is between 0.5 and 1, the envelope is considered to be pre-Rayleigh. This would correspond to the condition where the scatterer is randomly located in space but have random-scattering cross sections. In the case where the scatterer locations are random and the scatterers have nearly identical scattering cross sections, the Nakagami parameter approaches unity and the envelope is considered Rayleigh distributed. If the sample is made up of scatterers with some periodically located scatterers, then the Nakagami parameter is greater than unity and the distribution is considered Rician or post-Rayleigh. In the situation where there are few scatterers per resolution cell, the Nakagami parameter can take a value <0.5. In this scenario, the distribution is considered to be a Nakagami-Gamma. Therefore, the Nakagami distribution is useful for classifying tissues or samples by designating the type of scattering condition associated with the envelope distribution. Nakagami imaging has been explored for tissue characterization, where maps of the Nakagami parameter are constructed (Tsui et al. 2008; Li et al. 2010; Ho et al. 2012).

5.3.5   Homodyned K Distribution

The homodyned K distribution was first introduced by Jakeman (1980). Besides incorporating the capability of the K distribution to model situations with low effective scatterer number densities, the homodyned K distribution can also model situations where a coherent signal component exists due to periodically located scatterers. This makes the homodyned K distribution the most versatile of the models discussed, but also the most complicated (Destrempes and Cloutier 2010). The pdf of the homodyned K distribution does not have a closed-form expression; however, it can be expressed in terms of an improper integral as follows:

$$p_A(A) = A \int_0^\infty x J_0(sx) J_0(Ax) \left(1 + \frac{x^2 \sigma^2}{2\mu}\right)^{-\mu} dx, \tag{5.34}$$

where $J_0(\bullet)$ is the modified zeroth-order Bessel function of the first kind, s^2 is the coherent signal energy, σ^2 is the diffuse signal energy, and μ is the same as in the K distribution. The derived parameter $k = s/\sigma$ is the ratio of the coherent to diffuse signal energy and can be used to describe the level of structure or periodicity in scatterer locations.

To estimate the parameters of the homodyned K distribution, Hruska and Oelze (2009) developed an algorithm. The algorithm made use of fractional order moments of the envelope SNR (R), skewness (S), and kurtosis (K) to provide estimates of the k and μ parameters. The statistical parameters of the envelope are calculated as follows:

$$R = \frac{E[A^v]}{\left(E[A^{2v}] - E^2[A^v]\right)^{1/2}}, \tag{5.35}$$

$$S = \frac{E[A^{3v}] - 3E[A^v]E[A^{2v}] + 2E^3[A^v]}{\left(E[A^{2v}] - E^2[A^v]\right)^{3/2}}, \tag{5.36}$$

$$K = \frac{E[A^{4v}] - 4E[A^v]E[A^{3v}] + 6E[A^{2v}]E^2[A^v] - 3E^4[A^v]}{\left(E[A^{2v}] - E^2[A^v]\right)^2},$$

(5.37)

where v is an arbitrary moment order. A crucial observation is that if the amplitude is modeled according to the homodyned K distribution, then R, S, and K are functions of only k, μ, and v. Thus, for a given value of v, R, S, and K as defined in Equations 5.35 and 5.36 do not depend on the specific values of s and σ, only on μ and the ratio $k = s/\sigma$.

By estimating R, S, and K for a given value of v from ultrasonic echo data, it is possible to obtain estimates of k and μ. Rather than directly applying Equations 5.35 and 5.36 to data to obtain these estimates, standard statistical techniques are used to obtain unbiased estimates. A single R, S, or K estimate will not uniquely identify the desired parameters k and μ because there are usually infinitely many points in (k, μ) space where the theoretical value equals the estimate. In fact, the locus of such points forms a curve in (k, μ) space, which will be called a level curve. However, curves derived from the R, S, and K estimates will usually come close to intersecting at a point. Furthermore, level curves derived from different moment orders provide additional curves. By finding the point in (k, μ) space that is in some sense closest to the level curves, an estimate is obtained (Hruska 2009).

During estimation, R, S, and K are estimated for a data block. The estimated R, S, and K are used to generate level curves for two sets of fractional order moments for different values of k and μ. These curves are generated through comparison with a theoretical evaluation of R, S, and K. The location where the level curves intersect in the (k, μ) space provides the estimate of k and μ. By using six curves (three curves for R, S, and K for each fractional order moment used), an overdetermined system provides more robust estimates.

5.3.6 Estimating Envelope Statistics from Finite Samples

As in the spectral estimation techniques, an infinite amount of samples will not be available for estimation. Similar to the spectral estimation techniques, data blocks can be defined where a finite number of data samples can be analyzed to yield parameter estimated. In the case of the envelope statistics, the signal corresponding to the data blocks is processed to provide the envelope. The envelope of the RF signal can be detected through the Hilbert transform method. The amplitude of each envelope sample can be tallied in a histogram to examine its shape. Figure 5.9 shows an image of an envelope distribution constructed from ultrasound backscatter measurements. Along with the measured distribution, the Rayleigh pdf and the best-fit homodyned K distribution are shown. From the distribution of the envelope amplitudes, SNR, skewness, and kurtosis can be calculated for a particular data block. From these estimates, values of k and μ can be estimated for each data block. These data blocks can then be color coded to provide parametric images enhanced by information using the envelope statistics. Figure 5.10 shows an image of a tumor enhanced by information provided by the envelope statistics (homodyned K distribution).

The quality of estimates based on the envelope statistics depends on several factors. As for spectral-based estimates, the bias and the variance of estimates from the envelope statistics also depend on the number of samples available for estimation, that is, the

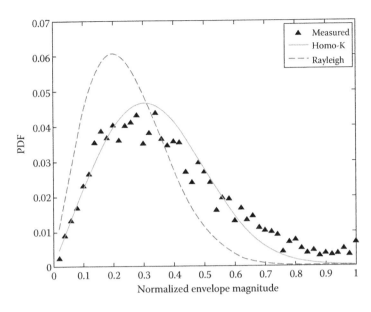

FIGURE 5.9 Distribution of ultrasound echo envelope values along with a best fit line using the homodyned K distribution and the Rayleigh pdf.

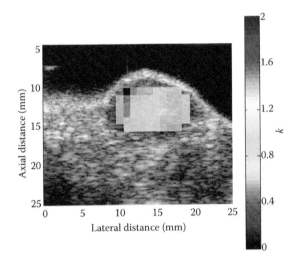

FIGURE 5.10 Parametric image of k parameter enhancing a B-mode image of a rodent tumor (MAT II-B).

larger the data block size, the better the bias and variance of the estimates (Hruska and Oelze 2009). In addition, when estimating the number of scatterers per resolution cell, μ, the accuracy of the estimate depends on the actual number of scatterers per resolution cell. When using the homodyned K distribution method and fractional order moments, μ is only accurate when the number of scatterers per resolution cell is less than 10. Other algorithms based on full order moments are even less sensitive to larger values of μ.

To convert the number of scatterers per resolution cell to the actual number density, the resolution cell of the ultrasonic imaging system at a location in space must be taken into account. The resolution cell size changes in the field of the source depending on the diffraction pattern, the sound speed profile of the sample, and the attenuation of the signal to the data block location. Attenuation downshifts the center frequency of the pulse and spreads the beam with distance (Zabzebski et al. 1999). Therefore, the resolution cell size will change depending on the attenuation of the medium. If attenuation is not correctly accounted for, then the estimates of μ and the conversion to the scatterer number density will be incorrect.

5.4   Applications of QUS

5.4.1   Diagnostic Applications

QUS techniques have successfully demonstrated the ability to uniquely classify different kinds of tissues for diagnostic purposes. QUS techniques offer imaging capabilities with new sources of image contrast as opposed to B-mode images where the contrast is based solely on small impedance mismatches between tissue structures. In addition, because QUS images provide specific numbers related to tissue structures and these techniques are approximately system and operator independent, QUS techniques can offer improved specificity for the diagnosis of disease. The following few sections provide a review of different successes related to the diagnostic applications of QUS imaging.

5.4.1.1   Ophthalmological Applications

Feleppa et al. (1986) found that the effective scatterer size in ocular tumors was a strong indicator of cancer. Larger scatterer sizes were observed in cancerous tumors as opposed to surrounding normal tissues. EAC was also integral to diagnostically distinguishing between ambiguous cases (Lizzi et al. 1987). For more than 20 years, Coleman and colleagues (Coleman and Lizzi 1983; Coleman et al. 1987, 1990, 1991, 2004; Silverman et al. 2001, 2003) strived to develop QUS techniques to diagnose/classify primary malignant melanoma of the chorioid and ciliary body to preserve vision without increasing the risk to the patient's life. They demonstrated that ultrasound backscatter properties were correlated with survival in patients with uveal malignant melanoma (Coleman et al. 1990, 1991). Recently, they successfully demonstrated that QUS parameters of extravascular matrix patterns (EMPs) correlated with histological EMP patterns to discriminate between lethal and less lethal tumors (Coleman et al. 2004).

5.4.1.2   Prostate Cancer

Currently, the definitive diagnosis of prostate cancer is achieved by core biopsy procedures. The standard means of guiding biopsy procedures is through conventional transrectal ultrasound (TRUS) imaging. However, conventional TRUS imaging cannot distinguish between normal prostate and diseased prostate. Therefore, to improve biopsy guidance and to possibly reduce the need for biopsy, QUS techniques have been examined for their ability to distinguish cancerous from non-cancerous regions in the prostate (Feleppa et al. 1996, 1997).

In this work, spectral-based QUS was used to identify regions of suspicion for cancer (Feleppa 2008). ESD, EAC, SS, SI, and MBF were examined for their capabilities to uniquely classify regions as cancer in the prostate. A receiver operator characteristic (ROC) analysis was conducted to quantify the ability of the classifier based on QUS parameters to predict cancer. In one study including 64 patients and 617 biopsies, a multilayer perceptron classifier (an artificial neural network [ANN] classifier) yielded an area under the curve (AUC) of 0.844 ± 0.018 using only the PSA level, SI, and MBF versus using B-mode imaging (which resulted in an AUC of 0.638 ± 0.031). In a subsequent study of the same data, four different classifiers were examined using the spectral-based parameters and the PSA level. The four classifiers were ANN, logitboost algorithms, support vector machines, and stacked, restricted Boltzmann machines (RBMs). Of the four classifiers, the RBM yielded the best performance with an AUC of 0.91 ± 0.04 (Feleppa et al. 2009). Therefore, the use of spectral-based features for detecting prostate cancer has demonstrated success.

5.4.1.3 Renal Studies

QUS scattering in renal tissues showed that changes in the scattering strength (acoustic concentration) were responsible for the anisotropy of backscatter and not changes in scatterer size and was thus an important parameter for characterizing the anisotropy of backscatter in tissues (Insana et al. 1991). The glomeruli ($\approx 200~\mu m$) and afferent and efferent arterioles ($\approx 50~\mu m$) were identified as the principal structures responsible for scattering. Scatterer size and sound speed were the most stable QUS parameters for characterizing the tissues. These early studies were the basis of investigations into the ability of QUS images using the scatterer properties to detect changes in renal microanatomy (Insana et al. 1992, 1993; Garra et al. 1994). QUS imaging techniques were demonstrated to be capable of differentiating among conditions that caused increased cortical echogenicity and structural changes such as glomerular hypertrophy, and QUS measurements agreed well with measurements of those structures in biopsy samples.

5.4.1.4 Mammary Tumors

Studies were conducted to determine if QUS could differentiate between benign tumors and malignant, and between different kinds of malignant tumors (Oelze et al. 2002, 2004; Oelze and Zachary 2006). QUS analysis was conducted over a broad ultrasonic frequency range of 5 to 25 MHz. The first tumors examined were spontaneous mammary fibroadenomas in rats (Oelze et al. 2002). The second kind of tumor examined was from a commercially available tumor cell line, the 4T1 MMT carcinoma for mice (ATCC, Manassas, VA) (Oelze et al. 2004). The carcinoma cells were cultured in medium and then injected subcutaneously into the fat pad of balb/c mice. Tumors were grown to a little over a centimeter in size and then examined using QUS techniques. The third kind of tumor examined was a mammary sarcoma (EHS sarcoma for mice; ATCC, Manassas, VA) (Oelze et al. 2002). Sarcoma cells were injected into mice (C57BL/6), and tumors were allowed to grow to a little over a centimeter in size before scanning.

The ESD and EAC estimates from the fibroadenomas and the malignant tumors revealed statistically significant differences using the spherical Gaussian model (SGM). The average ESD estimates were 105.9 ± 13.7, 30.0 ± 8.90, and $33.0 \pm 8.00~\mu m$ for the fibroadenomas, carcinomas, and sarcomas, respectively. The carcinoma tumors consisted of cellular hyperplasia that had undergone hypertrophy. Sizes of the cells were approximately the size of the ESD estimates from the tumors.

A distinct difference was observed between fibroadenomas with larger ESDs and carcinomas and sarcomas with the smaller ESDs. However, the QUS analysis using the SGM over the analysis bandwidth of 10 to 25 MHz was insufficient to separate the carcinoma from the sarcoma. Using the SGM, the carcinomas and sarcomas appeared almost identical.

Histological analysis of the tumors was then used to identify the dominant sources of scattering in the tumors. Figure 5.11a through c shows photomicrographs of the three kinds of solid tumors examined. In Figure 5.11a, the glandular acini were identified as the dominant source of scattering for the fibroadenomas. These glands are roughly spheroidal in shape and are characterized by a small band of epithelial cells surrounding a pocket of fluid. The average size of these acini was estimated to be around 100 μm in diameter, corresponding to the estimated ESD using QUS. In Figure 5.11b, the individual cells themselves were identified as the most probable source of dominant scattering in the tumors. The cells had an average nuclear diameter of 13 μm with the total cell size 50% to 200% larger than the nucleus.

Because the sarcoma ESD estimates were similar to the carcinoma estimates, it was expected that a similar type of structure would be observed in the sarcoma. However, as Figure 5.11c reveals, the structure of the sarcoma was vastly different than the structure of the carcinoma. The sarcoma was characterized by cells that were of the same size as the carcinoma cells, but instead of having a uniform distribution of cells, the sarcoma had a clumped distribution of cells. These clumped structures were as small as a single cell and sometimes larger than 100 μm. The clumps were surrounded by a fluid extracellular matrix consisting of laminin, collagen IV, entactin, and heparin sulfate proteoglycan (essentially a fluid).

QUS analysis using ESD estimates did not detect differences between carcinomas and sarcomas, although morphologies of the tumors were observed from optical photomicrographs to be distinctly different. *F*-tests of the normalized backscattered power spectra from the two kinds of malignant tumors revealed that statistically significant differences existed when the frequency of ultrasound was more than 16 MHz (Oelze and Zachary 2006). Therefore, differences that existed in the RF backscattered signature were not detected using the SGM over the initial analysis bandwidth of 10 to 25 MHz. However, when the ESD was estimated over the analysis bandwidth of 16 to 25 MHz, significant differences were observed between sarcomas (32.1 ± 3.81) and carcinomas (42.1 ± 4.01).

From the estimates of the scatterer properties, images that display the scatterer property information were constructed. These images are called QUS images or ultrasound

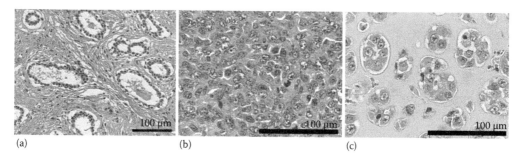

(a)   (b)   (c)

FIGURE 5.11   Photomicrographs of (a) rat fibroadenoma, (b) mouse mammary carcinoma, and (c) mouse mammary sarcoma.

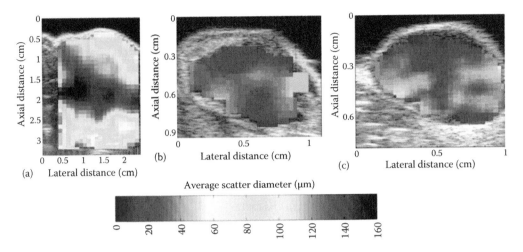

FIGURE 5.12   ESD images of mammary tumors: rat fibroadenoma (5–12 MHz) (a), mouse mammary carcinoma (10–25 MHz) (b), and mouse mammary sarcoma (10–25 MHz) (c) tumors.

parametric images. Figure 5.12 shows images of the three kinds of tumors that were examined. The images are useful in that they can provide an immediate understanding of the underlying structure based on the scatter property estimates. For example, the images show a clear distinction between fibroadenomas and other tumors. When using the higher analysis bandwidth, differences were observed between carcinomas and sarcomas, as shown in Figure 5.13.

5.4.1.5   Detection of Micrometastases in Lymph Nodes

High-frequency ultrasound (>20 MHz) and QUS techniques have been used to detect and localize small metastatic foci in human lymph nodes (Mamou et al. 2010, 2011). The detection of micrometastases from histology is a significant medical problem because lymph nodes are typically cut into small slices of a millimeter or larger for examination. If the micrometastases are not within the slice region, they can be easily missed. Therefore, there is a sampling problem for pathologists to find these micrometastases.

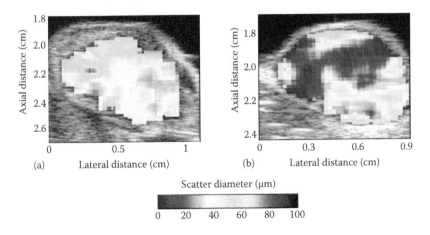

FIGURE 5.13   ESD images of a carcinoma (a) and sarcoma (b) using analysis bandwidth of 16 to 25 MHz.

Chapter 5

To provide pathologists with improved sampling of the lymph nodes for the detection of micrometastases, freshly excised lymph nodes from cancer patients undergoing lymphadectomy were examined using QUS with both spectral-based parameters and envelope statistics parameters. Three-dimensional volumes were scanned with a transducer having a center frequency of 25.6 MHz. After ultrasound scanning, each lymph node was inked to recover the orientation for later comparison to QUS images. The lymph nodes were then cut in half, embedded, fixed, sectioned in 65 μm intervals, stained with hematoxylin and eosin, and imaged optically with a high quality slide scanner. Metastatic regions were highlighted in each region by a pathologist.

To obtain the QUS parameter estimates, data blocks were defined as cylindrical volumes, and the signal from the cylindrical volume was gated and processed for QUS. The cylindrical volumes had a 1 mm diameter and a 1 mm length. From the normalized power spectrum, ESD, EAC, SS, SI, and MBF were calculated. From the envelope statistics, k and μ parameters were estimated from the homodyned K distribution, and m and Ω parameters were estimated from the Nakagami distribution. QUS images were constructed from voxels corresponding to the cylindrical data blocks.

In one study of 112 lymph nodes, a total of 13 combinations of the parameters were examined to build a classifier (Mamou et al. 2011). From this analysis, it was observed that the ESD alone yielded a sensitivity and specificity of 95% and 91.3%, respectively. ROC curves were also generated, and an area under the ROC curve for the ESD parameter was found to be 0.986. The best classification was observed when combing the ESD with the k parameter with a 95%, 95.7%, and 0.996 sensitivity, specificity, and AUC, respectively. Therefore, QUS has demonstrated success at detecting micrometastases in human lymph nodes.

5.4.1.6 Fatty Liver Disease

Multiple QUS parameters have been examined for liver disease diagnosis, including attenuation, envelope statistics (first-order statistics), and RF spectral analysis (second-order statistics) (Oosterveld et al. 1993a,b; Sasso et al. 2010). These studies used the QUS parameters to detect the classes of diffuse liver disease against a population of normal livers. The researchers found that attenuation, envelope statistics, and spectral parameters were highly important parameters for discriminating the different disease states. Similarly, Lizzi et al. (1988) examined the use of scattering models and spectral analysis from backscattered ultrasound to interpret liver state. By providing parameters related to the tissue microstructure, they argued that better interpretation and insights into disease states could be obtained.

Studies were conducted on the livers of New Zealand white rabbits, which examined the ability of QUS parameters (ESD, EAC, k parameter, and μ parameter) to characterize fatty liver. The rabbits had been on a normal diet, an atherogenic diet for 3 weeks, or an atherogenic diet for 6 weeks. The lipid levels in the livers were quantified by estimating the mass of lipids in the liver versus the total liver mass and were recorded for each group. From the ultrasonic backscatter from the livers, ESD and EAC were estimated. From the envelope of the backscattered signals, k and μ parameters were estimated using the homodyned K distribution (Hruska and Oelze 2009). Table 5.1 lists QUS parameter estimates from rabbit livers for each group along with the corresponding lipid levels. Each of the QUS parameters is correlated with the degree of lipid level content. The

Table 5.1   QUS Estimates from Rabbit Livers versus Duration on Atherogenic Diet and Liver Lipid Content

Weeks on Diet	ESD (μm)	EAC (dB)	k Parameter	μ Parameter	Concentration of Lipids (mg/g)
0 weeks	57.4 ± 5.63	29.2 ± 8.34	0.68 ± 0.11	0.30 ± 0.16	15 ± 9.0
3 weeks	28.2 ± 0.78	53.3 ± 1.29	0.40 ± 0.12	1.57 ± 0.38	56 ± 22
6 weeks	21.4 ± 3.00	61.5 ± 6.44	0.39 ± 0.09	1.99 ± 0.92	139 ± 28

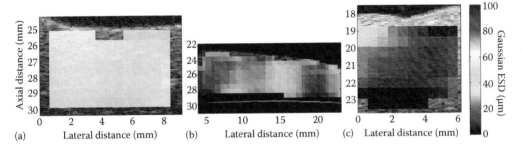

FIGURE 5.14   QUS images of normal and fatty liver with (a) normal diet, (b) 3 weeks on fatty diet, and (c) 6 weeks on fatty diet. The upper panels are ESD images and the lower panels are EAC images.

correspondence of lipid level increases with changes in QUS parameters, suggesting that QUS has the potential to differentiate between normal and fatty liver and to grade the degree of fatty liver. Figure 5.14 shows QUS (ESD and EAC) images of a normal liver, a liver from a rabbit on a fatty diet for 3 weeks, and a liver from a rabbit on a fatty diet for 6 weeks. From the images, it can be observed that the QUS parameters were able to provide new contrast to differentiate and classify the liver state (fatty versus nonfatty).

5.4.2   Monitoring Therapy

Because QUS is hypothesized to be sensitive to tissue microstructure and changes in tissue microstructure, QUS techniques have been used to monitor and assess therapeutic response. The following sections describe some efforts in using QUS for therapy assessment and monitoring.

5.4.2.1   Detection of Apoptosis

QUS has been used to monitor the response of tumors to different kinds of therapy. QUS techniques have been used to monitor the efficacy of cancer treatment in animal models and cell cultures. Studies have used ultrasound backscatter microscopy (40 MHz) to examine the differences between pellets of viable, dead, and apoptotic cells in vitro (Czarnota et al. 1997, 1999; Kolios et al. 2002; Vlad et al. 2009). Czarnota et al. (1999) examined cancer cells in culture using QUS techniques when undergoing different treatment protocols. They found that therapy-induced apoptosis could be detected and quantified in cell cultures using QUS. Specifically, results from the study indicated a twofold increase in backscatter intensity from apoptotic cells over viable cells. Cells that

Chapter 5

had been killed by heat exhibited an intermediate level of backscatter intensity. The ESD of the cells was estimated and observed to decrease when cells underwent apoptosis. In this study, it was conjectured that the nucleus was a major contributor to cellular scattering.

In addition, QUS was successfully used to assess and monitor therapy (i.e., chemotherapy and photodynamic therapy) in animal models of cancer (Czarnota et al. 1999). In one study, tumor-bearing mice were treated with radiation, and the changes in the tumor structure were examined using the spectral-based QUS parameters of SS, SI, and integrated backscatter (IB) (Vlad et al. 2009). SS and IB were observed to increase in treated mice compared with control animals. Parametric images were constructed, and results correlated with histology. Parametric images of the IB and SS were able to demarcate regions of therapeutic response from regions where no response was observed.

5.5 Conclusion

QUS techniques have been developing for more than 30 years. Early successes were demonstrated in ocular tumors. Since then, QUS techniques have been applied to improving diagnostics in cancer detection by increasing the specificity of imaging and by offering new sources of image contrast. QUS techniques have also been successfully demonstrated to be sensitive to tissue structural changes associated with cell death and therapeutic response. QUS will continue to be a significant ultrasonic imaging mode for improved diagnostics and capabilities for ultrasound imaging.

References

Abeyratne UR, Petropulu AP, and Reid JM. On modeling the tissue response from ultrasonic b-scan images. *IEEE Transactions on Medical Imaging*, 15, 479–490 (1996).

Baldeweck T, Laugier P et al. Application of autoregressive spectral analysis for ultrasound attenuation estimation: Interest in highly attenuating medium. *IEEE Transactions on Ultrasonics, Ferroelectrics and Frequency Control*, 42, 99–109 (1995).

Bigelow T. Estimating the total ultrasound attenuation along the propagation path by applying multiple filters to backscattered echoes from a single spherically focused source. *IEEE Transactions on Ultrasonics, Ferroelectrics and Frequency Control*, 57, 900–907 (2010).

Chaturvedi P and Insana MF. Error bounds on ultrasonic scatterer size estimates. *Journal of the Acoustical Society of America*, 100, 392–399 (1996).

Coleman DJ and Lizzi FL. Computerized ultrasonic tissue characterization of ocular tumors. *American Journal of Ophthalmology*, 96, 165–175 (1983).

Coleman DJ, Rondeau MJ et al. Computerized ultrasonic biometry and imaging of intraocular tumors for monitoring of therapy. *Transactions of the American Ophthalmological Society*, 85, 48–81 (1987).

Coleman DJ, Silverman RH et al. Correlations of acoustic tissue typing of malignant-melanoma and histopathologic features as a predictor of death. *American Journal of Ophthalmology*, 110, 380–388 (1990).

Coleman DJ, Silverman RH et al. Ultrasonic tissue characterization of uveal melanoma and prediction of patient survival after enucleation and brachytherapy. *American Journal of Ophthalmology*, 112, 682–688 (1991).

Coleman DJ, Silverman RH et al. Noninvasive in vivo detection of prognostic indicators for high-risk uveal melanoma: Ultrasound parameter imaging. *Ophthalmology*, 111, 558–564 (2004).

Czarnota G J, Kolios MC et al. Ultrasonic biomicroscopy of viable, dead and apoptotic cells. *Ultrasound in Medicine and Biology*, 23, 961–965 (1997).

Czarnota GJ, Kolios MC et al. Ultrasound imaging of apoptosis: High-resolution non-invasive monitoring of programmed cell death in vitro, in situ and in vivo. *British Journal of Cancer*, 81, 520–527 (1999).

Destrempes F and Cloutier G. A critical review and uniformized representation of statistical distributions modeling the ultrasound echo envelope. *Ultrasound in Medicine and Biology*, 36, 1037–1051 (2010).

Donohue KD, Huang L et al. Tissue classification with generalized spectrum parameters. *Ultrasound in Medicine and Biology*, 27, 1505–1514 (2001).

Dutt V and Greenleaf JF. K distribution model of ultrasound speckle: Fractional order SNRs and log compression variance. *Proceedings of the IEEE Ultrasonics Symposium*, 1375–1378 (1995).

Dutt V and Greenleaf JF. Ultrasound echo envelope analysis using a homodyned K distribution signal model. *Ultrasonic Imaging*, 16, 265–287 (1994).

Feleppa EJ, Lizzi FL et al. Diagnostic spectrum analysis in ophthalmology: A physical perspective. *Ultrasound in Medicine and Biology*, 12, 623–631 (1986).

Feleppa EJ, Kalisz A et al. Typing of prostate tissue by ultrasonic spectrum analysis. *IEEE Transactions on Ultrasonics, Ferroelectrics and Frequency Control*, 43, 609–619 (1996).

Feleppa EJ, Liu T et al. Ultrasonic spectral-parameter imaging of the prostate. *International Journal of Imaging Systems and Technology*, 8, 11–25 (1997).

Feleppa EJ. Ultrasonic tissue-type imaging of the prostate: Implications for biopsy and treatment guidance. *Cancer Biomarkers*, 4, 201–212 (2008).

Feleppa EJ, Rondeau MJ et al. Prostate-cancer imaging using machine-learning classifiers: Potential value for guiding biopsies, targeting therapy, and monitoring treatment. *Proceedings IEEE Ultrasonics Symposium*, Rome, Italy, 527–529 (2009).

Fellingham LL and Sommer FF. Ultrasonic characterization of tissue structure in the in vivo human liver and spleen. *IEEE Transactions on Sonics and Ultrasonics*, SU-31, 418–428 (1984).

Fink M, Hottier F, and Cardoso JF. Ultrasonic signal processing for in vivo attenuation measurement: Short time Fourier analysis. *Ultrasonic Imaging*, 5, 117–135 (1983).

Garra BS, Insana MF et al. Quantitative ultrasonic detection of parenchymal structural change in diffuse renal disease. *Investigative Radiology*, 29, 134–140 (1994).

Gerig AL, Varghese T, and Zagzebski JA. Improved parametric imaging of scatterer size estimates using angular compounding. *IEEE Transactions on Ultrasonics, Ferroelectrics and Frequency Control*, 51, 708–715 (2004).

Ghoshal G and Oelze ML. Improved scatterer property estimates from ultrasound backscatter using gate-edge correction and a pseudo-Welch technique. *IEEE Transactions on Ultrasonics, Ferroelectrics and Frequency Control*, 57, 2828–2832 (2010).

Ho MC, Lin JJ et al. Using ultrasound Nakagami imaging to assess liver fibrosis in rats. *Ultrasonics*, 52, 215–222 (2012).

Hruska DP. Improved techniques for statistical analysis of the envelope of backscattered ultrasound using the homodyned K distribution. MS thesis, University of Illinois at Urbana-Champaign (2009).

Hruska DP and Oelze ML. Improved parameter estimates based on the homodyned K distribution. *IEEE Transactions on Ultrasonics, Ferroelectrics and Frequency Control*, 56, 2471–2481 (2009).

Huisman HJ and Thijssen JM. Precision and accuracy of acoustospectrographic parameters. *Ultrasound in Medicine and Biology*, 22, 855–871 (1996).

Insana MF. Modeling acoustic backscatter from kidney microstructure using and anisotropic correlation function. *Journal of the Acoustical Society of America*, 97, 649–655 (1995).

Insana MF and Brown DG. Acoustic scattering theory applied to soft biological tissues. In *Ultrasonic Scattering in Biological Tissues*, edited by K. Shung and G. Thieme. CRC Press, Boca Raton, FL (1991).

Insana MF and TJ Hall. Parametric ultrasonic imaging from backscatter coefficient measurements—Image formation and interpretation. *Ultrasonic Imaging*, 12, 245–267 (1990).

Insana MF, Garra BS et al. Analysis of ultrasound image texture via generalized Rician statistics. *Optical Engineer*, 25, 743 (1986).

Insana MF, Wagner RF et al. Describing small-scale structure in random media using pulse-echo ultrasound. *Journal of the Acoustical Society of America*, 87, 179–192 (1990).

Insana MF, Hall TJ, and Fishback JL. Identifying acoustic scattering sources in normal renal parenchyma from the anisotropy in acoustic properties. *Ultrasound in Medicine and Biology*, 17, 613–626 (1991).

Insana MF, Wood JG, and Hall TJ. Identifying acoustic scattering sources in normal renal parenchyma in vivo by varying arterial and ureteral pressures. *Ultrasound in Medicine and Biology*, 18, 587–599 (1992).

Insana MF, Hall TJ et al. Renal ultrasound using parametric imaging techniques to detect changes in microstructure and function. *Investigative Radiology*, 28, 720–725 (1993).

Jakeman E. On the statistics of K-distributed noise. *Journal of Physics A: Mathematical and General*, 13, 31–48 (1980).

Jakeman E and Pusey PN. A model for non-Rayleigh sea echo. *IEEE Transactions on Antennas and Propagation*, 24, 806–814 (1976).

Kim H and Varghese T. Hybrid spectral domain method for attenuation slope estimation. *Ultrasound in Medicine and Biology*, 34, 1808–1819 (2008).

Kolios MC, Czarnota GJ et al. Ultrasonic spectral parameter characterization of apoptosis. *Ultrasound in Medicine and Biology*, 28, 589–597 (2002).

Kuc R and Li H. Reduced-order autoregressive modeling for center-frequency estimation. *Ultrasonic Imaging*, 7, 244–251 (1985).

Lavarello RJ, Oelze ML et al. Implementation of scatterer size imaging on an ultrasonic breast tomography scanner. *Proceedings of the 2009 IEEE Ultrasonics Symposium*, Rome, Italy, 305–308 (2009).

Lavarello RJ and Oelze ML. Quantitative ultrasound estimates from populations of scatterers with continuous size distributions. *IEEE Transactions on Ultrasonics, Ferroelectrics and Frequency Control*, 58, 744–749 (2011).

Lavarello RJ and Oelze ML. Quantitative ultrasound estimates from populations of scatterers with continuous size distributions: Effects of the size estimator algorithm. *IEEE Transactions on Ultrasonics, Ferroelectrics and Frequency Control*, 59, 2066–2076 (2012).

Li ML, Li DW et al. Ultrasonic Nakagami visualization of HIFU-induced thermal lesions. *Proceedings of the 2010 IEEE Ultrasonic Symposium*, San Diego, CA, 2251–2253 (2010).

Liu W and Zabsebski JA. Trade-offs in data acquisition and processing parameters for backscatter and scatterer size estimations. *IEEE Transactions on Ultrasonics, Ferroelectrics and Frequency Control*, 57, 340–352 (2010).

Lizzi FL, Greenebaum M et al. Theoretical framework for spectrum analysis in ultrasonic tissue characterization. *Journal of the Acoustical Society of America*, 73, 1366–1373 (1983).

Lizzi FL, Ostromogilsky M et al. Relationship of ultrasonic spectral parameters to features of tissue microstructure. *IEEE Transactions on Ultrasonics, Ferroelectrics and Frequency Control*, 34, 319–329 (1987).

Lizzi FL, King DL et al. Comparison of theoretical scattering results and ultrasonic data from clinical liver examinations. *Ultrasound in Medicine and Biology*, 14, 377–385 (1988).

Lizzi FL, Astor M et al. Statistical framework for ultrasonic spectral parameter imaging. *Ultrasound in Medicine and Biology*, 23, 1371–1382 (1997).

Lu ZF, Zagzebski JA et al. A method for estimating an overlying layer correction in quantitative ultrasound imaging. *Ultrasonic Imaging*, 17, 269–290 (1995).

Luchies A. Quantitative ultrasonic characterization of diffuse scatterers in the presence of structures that produce coherent echoes. MS Thesis, University of Illinois at Urbana-Champaign (2011).

Luchies A, Ghoshal G et al. Reducing the effects of specular scatterers on QUS imaging using the generalized spectrum. *Proceedings of the 2010 IEEE Ultrasonics Symposium*, San Diego, CA, 728–731 (2010).

Madsen EL, Insana MF, and Zagzebski JA. Method of data reduction for accurate determination of acoustic backscatter coefficients. *Journal of the Acoustical Society of America*, 76, 913–923 (1984).

Mamou J, Coron A et al. Three-dimensional high-frequency characterization of cancerous lymph nodes. *Ultrasound in Medicine and Biology*, 36, 361–375 (2010).

Mamou J, Coron A et al. Three-dimensional high-frequency backscatter and envelope quantification of cancerous human lymph nodes. *Ultrasound in Medicine and Biology*, 37, 345–357 (2011).

Nakagami M. *The m-Distribution, a General Formula of Intensity Distribution of Rapid Fading*. Pergamon Press, Oxford, UK (1960).

Nam K, Zagzebski JA, and Hall TJ. Simultaneous backscatter and attenuation estimation using a least squares method with constraints. *Ultrasound in Medicine and Biology*, 37, 2096–2104 (2011).

O'Donnell M and Miller JG. Quantitative broadband ultrasonic backscatter: An approach to nondestructive evaluation in acoustically inhomogeneous materials. *Journal of Applied Physics*, 52, 1056–1065 (1981).

Oelze ML and O'Brien, Jr. WD. Frequency-dependent attenuation compensation functions for ultrasonic signals backscattered from random media. *Journal of the Acoustical Society of America*, 111, 2308–2319 (2002).

Oelze ML and O'Brien, Jr. WD. Defining optimal axial and lateral resolution for estimating scatterer properties from volumes using ultrasound backscatter. *Journal of the Acoustical Society of America*, 115, 3226–3234 (2004a).

Oelze ML and O'Brien, Jr. WD. Improved scatterer property estimates from ultrasound backscatter for small gate lengths using a gate-edge correction factor. *Journal of the Acoustical Society of America*, 116, 3212–3223 (2004b).

Oelze ML and Zachary JF. Examination of cancer in mouse models using quantitative ultrasound. *Ultrasound in Medicine and Biology*, 32, 1639–1648 (2006).

Oelze ML, Zachary JF, and O'Brien, Jr. WD. Characterization of tissue microstructure using ultrasonic backscatter: Theory and technique optimization using a Gaussian form factor. *Journal of the Acoustical Society of America*, 112, 1202–1211 (2002).

Oelze ML, Zachary JF, and O'Brien, Jr. WD. Differentiation and characterization of mammary fibroadenomas and 4T1 carcinomas using ultrasound parametric imaging. *IEEE Transactions on Medical Imaging*, 23, 764–771 (2004).

Oosterveld BJ, Thijssen JM et al. Detection of diffuse liver disease by quantitative echography: Dependence on a priori choice of parameters. *Ultrasound in Medicine and Biology*, 19, 21–25 (1993a).

Oosterveld BJ, Thijssen JM et al. Correlations between acoustic and texture parameters from RF and B-mode liver echograms. *Ultrasound in Medicine and Biology*, 19, 13–20 (1993b).

Oppenheim A and Schafer R. *Digital Signal Processing*. Prentice Hall, Englewood Cliffs, NJ (1975).

Parker KJ, Lerner RM, and Waag RC. Comparison of techniques for in vivo attenuation measurements. *IEEE Transactions on Biomedical Engineering*, 35, 1064–1068 (1988).

Parker KJ and Waag RC. Measurement of ultrasonic attenuation within regions selected from B-scan images. *IEEE Transactions on Biomedical Engineering*, 30, 431–437 (1983).

Sanchez JR, Pocci D, and Oelze ML. A novel coded excitation scheme to improve spatial and contrast resolution of quantitative ultrasound imaging. *IEEE Transactions on Ultrasonics, Ferroelectrics and Frequency Control*, 56, 2111–2123 (2009).

Sasso M, Beaugrand M et al. Controlled attenuation parameter (CAP): A novel VCTE™ guided ultrasonic attenuation measurement for the evaluation of hepatic steatosis: Preliminary study and validation in a cohort of patients with chronic liver disease from various causes. *Ultrasound in Medicine and Biology*, 36, 1825–1835 (2010).

Sigelmann RA and Reid JM. Analysis and measurement of ultrasound backscattering from an ensemble of scatterers excited by sine-wave bursts. *Journal of the Acoustical Society of America*, 53, 1351–1355 (1973).

Silverman RH, Lizzi FL et al. High-resolution ultrasonic imaging and characterization of the ciliary body. *Investigative Ophthalmology and Visual Science*, 42, 885–894 (2001).

Silverman RH, Folberg R et al. Spectral parameter imaging for detection of prognostically significant histologic features in uveal melanoma. *Ultrasound in Medicine and Biology*, 29, 951–959 (2003).

Suzuki H. A statistical model for urban radio propagation. *IEEE Transactions on Communications*, 7, 673–680 (1977).

Tsui PH, Yeh CK et al. Classification of breast masses by ultrasonic Nakagami imaging: A feasibility study. *Physics in Medicine and Biology*, 53, 6027–6044 (2008).

Vlad RM, Brand S et al. Quantitative ultrasound characterization of responses to radiotherapy in cancer mouse models. *Clinical Cancer Research*, 15, 2067–2075 (2009).

Wear KA, Wagner RF et al. Application of autoregressive spectral analysis to cepstral estimation of mean scatterer spacing. *IEEE Transactions on Ultrasonics, Ferroelectrics and Frequency Control*, 40, 50–58 (1993).

Yao LX, Zagzebski JA, and Madsen EL. Backscatter coefficient measurements using a reference phantom to extract depth-dependent instrumentation factors. *Ultrasonic Imaging*, 12, 58–70 (1990).

Zabzebski JA, Chen JF et al. Intervening attenuation affect first-order statistical properties of ultrasound echo signals. *IEEE Transactions on Ultrasonics, Ferroelectrics and Frequency Control*, 46, 35–40 (1999).

Chapter 5

6. Ultrasound Tomography
A Decade-Long Journey from the Laboratory to the Clinic

Neb Duric, Peter J. Littrup, Cuiping Li, Olivier Roy, and Steve Schmidt

Ultrasound Imaging and Therapy. Edited by Aaron Fenster and James C. Lacefield © 2015 CRC Press/Taylor & Francis Group, LLC. ISBN: 978-1-4398-6628-3

Chapter 6

6.1 Introduction

Ultrasound tomography (UST) is an imaging technique that combines sonography with computed tomography (CT) methods to solve an inverse problem. It is well suited for inferring biomechanical properties of a volume of tissue from measurements made along a surface surrounding the tissue. One clinically relevant application is the detection of breast cancer, a major subject of this review.

In x-ray imaging, CT is simplified by the assumption that x-rays travel in straight lines. This assumption is reasonable because x-rays are photons with sufficiently high energies to behave as particles in this regard. Consequently, x-ray image reconstructions are relatively straightforward, relying on methods that are variants of the radon transform. By contrast, UST uses sound waves, which unlike x-rays are purely wavelike and therefore do not travel as particles. Consequently, UST must take into account wave propagation phenomena such as reflection, refraction, and even diffraction. In an inhomogeneous medium, ultrasound pulses do not travel in straight lines, thereby complicating the tomographic inversion and placing extra burden on the computational requirements. The need for a high level of computing power and associated data processing has been a major historical factor in limiting the development of UST compared with CT and other tomographic methods.

Despite this limitation, UST has been under development for more than 30 years, motivated by many potential advantages over x-ray CT in the area of medical imaging. At diagnostic levels, sound waves do not appreciably heat tissue and, unlike x-rays, do not damage tissue through the process of ionization. With mounting concerns over radiation exposure, UST offers a nonsignificant risk alternative for medical imaging. In the area of breast imaging, UST offers the possibility of a comfortable alternative to mammography, which requires substantial compression that many women find uncomfortable and some even painful. Furthermore, UST is poised to address limitations associated with current clinical breast imaging.

Mammography is the current gold standard for breast cancer screening, and although it has been shown to reduce the mortality rate in multiple screening trials,[1] it generates many abnormal findings not related to cancer, which leads to additional costly imaging procedures and biopsies.[2–5] Magnetic resonance imaging (MRI) is making increasing inroads into diagnostic breast imaging by virtue of its high sensitivity and operator independence. Consequently, for high-risk women, MRI is now viewed as the highest standard for breast cancer early detection and screening.[6,7] However, MRI can have a high false-positive rate and requires contrast injection, and the examinations can be both relatively long and costly.

Recent studies have demonstrated the effectiveness of ultrasound imaging in detecting breast cancer,[8–10] particularly for women with dense breasts. The ACRIN 6666 study, funded by the Avon Foundation and the National Cancer Institute, represents a definitive trial evaluating the potential of ultrasound as a screening tool.[9] The latest reports have shown the potential to screen for small breast masses otherwise missed by mammography.[10] Despite these successes, ultrasound is unlikely to fill the gap between the cost-effectiveness of mammography and the imaging quality of MRI for the following reasons:

i. The added cancers found by ultrasound are offset by an increased burden of false positives,[10] indicating that the characterization of small masses by conventional ultrasound is perhaps even more limited than that of MRI.
ii. The operator-dependent nature of ultrasound will prevent uniform replication of results.
iii. The associated small aperture imaging leads to long examination times and the need to "stitch" the localized images into a view of the entire breast.

Addressing the problems of current ultrasound to give it the volumetric image acquisition and operator independence of MRI would be therefore a major advance toward a clinically successful screening option. We, along with other groups, are motivated by the possibility that UST could provide such an option.

Driven by ongoing research in this area, UST is riding Moore's law into relevancy. This review will attempt to capture the UST concept and its history, ongoing status, and future status as a viable medical imaging technology.

6.2 UST as an Inverse Problem

In an inverse problem, one wishes to find m such that

$$d = G(m),$$

where G is an operator describing the explicit relationship between the data d and the model parameter m, and it is a representation of the physical system. The operator G is often called the *forward operator*.

In most practical applications, the object can be confined to either a 2-D plane or a 3-D volume while the measurements are made in 1-D or 2-D space, respectively (Figure 6.1). In

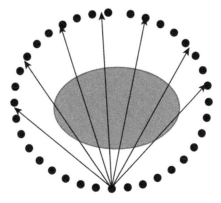

FIGURE 6.1 A series of emitters and receivers are used to probe the object of interest (shaded). A single transmit is accompanied by multiple receivers.

Chapter 6

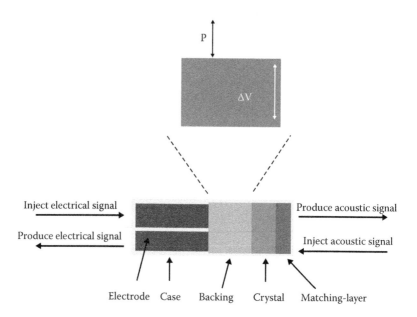

FIGURE 6.2 A transducer converts mechanical energy into an electrical signal when receiving and performs the opposite function when transmitting.

the case of UST, a typical setup would involve either surrounding the object with an array of transducer elements or rotating a transducer around the object to probe the object with sound waves and to measure the resulting interaction between the sound waves and the object. The measurements would then be recorded and used to construct an image tomographically.

Generally, UST sensors consist of ultrasound transducers that are made of piezoelectric materials (PZT) because they are highly efficient at converting electrical energy into mechanical energy and vice versa (Figure 6.2). Thus, PZT transducers can be used as either receivers or emitters, and the same transducer can be switched between the two modes.

A transducer can be mechanically rotated around the object to provide many points of insonification and measurement, or a transducer consisting of an array of elements can be placed around the object (Figure 6.1). The advantage of the former technique (mechanical multiplexing) is its simplicity and possible lower cost. An array of elements is generally more expensive to build, but it offers the possibility of either an electronic multiplexing or a parallelized system that provides data channels for many or all elements, thereby greatly accelerating the data acquisition process. There is therefore a trade-off between the cost and the speed associated with any UST implementation. The exact amount of trade-off is governed by the application and favors high-speed implementations in the case of clinical imaging, as described later.

6.2.1   Solving the UST Inverse Problem

The algorithms used to reconstruct images from the acquired data are derived from their counterparts in the seismic community. Typically, seismic studies involve systems many kilometers in size using sound waves with wavelengths of up to 1 km, whereas clinical studies involve objects many millimeters in size using sound waves with wavelengths of up to 1 mm (Figure 6.3). In essence, the two systems are of similar size when counted in wavelengths of the probing signal.

Consequently, algorithms used in the seismic community can be used almost directly for clinical applications. One example is the Kirchoff migration algorithm, long used in seismology for reconstructing images from reflected signals in support of oil exploration. Figure 6.4 sketches the concept of how Kirchhoff migration constructs an echo intensity value at a specific image pixel from a set of measurements, as defined by the inverse problem. Kirchhoff migration can be further simplified to a method known as *synthetic aperture formation technique*—a process of delaying and summing signals to reconstruct the amplitude of the reflected waveform at particular time instances in

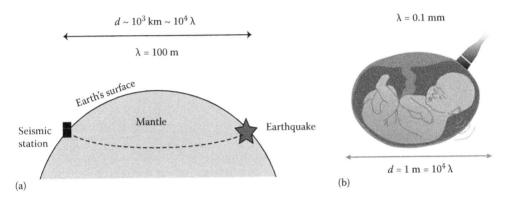

FIGURE 6.3 A typical seismic measurement system has similar dimensions as medical ultrasound when measured in wavelengths.

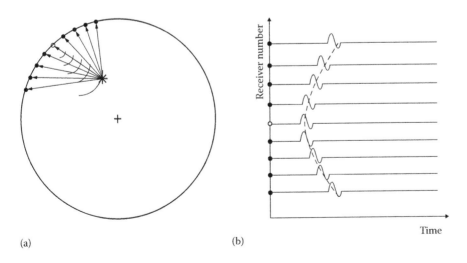

FIGURE 6.4 (a) For a given transmitter (open circle), a scatterer will radiate acoustic energy toward the receivers (filled circles). (b) The received signals arrive at different times, and the locus of such signals is related to the position of the scatterer.

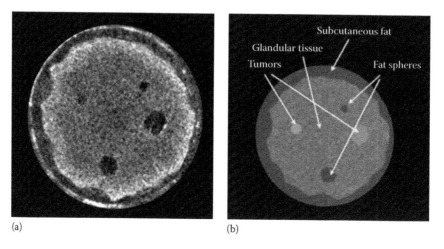

Subcutaneous fat

Glandular tissue

Tumors

Fat spheres

(a)   (b)

FIGURE 6.5 (a) A reflection image constructed using the method illustrated in Figure 6.4. (b) The truth image is represented by a CT scan of an anthropomorphic breast phantom.

the imaging region. An example of an image constructed in this manner is shown in Figure 6.5.

The second example is the reconstruction of the speed of sound images from signals transmitted through the object (as in the entire earth seismology from the sound generated by earthquakes). Sometimes called *time-of-flight* (TOF) tomography, the reconstruction relies on the measurements of the first arrival of the signal. Finally, the third type of reconstruction based on the amplitude changes of the first arriving transmitted signal is analogous to CT and yields attenuation images.

The quality of the reconstructed image depends on the quality of the signals acquired by the UST system and by the sophistication of the reconstruction algorithm. The latter is defined by how well the physics of the sound propagation are modeled. Generally, the simpler the wave-based assumptions, the faster an algorithm can run but the lower the quality of the final image. Therefore, a trade-off exists between reconstruction speeds and image quality. This trade-off can be understood by discussing the wave propagation theory and its computational implementations, as summarized in the following paragraph.

Sound propagates according to the acoustic wave equation as

$$\nabla^2 p - \frac{1}{c^2}\frac{\partial^2 p}{\partial t^2} = 0,$$

where ∇^2 is the Laplace operator, p is the acoustic pressure (the local deviation from the ambient pressure), and c is the speed of sound. The latter can also be expressed as

$$c^2 = \frac{K}{\rho},$$

where ρ is the material density and K is the compressibility constant. The solution for a spherical wave in a homogeneous medium is given as follows:

$$p(r,k) = \frac{A}{r} e^{\pm ikr},$$

where k is the wave number and A is the amplitude of the wave, which falls off with the radial distance traveled, r. It is evident that the propagation of the acoustic wave is sensitive to the material properties of density and stiffness. Therefore, for a heterogeneous medium, the solution is much more complex and requires algorithmic computations. Furthermore, both transverse and longitudinal waves are supported in any actual system. However, all UST implementations rely on the measurements of the longitudinal wave because it propagates much more rapidly and decays relatively slowly. The longitudinal waves can also undergo mode conversions creating surface waves, such as those called the *whispering gallery*. A full solution to the wave equation is therefore computationally daunting, given the complexity of the physics being described. Most reconstruction methods therefore make simplifying assumptions to make the problem tractable.

6.2.2 Ray Tomography

For finite bandwidth sound waves used in ultrasound imaging (1–20 MHz), energy travels from transmitter to receiver along a hollow banana-shaped volume, which can be represented as a "banana–donut,"[11–16] as shown in Figure 6.6. The center width of the "banana–donut" for dominant frequency is the width of the first Fresnel zone, $\sqrt{\lambda L}$, where λ is the wavelength and L is the distance between transmitter and receiver. In ray theory, this volume is collapsed into an infinitesimal line (ray path) by assuming the infinite frequency,[15] similar to what is assumed for geometrical optics.

6.2.2.1 Straight Ray Approximation

The straight ray approximation is similar to the assumption made for CT reconstructions, which assume x-rays travel in straight lines. Thus, every transmitter is connected to every receiver by a straight line. An example of an image made this way is shown in Figure 6.7, using a numerical phantom that mimics the overall structure of *in vitro* phantoms but enables a robust testing of multiple algorithms.

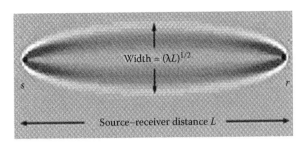

FIGURE 6.6 The banana–donut concept.

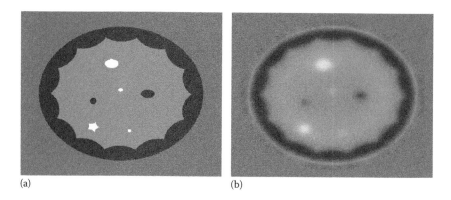

(a)   (b)

FIGURE 6.7   (a) A numerical breast phantom and (b) its straight ray reconstruction.

The spatial resolution of the reconstructed images is poor because straight ray approximation does not take into account the refraction of the waves as they pass through an inhomogeneous tissue. This blurring can be reduced by taking refraction into account when reconstructing the images by allowing rays to bend as they propagate from transmitter to receiver.

6.2.2.2   Bent Ray Tomography

Bent ray tomography relies on the knowledge that refraction is governed by changes in sound speed. The initial model of sound speed can be homogeneous or heterogeneous. The model is used to bend the rays as they propagate from one pixel to the next. The traced ray path and predicted arrival times are used to generate the next sound speed model, which enables more accurate bending. The process is repeated until convergence is achieved. The net effect of bending the rays is to compensate for the refractive effects and thereby reduce the amount of biological blurring. Figure 6.8 shows an image reconstructed with bent ray tomography. It is evident that a significant improvement in spatial resolution is achieved. More studies on bent-ray tomography can be found.[17-26]

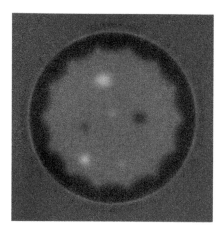

FIGURE 6.8   Image of the same numerical breast phantom using bent ray reconstruction. Better contrast is achieved compared with the straight ray model. The star-shaped boundary of the subcutaneous fat is better resolved.

6.2.2.3   Ray-Based Transmission Algorithm

A typical transmission algorithm has three components:

1. Data processing

     Before performing ray-based sound speed tomography on the acquired ultrasound data, the TOF for each received waveform needs to be picked. In other words, the onset time of ultrasound signal arriving at receiver needs to be determined. The picked TOFs are used to reconstruct sound speed images. To determine the TOF for each waveform, either manual picking or some forms of automatic picking can be exploited. Automatic pickers are required when large volumes of data are collected, as in medical imaging.[27–36]

     Before performing a ray-based attenuation tomography on the acquired ultrasound data, the attenuation data for each received waveform need to be calculated. Various authors[37–42] present different ways to determine attenuation data from the received waveforms.

2. Forward model

     In straight ray tomography, the propagation paths of ultrasound wave are assumed to be straight lines.

     In bent ray tomography, the 2-D ultrasound wave propagation is governed by the eikonal equation, as follows:

$$(\nabla E)^2 = (\partial T/\partial x)^2 + (\partial T/\partial y)^2 = (1/v)^2 = \left(s_x^2 + s_y^2\right)^2, \tag{6.1}$$

where T is the travel time, v is the sound speed, and (S_x, S_y) is the slowness vector of the ultrasound wave that is defined as the inverse of sound speed. In Equation 6.1, E = const describes the "wavefronts," and "rays" are defined as the orthogonal trajectories of these wavefronts. The eikonal equation can be obtained from the wave equation in the limit of infinite frequency.

     To solve the UST problem, a regular rectangular grid model is created on the image plane, whose boundaries enclose the acquisition geometry. Equation 6.1 is solved to obtain a travel-time map for each source (transmitter) position, which is later used to calculate the travel-time gradient for ray tracing. An ultrasound ray is back-propagated from receiver to transmitter based on either the straight ray path (straight ray tomography) or the travel-time gradient method (bent ray tomography) [16]. The traced ray paths serve as a sensitivity matrix in the inverse process.

3. Inversion

     We take sound speed inversion as an example to explain the inverse process of ray tomography. The procedures can be easily modified for attenuation inversion.

Let Δt_i be the difference between the ith picked TOF for the recorded ultrasound data and the ith calculated TOF based on the current sound speed model. The inverse problem can be described as follows:

$$\sum_{j}^{M} l_{ij}\Delta s_j = \Delta t_i, \tag{6.2}$$

Chapter 6

where Δs_j is the slowness (inverse of sound speed) perturbation for the jth grid cell, which needs to be inverted, and l_{ij} is the traced ray length of the ith ray within the jth cell. Equation 6.2 can be expressed in matrix form as follows:

$$L\Delta S = \Delta T, \tag{6.3}$$

where the ray matrix L is the so-called sensitivity matrix.

This is a linear problem for straight ray tomography because matrix L is fixed through the entire inverse process. The problem becomes nonlinear when we take the ray bending into consideration, in which case the matrix L depends on the current sound speed model.

The cost function for the inverse problem can be described as in the following equation:

$$f = \arg\min_{\Delta S}\left(\| L\Delta S_\lambda - \Delta T \|^2 + \lambda \mathrm{REG}(S_\lambda)\right). \tag{6.4}$$

The extra term $\lambda\mathrm{REG}(S_\lambda)$ in Equation 6.4 can be any form of regularization whose contribution to the total cost function is balanced by the trade-off parameter λ.

Starting with a homogeneous sound speed model, the optimization problem in Equation 6.4 is solved iteratively. For bent ray tomography, ray paths are traced on the updated sound speed model after each iteration. The iteration continues until the solutions converge. The techniques to determine trade-off parameter λ include the L-curve method[43,44] and the generalized cross-validation method.[44] There are excellent discussions in Refs. 45–47 on convergence rate and stopping criteria for the iterative inverse process. A simple stopping criterion is that the cost function for the current iteration is not significantly improved from the previous iteration.

6.2.3   Diffraction Tomography

As noted earlier, waveform tomography is computationally intensive, whereas ray tomography is fast but provides inferior spatial resolution. Investigators have sought simplified forms of the wave equation to reduce the computational burden while avoiding the ray approximation. The most common simplifying assumptions are known as the *first Born approximation* and the *first Rytov approximation*.[48–54]

6.2.3.1   First Born Approximation

In this approach, the wave equation is simplified by assuming that scattering is weak and there is no multiple scattering as the wave propagates from the transmitter to the receiver. The first Born approximation assumes the heterogeneity in the propagating medium perturbs the total wave field. It consists of taking the incident wave field in place of the total wave field as the driving wave field at each scatterer. This approximation is accurate enough if the scattered wave field is small compared with the incident wave field. It will break down if the scattered wave field becomes large relative to the reference wave field. Consequently, this method achieves high resolution but fails to properly reconstruct images with more than a few percent contrast differences. In most clinical applications, it is tantamount to assuming that the object being imaged can be inhomogeneous but with very small contrast variations.

6.2.3.2 Distorted Born Method

The distorted Born method is a high-order Born approximation. It computes iterative solutions to nonlinear inverse scattering problems through successive linear approximations. By decomposing the scattered field into a superposition of scattering by an inhomogeneous background and by a material perturbation, large or high-contrast variations in medium properties can be imaged through iterations that are each subject to the distorted Born approximation. However, the repeated numerical computation of forward solutions (Green's function) imposes a very heavy computational burden, which limits its clinical application.

6.2.3.3 First Rytov Approximation

The first Rytov approximation starts by assuming the heterogeneity in the medium perturbs the phase of the scattered wave field. This approximation is valid under a less restrictive set of conditions than the first Born approximation.[48,49] The validity of the first Rytov approximation is governed by the change in the scattered phase over one wavelength, not the total phase. In other words, the first Rytov approximation is valid when the phase change over a single wavelength is small (a few percent). However, most clinical applications violate this assumption.

6.2.3.4 Hybrid Method (Wave Path TOF Tomography)

Diffraction tomography has usually been presented in the Fourier domain, for a single-frequency source.[51–53] Woodward formulated diffraction tomography as a multifrequency inverse problem in the space domain and proposed the wave path concept for wave equation tomography to account for the finite frequency effects.[11] For band-limited wave propagation through a nondispersive medium, the phase shifts experienced by each frequency are equivalent. For the Rytov approximation, which naturally separates the amplitude (real part) from the phase (imaginary part) of wave, this means that the imaginary part of the Rytov wave paths can be summed over all frequencies without loss of information.[11] This summation over frequencies yields a narrow wave path resembling a volume like a banana-donut that runs from a source to a receiver (Figure 6.6). The TOF perturbation of a sound wave is linearly related to its phase shift that is closely related to the medium sound speed. This linear relationship provides a natural way to reconstruct the medium sound speed by applying the wave path TOF tomography (WTFT) technique. WTFT has been investigated in geophysics but, to the best of our knowledge, not yet in medical imaging.[10–16]

6.2.3.5 Reflection Tomography

In contrast to the use of transmitted signals, reflection tomography relies on the reflected echoes to construct images of relative echo amplitudes. Although the resulting images are somewhat analogous to B-mode clinical ultrasound, there are significant differences. Because UST data do not generally use beam forming on the front end, the Kirchoff migration technique does so "after the fact," in other words, after all the data have been gathered. Consequently, there is a great deal of flexibility on how the data are used.

Chapter 6

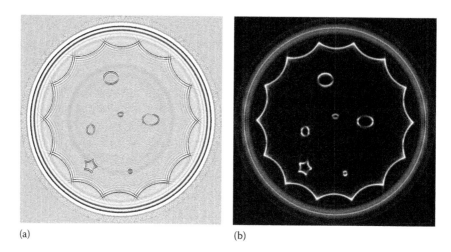

(a)   (b)

FIGURE 6.9   A reflection image from synthetic data shows fine details in this numerical phantom when the raw RF signal is used (a). With envelope fitting the image assumes a more conventional look (analogous to B-mode).

In particular, it is possible to use apertures of arbitrary size to reconstruct a reflection image. Unlike B-mode imaging, this method enables the reception of aperture sizes that are limited only by the overall locus of the receiving transducer elements. Such flexibility enables higher resolution and contrast relative to B-mode imaging. Finally, because the Kirchhoff migration method operates on raw data, there is more flexibility in terms of how the signals are processed before and after the migration process is initiated. Figure 6.9 shows reflection images made with and without envelope fitting, for example.

Generally, migration methods rely on the assumptions that every point in the object is a scattering object, independent of one another, and that only one scatter occurs on a path that connects that point to an emitter–receiver pair. Consequently, traditional migration methods do not take into account diffraction and other wave properties. Recent developments in wave-based migration methods promise to overcome these barriers.[55] Some migration methods have been adapted to correct for refraction and for attenuation. Using the reconstructed data sets of other modalities, sound speed and attenuation, enables proper time delays to be calculated for signal alignment and variable signal amplification to correct for inhomogeneous energy loss. The resulting effects to the image include sharper boundaries, higher contrast, less background noise, and, in some cases, the resolution of objects otherwise lost because of constant media assumptions. The ease of implementing these corrections is unique to the data acquisition method and the geometry of the problem. Examples are shown in Figure 6.10.

6.2.4   Waveform Tomography

With ever-increasing computational power, the ability to solve the wave equation is being realized. Even in this case, however, it is necessary to make some assumptions. The major assumption is that the density of the object being imaged is constant. Solutions based on this assumption are now possible, for both sound speed and attenuation.[56] The advantage of this approach in light of the computational burden is that it enables

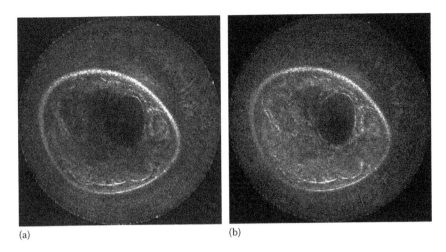

(a)   (b)

FIGURE 6.10   (a) An in vivo reflection image of a cyst. (b) Same object with refraction correction applied.

diffraction as well as a better correction for refractive effects. Furthermore, by using all of the recorded wave information (as opposed to the arrival time of the signal), the method has the potential to increase image contrast while suppressing artifacts. The limiting resolution of $\lambda/2$ is up to an order of magnitude better than ray tomography. Figure 6.11 shows a waveform reconstruction of the numerical breast phantom. We can observe that the quality of the reconstruction is significantly enhanced compared with ray-based reconstructions. In particular, the boundary of the star-shaped mass is well resolved.

Waveform tomography reconstruction methods have been formulated in the time domain[57,58] and in the frequency domain.[59] The latter usually enables a simpler formulation of the problem because convolution and differential operators are mapped to multiplications. The reconstruction process is similar to ray tomography. We start from an initial model of the unknown parameters (sound speed and attenuation) and solve a forward problem. The solution of this forward problem is a set of simulated waveforms recorded at the transducer locations. The residual between the recorded waveforms and

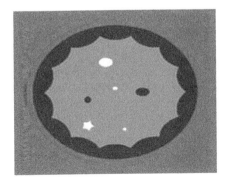

FIGURE 6.11   Reconstruction of the numerical breast phantom using waveform tomography. Compared with ray tomography, the resolution and the contrast are significantly better, and the boundaries of the masses are reconstructed with much higher accuracy.

the measured ones is then used to update iteratively the unknown parameters until convergence. This approach faces two major challenges.

The first one relates to the complexity of modeling the forward operator. The forward operator in sound speed waveform tomography must predict the waveform recorded at the transducer positions for a given sound speed model. This is in contrast to ray tomography where only the arrival time of the waveform needs to be estimated. Forward modeling is usually achieved by means of finite difference or finite element methods. These methods must be accurate enough to avoid numerical dispersion and to properly account for the boundaries of the simulation area (e.g., absorbing boundary conditions).[60,61] For a given accuracy, the size of the model typically scales linearly with the frequency of the probing pulse. Complexity can be lowered using approximations. It can also be addressed by means of efficient parallel implementations.[62,63] However, this issue remains a challenging one, especially in medical imaging applications where reconstruction time must be maintained at a minimum to maintain a high patient throughput.

The second challenge with waveform tomography pertains to the convergence of the iterative reconstruction process. Although ray tomography only matches travel times, its waveform counterpart attempts to minimize a residual computed over (a portion of) the entire waveform. For a single frequency, this process can be thought of as finding sound speed and attenuation models such that harmonic components match in phase and amplitude. Convergence to the correct cycle of the waveform thus requires an accurate initial model, especially at high frequencies. One approach is to start from an initial model obtained using ray tomography and to sequentially drive the iterative algorithm using waveform components from low to high frequencies.

6.3   History of Implementation

The idea of solving acoustic inverse problems in medicine can be traced back to the work of Wilde and Reid[64] and Howry and Bliss[65] in the 1950s. At that time, the systems used were crude mechanical scanners using a single transducer that rotates on an arm and collects reflected signals using the pulse-echo technique. The first cross-sectional breast tissue images were made at that time. Figure 6.12 shows one such scanner along with the first breast images.

However, the lack of computational power, combined with the slow rotation, made it impossible to apply this technique clinically. These early methods did give birth to what is now known as *B-mode clinical ultrasound*. However, the tomographic aspect had to wait almost 30 years before the concept of UST was seriously revisited.

Historically, two general approaches have been used to advance operator-independent sonography. One approach has been based on improving the current ultrasound devices and techniques that rely on reflection (or B-mode) imaging, whereas the second approach uses transmission imaging to characterize masses.

Stavros et al.[66] proposed that the analysis of mass margins, shape, and echo-properties based on conventional, reflection ultrasound images could lead to a highly accurate differentiation of benign masses from cancer. These observations led to the development of the "Stavros criteria," which evolved into the Breast Imaging Reporting and Data System (BIRADS) for ultrasound. To implement this analysis into a screening scenario,

Carcinoma

FIGURE 6.12   Early mechanical scanners of Howry et al. A scan of a cadaveric breast is shown on the right. (From Howry, D. H. and Bliss, W. R., *Journal of Laboratory and Clinical Medicine*, 40:579–592, 1952.)

attempts have been made to construct operator-independent scanners that image the entire breast (see Norton and Linzer[67] for an early example). The only commercial device to achieve any formal clinical acceptance thus far has been that of U-Systems.[68]

In 1976, Greenleaf et al.[69] made the seminal observation that acoustic measurements of breast tissue samples, made with transmission ultrasound, could be used to characterize breast tissue. On the basis of these studies, they concluded that using the imaging parameters of sound speed and attenuation (henceforth the Greenleaf criteria) could help differentiate benign masses from cancer.

Figure 6.13 shows the first renderings of sound speed images of a cadaveric breast from the work of Greenleaf and Johnson.[70] This work represents a big step from laboratory measurements of small tissue samples to cross-sectional imaging of the entire breast. The scan method consisted of rotating an opposing pair of simple transducers. The algorithms were based on the straight ray approximation.

As a direct result of this and other similar studies, several investigators developed operator-independent ultrasound scanners based on the principles of UST in an attempt to measure the Greenleaf criteria with in vivo scans.[71–75] Clinical examples include the work of Carson et al. (U. Michigan),[71] Andre et al. (UCSD),[72] Johnson et al. (TechniScan Medical Systems),[73] Marmarelis et al. (USC),[74] Liu and Waag (U. Rochester),[75] and Duric and Littrup et al. (KCI).[23,24] More recently, Ruiter et al.[76] have reported progress on a true 3-D scanner using a hemispherical array of transducers. Although no clinical results have been reported to date, clinical studies are currently being planned.[76] Although it would be difficult to review all UST projects carried out to date, we will instead highlight specific developments that represent key steps in the historical evolution of UST. The clinical systems developed by these groups used similar patient positioning systems. Patients were positioned in a prone position on a flat table with breasts suspended through a hole in the table in a water bath lying just below the table surface. The water

Chapter 6

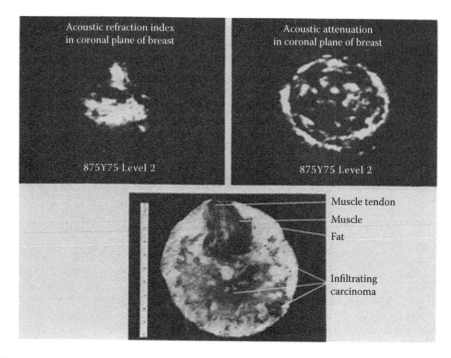

FIGURE 6.13 Cross-sectional images of sound speed and attenuation of a cadaveric breast.

bath is a requirement that ensures minimal distortion of the breast while allowing strong coupling of acoustic waves to the tissue.

The first successful in vivo transmission images of breast were reported by Glover et al.,[77] Greenleaf et al.,[78] and Carson et al.[71] Consisting of rotating transducers, the systems were sufficiently fast to complete one cross-sectional scan with little motion artifact. In the case of Carson et al., the system also performed reflection tomography to complement the transmission images, another first. Figure 6.14 shows reflection, sound speed, and attenuation images of the same breast cross section. As in the case of Greenleaf and

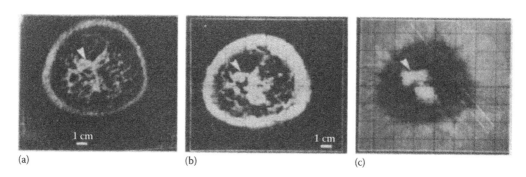

FIGURE 6.14 Reflection and transmission images from Carson et al.

Johnson, the transmission images are of limited resolution, primarily because of the straight ray approximation used by both groups. Although an in vivo first, it was still not practical to image an entire breast because of the need for mechanical rotation.

The first practical in vivo scans of an entire breast were carried out by Andre et al.[72] using a radically different concept. Instead of mechanical rotation, the system used a fixed, ring-shaped array of 1024 individual transducers that encircled the breast. The absence of mechanical rotation combined with a multichannel data acquisition allowed rapid scanning (minutes) of the entire breast under in vivo conditions. Another innovation was the use of diffraction tomography in place of ray-based reconstructions. Because diffraction tomography represents wave-based reconstructions in the Borne limit approximation, the resulting sound speed images were of much higher spatial resolution, as illustrated in Figure 6.15.

(a)

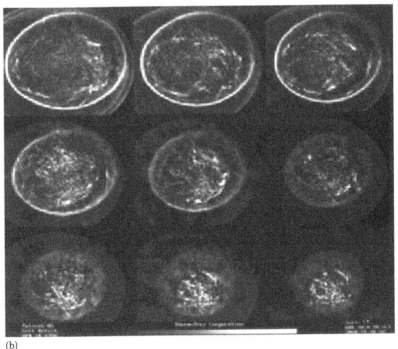

(b)

FIGURE 6.15 (a) Diffraction tomography setup of Andre et al. (b) Reconstructed in vivo cross-sectional images.

Chapter 6

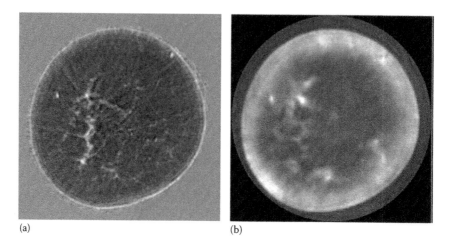

(a)   (b)

FIGURE 6.16   Sound speed and attenuation images from TechniScan group. These in vivo images were obtained with an inverse scattering algorithm.

In parallel with these first efforts to use wave-based imaging, investigators at TechniScan Inc. used a more robust approach using an algorithm based on inverse scattering theory.[73] Although this group uses mechanical multiplexing, their use of transducer arrays and multichannel data acquisition allowed them to perform in vivo scans of the entire breast. Examples of sound speed and attenuation images are shown in Figure 6.16. The images have superb spatial resolution and contrast, and they represent the gold standard in transmission UST at the moment.

In our laboratories, at the Karmanos Cancer Institute, our group has also focused on the development of UST for breast imaging. To that end, we have been developing and testing a clinical prototype in KCI's breast center. The continuing development of the prototype and its associated UST methodology have been guided by clinical feedback from these studies and have led to continuing evolution in imaging performance, leading to increasingly greater clinical relevance. This water bath system uses a solid-state ring array transducer consisting of 256 elements that encircle the breast in a similar manner to that of Andre et al. However, it differs in two important respects. It uses a 256-channel data acquisition system that enables single-slice acquisitions in approximately 30 ms, resulting in scans of the entire breast for 1 minute or less. Furthermore, it uses bent-ray reconstructions for imaging. Although such images have lower spatial resolution compared with wave-based approaches, they can be run fast, in keeping with the goal of a clinically fast system. The clinical prototype is currently being upgraded into a commercial system named SoftVue, through the start-up company Delphinus Medical Technologies. The new system will have 1024 active elements and use a 512-channel data acquisition system. Table 6.1 lists the major past and ongoing attempts to build and implement UST scanners.

6.4   Clinical Motivation for Breast Imaging

According to SEER statistics, breast cancer incidence varies with the stage of the disease.[79] Approximately 61% of breast cancers are localized, and approximately 31% are

Table 6.1 Major Distinguishing Attributes of Past and Ongoing Attempts to Build and Implement UST Scanners

Group	Transducer Type	DAQ Type	Entire Breast Scan Time	Dimensionality	Images Output R = Reflection S = Sound Speed A = Attenuation	Algorithm Types
Wild and Reid (1953)	Freehand, single	Single channel	Hours	2-D	R	B-mode
Howry et al. (1953)	Rotating, single	Single channel	Hours	2-D	R	B-mode
Greenleaf and Johnson (1978)	Rotating pair	Single channel	Hours	2-D	S, A	Straight ray
Carson et al. (1981)	Rotating pair	Single channel	Hours	2-D	R, S, A	Straight ray and B-mode
Andre et al. (1997)	Fixed ring array	Multichannel	Minutes	2.5-D	S, A	Diffraction
Johnson et al. (2007)	Rotating linear array	Multichannel	<1 hour	2.5-D–3-D	R, S, A	Inverse scattering
Duric and Littrup (2007)	Fixed ring array	Multichannel	~1 minute	2.5-D	R, S, A	Bent ray, migration

regional at the time of diagnosis. Another 5% are diagnosed with distant metastases, and approximately 3% are unstaged. The 5-year survival rate varies strongly with stage. The 5-year survival rate for women with localized and regional cancer is 98% and 84%, respectively. In the case of distant stage, the survival rate drops dramatically to 23%. For unstaged cancers, it is approximately 58%. The survival numbers decline further when periods longer than 5 years are considered.[79] The combination of these trends applied to the population of the United States results in approximately 190,000 new cancer cases diagnosed each year, with a corresponding mortality rate of 40,000/year.[79] These numbers suggest that many cancers are not detected at an earlier stage when they are more treatable. There are many reasons why cancers are not detected early, but some of the major factors are related to limited participation in breast screening and the performance of screening mammography.

6.4.1   Limited Participation in Screening

National cancer screening statistics indicate that only 51% of eligible women undergo annual mammogram.[80] Limited access, fear of radiation, and discomfort are among the factors cited explaining the low participation rate. Greater participation would lead to the detection of cancer at an earlier stage leading to a greater survival rate. Increased participation and improved breast cancer detection would have the greatest effect on nearly one in three women who are diagnosed each year with later stage (regional or greater) breast cancer, totaling approximately 60,000 women per year in the United States. The net effect would be an increase in survival time and a corresponding decrease in mortality rates, particularly among women who have been reluctant toward screening mammography.

6.4.2   Limited Performance of Mammography

For women with dense breast tissue, who are at the highest risk for developing breast cancer,[81–83] the performance of mammography is at its worst.[84] Consequently, up to 50% of cancers are missed at their earliest stages when they are the most treatable. Improved cancer detection for women with denser breasts would decrease the percentage of breast cancer diagnosed at later stages, which would result in lower mortality rates.

6.4.3   Role of UST Imaging

Although tomosynthesis may improve some of the limitations of standard mammography, it is unlikely to create a paradigm shift in performance and still uses ionizing radiation. By contrast, MR provides volumetric, radiation-free imaging with high sensitivity. Studies have shown that MR can affect a large swath of the breast management continuum ranging from risk assessment to diagnosis and treatment monitoring.[85–92] However, MR requires relatively long examination times and the use of contrast agents. Furthermore, MR has long been prohibitively expensive for routine use, and there is a need for a low-cost equivalent alternative. Positron emission tomography is also limited by cost. Conventional sonography, which is inexpensive, comfortable, and radiation free, continues to only play an adjunctive role in breast imaging because of its operator dependence and the time needed to scan the entire breast.[10] The lack of an alternative that balances between the cost-effectiveness of mammography and the imaging performance of MR is a barrier dramatically affecting mortality and morbidity through improved screening.

UST has the potential to eliminate this trade-off by combining the low-cost advantage of mammography with the superior imaging performance of MR. Furthermore, UST offers comfort, operator independence and low-installation and maintenance costs. These advantages have the potential to support early breast cancer detection and to improve participation in breast cancer screening programs. When fully realized, UST may have a positive societal impact because the accurate identification of cancer, at relatively low cost, would enable widespread access in clinics and a reduction in death rates from breast cancer.

6.4.4   Potential Advantages of UST for Breast Imaging

Current clinical practice is based on screening mammography with diagnostic follow-up and biopsy. UST has the potential to streamline this process by combining screening and diagnosis and by dramatically reducing the biopsy rate. Such streamlining is made possible by the fact that UST provides volumetric imaging of the entire breast, which has the same performance regardless of whether it is used for screening or diagnostics. Furthermore, the paradigm of breast compression and use of ionizing radiation is directly averted. The former is obviated by the use of a water bath, which eliminates probe contact with the breast, whereas the latter is eliminated by using sonic energy.

6.4.4.1   Safety
As noted earlier, clinical ultrasound is a nonsignificant risk imaging modality. UST uses acoustic energies that can be actually lower than conventional ultrasound. The reason for this is that many currently developing UST systems use unfocused beams from small aperture transducer elements (i.e., no beamforming) that are generally fired sequentially or are beamformed to launch plane waves. In either case, the instantaneous energy density and the peak power level are well below B-mode ultrasound (typically ~100 mW peak power).

6.4.4.2   Cost-Effectiveness
Compared with MRI and mammography, ultrasound is inexpensive by the nature of its components and by the fact that no room shielding or special installation is required. The transducers used in ultrasound equipment are much cheaper than x-ray tubes and far cheaper than the large magnets required by MRI. Furthermore, the electronics in ultrasound system handle power signals, adding to the cost advantage.

6.4.4.3   Patient Comfort
In a typical UST setup (e.g., Figure 6.17), the transducer signals are coupled to the breast tissue via a water bath. Consequently, the sensor does not touch the breast but stays immersed in the water.

6.4.5   Potential Diagnostic Value

Conventional reflection ultrasound exploits differences in acoustic impedance between tissue types to provide anatomical images of breast tumors. However, reflection is just

Chapter 6

FIGURE 6.17 The ring transducer used at the Karmanos Cancer Institute. The ring's 256 elements surround the breast (a breast phantom is shown).

one aspect of a multifaceted set of acoustic signatures associated with the biomechanical properties of tissue. Unlike palpation, for example, which relies on stiffness to identify suspicious masses, conventional ultrasound does not provide information on the elastic properties of tissue. UST records a greater set of acoustic properties to paint a more complete picture of the biomechanical properties of breast tissue. By merging reflection images with images of the bioacoustic parameters of SS and AT, UST offers the possibility of exploiting differences in anatomical and elastic properties of tissue to accurately differentiate cancer from normal tissue or benign disease.

The diagnostic value of UST is derived from its ability to probe the biomechanical properties of tissue. As noted in Section 6.2, a sound wave interacts strongly with tissue, thereby allowing the measurement of parameters such as reflectivity, sound speed, and attenuation in the form of images.

Reflectivity is determined by gradients in the acoustic impedance. In a simple example, where an acoustic signals travel between regions of distinctly different impedance (Z), at normal incidence,

$$R = \frac{Z_1 - Z_2}{Z_1 + Z_2}, \quad Z = \rho c$$

where ρ is the material density and c is the speed of sound.

This parameter is well suited for imaging boundaries that separate different tissue types. The currently practiced ultrasound BIRADS criteria are based on how tumor shape and margins are visualized, including echo-texture differences. This approach has been championed by Stavros to improve the specificity of diagnostic ultrasound examinations. The use of the Stavros criteria can be easily extended to UST images where they can be applied to the entire volume of the breast. Recent results with the U-Systems whole-breast scanner have already shown that the application of the Stavros criteria in the coronal plane can identify architectural distortion, a key element in diagnosing breast cancer.

The speed of sound in tissue varies with the physical parameters of density and stiffness, according to equation. It has been shown in the laboratory[93–95] that that stiffness and density are not independent variables and that K $\propto$ ρ^3 in human tissue. This approximate dependence suggests that sound speed is linearly proportional to density and varies as the cube root of the stiffness, on at least some spatial scales of the tissue. It is therefore a measure of both tissue properties, but with much greater relative sensitivity to density. Sound speed is an absolute quantity with units of kilometers per second, which enables measurements to be compared at different times or between patients. It is therefore a quantitative and virtual analog of manual palpation, which recognizes that "hard masses" have a different feel relative to normal soft tissue. The typical range of sound speed in soft human tissue is ~1.35 to 1.6 km/s.

The amplitude of a propagating longitudinal wave changes because of the two dominant processes of scattering and absorption. The former is an elastic process that simply redistributed the acoustic energy through the tissue, whereas the latter converts sound energy into heat. The two processes act together to define the attenuation coefficient a such that the decay of the signal amplitude is given as follows:

$$A(r) = A_0 e^{-\alpha r}.$$

The two processes have strong but different dependencies on frequency such that absorption is relatively important at high frequencies (>10 MHz) while scattering dominates at the lower frequencies. Most UST scanners being developed today use frequencies less than 10 MHz so that attenuation measurements can be interpreted as a combination of absorption and scatter. Scattering is strongly dependent on the cellular structure of tissue as well as the macroscopic structures smaller than a wavelength in size. It is the scattering of such structures (nonspecular reflections) that gives rise to speckle in conventional ultrasound images. Cancer is known to alter tissue on cellular and macroscopic scales and even its immediate environment, which in turn yields an attenuation signature. As Greenleaf and Johnson[70] showed, cancers tend to have higher levels of attenuation relative to normal tissue, a conclusion that can also be inferred from shadowing caused by cancer on conventional ultrasound images. Attenuation is strongly frequency dependent and has a value of ~0.5 dB/cm/MHz in tissue.

The diagnostic value of these UST parameters is that they can be combined to characterize breast tissue and to predict the presence of cancer as discussed in the next section.

6.4.5.1 Composition of Breast

Breast tissue is composed of fat and fibroglandular tissue. Fat is relatively anhydrous and has lower density and lower average atomic weight (Z number) compared with fibroglandular tissue. In mammography, these differences explain why fat has lower x-ray absorption than fibroglandular tissue. Similarly, the water content differences on MR images explain why fat and glandular tissue have lower and higher signal intensities, respectively. For UST, the differences are biomechanical in nature. Thus, fat has low sound speed and attenuation compared with fibroglandular tissue. Figure 6.18 illustrates how UST can separate fibroglandular tissue from fat.

6.4.5.2 Characterizing Lesions

As an extension of the previous discussion, we expect that we can exploit differences in biomechanical properties of lesions to differentiate between benign masses and cancer. The analysis of the images acquired in the manner described previously suggests that we

Chapter 6

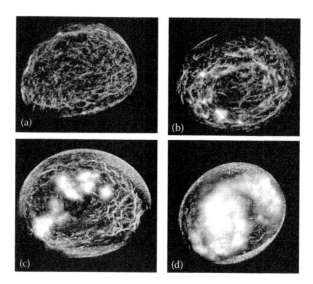

FIGURE 6.18   (a–d) Cross-sectional breast images corresponding to increasing breast density.

can detect the variety of mass attributes noted by current ultrasound BIRADS criteria, such as mass shape, acoustic mass properties, and architecture of the tumor environment. These attributes help quantify current BIRADS criteria (e.g., "shadowing" or high attenuation) and provide greater possibilities for defining a unique signature of cancer listed as follows:

1. Irregular margins: spiculated, microlobulated, or ill-defined margins are more suspicious than thin and smooth margins.
2. Architectural distortion: surrounding tissue shows altered anatomy (e.g., mass effect and/or retraction).
3. Elevated sound speed: higher sound speed than surrounding tissue is noted within the mass. Typically, the sound speed is elevated by 50 to 150 m/s relative to fat.
4. Elevated attenuation: higher attenuation than surrounding tissue is noted within the mass. The amount of enhancement varies but is approximately 2.5 to 5 dB/cm relative to fat at 2 MHz.

The first two attributes are linked to the acoustic shape of the mass as defined by the appearance of the mass in the reflection images. They represent straightforward applications of the reflection criteria of ultrasound BIRADS. The third and the fourth attributes are unique to transmission ultrasound, as first defined by Greenleaf. They represent the internal acoustic properties of the mass that can be measured quantitatively in the sound speed and attenuation images. The above attributes are defined such that the probability of cancer increases with the number of attributes that are present.

6.5   Clinical Results

As UST has matured, its clinical relevance has begun to be tested on the clinical stage. In recent years, an increasing number of studies have tested the technology under real-world

Table 6.2 Summary of Imaging Characteristics of the Prototype Scanner Used to Collect Data

Imaging Mode	Spatial Resolution[a]	Inversion Type	Tissue Properties Measured
Reflection	0.5 mm	Refraction corrected migration	Fibrous stroma (specular reflectors) and glandular tissue (nonspecular)
Sound speed	2 mm	Refraction corrected (bent ray) tomography	Differentiate tissue types based on density and compressibility
Attenuation	2 mm	Refraction corrected (bent ray) tomography	Differentiate tissue types based on tissue scattering properties

[a] Slice thickness, ~5 mm.

clinical conditions. TechniScan Inc. now has two active studies, one in Freiberg, Germany, the other at the University of California at San Diego. These studies have recruited more than 100 patients. For the past 6 years, our team at the Karmanos Cancer Institute in Detroit has undertaken multiple studies in support of scanner development. To date, more than 600 patient scans have been completed. A spin-off company from this project, Delphinus Medical Technologies, is currently assembling a new generation of scanner that will also be used in international multicenter trials. The outcome of these multicenter trials will provide the first definitive assessment of UST efficacy by comparing its performance against mammography and MRI. In this section, we present representative images from our group's work to illustrate the potential clinical relevance of UST (Table 6.2).

6.5.1 Breast Architecture

We performed two independent studies to assess UST's ability to record the major tissue components of the breast.

It is difficult to directly compare UST and mammographic images because the former measures the volume of the breast whereas the latter measures the area corresponding to the compressed breast. We therefore chose to compare the UST images with MRI because the latter also measure the volume of the breast. We performed a study of 36 patients to first determine whether the scanner is sensitive to similar breast structures as MRI. The initial focus of the study was to determine how reliably and accurately the breast architecture could be measured. Figure 6.19 illustrates a comparison of anatomy visualized by the prototype compared with that of MRI for the same patient. During the examination, the breast is less distorted by gravity because it is surrounded by water, whereas in MRI, the breast is also pendulant but surrounded by air. Apart from these differences, these results demonstrated that the prototype can accurately map breast anatomy, thereby allowing direct volumetric comparisons to MRI.

The similarity between the MRI and the UST images may appear surprising given that MRI measures water content of tissue using magnetic resonance whereas UST measures biomechanical properties using acoustics. A likely explanation is that both modalities

Chapter 6

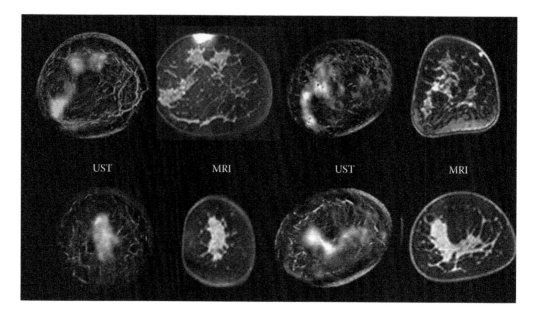

FIGURE 6.19 A comparison of cross-sectional UST breast images with MRI for four patients. Light regions indicate dense fibroglandular tissue, and dark areas represent fat. The enhanced MRI image (right images, fat subtracted and T1 weighted) shows the presence of fatty tissue (dark gray), parenchyma (light gray), and fibrous stroma (light bands). The corresponding ultrasound tomography images (left) show fatty tissue (dark), parenchyma (light gray), and fibrous stroma (white bands).

trace similar structures because both water content and sound speed increase with tissue density[96–98] as we now discuss.

6.5.1.1   Bioacoustic Properties of the Breast

The properties of breast tissue that are relevant to understanding their interaction with acoustic waves are their density, compressibility, and scattering properties. As in conventional sonography, reflections arise at interfaces between tissues of differing density and stiffness, which leads to differences in sound speed and acoustic impedance. Such reflections are said to be specular when the reflecting surface is smooth on scales $\gg l$, where l is the wavelength of the acoustic signal. By contrast, nonspecular reflections occur when acoustic waves interact with tissue structures that are smaller than l. These interactions lead to the speckle pattern that characterizes conventional US images. They also lead to attenuation because acoustic energy is scattered as a result of these interactions. At very low ultrasound frequencies, such as those used with this scanner (e.g., ~2 MHz), attenuation is dominated by this process. Dense tissue is generally accompanied by greater attenuation and sound speed with increased speckle content, whereas fibrous stroma, which represent large discontinuous changes in acoustic impedance, are manifested as bright specular reflectors. The development of thresholded sound speed and attenuation images overlaid upon reflection is shown in Figure 6.20.

6.5.1.2   Magnetic Resonance Properties of Breast Tissue

MRI operates on a fundamentally different principle. A strong magnetic field, in combination with RF signals, is used to manipulate the magnetic moments of nuclei that make up biological tissue. The signals detected are predominantly from hydrogen (H)

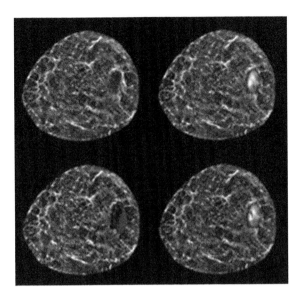

FIGURE 6.20   UST images showing reflection image (top left), addition of thresholded sound speed only (top right), addition of thresholded attenuation only (bottom left), and fusion image (bottom right) showing pixels that satisfy both thresholds. This scheme for depicting cancers also suggested the possibility of a graded differentiation scheme, whereby fibroadenomas would only appear yellow because they typically do not have high attenuation.

protons because imaging differences between biological tissues rely on relative water and/or blood content. The relative free-water content, or unbound H, is best noted on T2-weighted (spin–spin) sequences showing cysts with a very bright signal. Alternatively, the relative vascularity of tissues (i.e., blood flow) are best noted as enhancement differences caused by paramagnetic effects of gadolinium on T1-weighted images, whereby T1 signal (i.e., spin-lattice) is increased within parenchymal tissue and most tumors because of higher relative blood flow than breast fat.

The similarity of the MRI and UST images likely results from the importance of density in both types of interactions. Density drives signal strength from proton flips in the case of MRI, whereas it drives sound speed, attenuation, and acoustic impedance changes from proton flips in the case of UST. The high degree of spatial correlation of MR and sound speed images is therefore largely driven by similar sensitivity to changes in tissue density.

6.5.2   Breast Cancer Detection

Three types of images are produced from the raw data using previously described tomographic reconstruction algorithms: (i) sound speed, (ii) attenuation, and (iii) reflection.[25,36,42] Sound speed images are based on the arrival times of acoustic signals. Previous studies have shown that cancerous tumors have enhanced sound speed relative to normal breast tissue,[99–102] a characteristic that can aid the differentiation of masses, normal tissue, and fat. Attenuation images are tomographic reconstructions based on acoustic wave amplitude changes. Higher attenuation in cancer causes greater scatter of the ultrasound (US) wave, so attenuation data in conjunction with sound speed

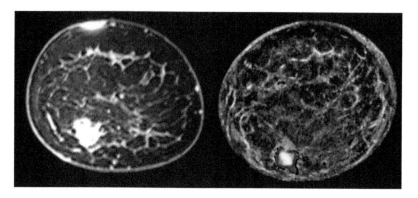

FIGURE 6.21   (Left) MRI cross-sectional image showing a cancer at seven o'clock. (Right) A fused UST image showing the same region.

provide a potentially effective means for determining malignancy as illustrated visually in Figure 6.20. Reflection images, derived from changes of acoustic impedance, provide echo-texture data and anatomical detail for the entire breast. Reflection images are valuable for defining tumor margins, which can be used to characterize lesions through the BIRADS criteria.[66] These three types of images can be combined without geometric discrepancy by means of image fusion, enabling a multiparameter visual and quantitative characterization of masses.[101,102]

In the comparative study with MRI discussed earlier, UST-fused images were compared with contrast-enhanced MRI images. Figure 6.21 shows an example of such a comparison. The mass at seven o'clock appears distinct in both the UST and the MRI images, indicating that the fusion technique is a potential tool for identifying and visualizing breast lesions from UST data. A separate study exploring this possibility is being carried out.

The ability to image both the US-BIRADS and the Greenleaf criteria suggests that it may be possible to improve diagnostic accuracy by (i) combining these criteria to better quantify US-BIRADS criteria, such as using decibels per centimeters for attenuation instead of "shadowing," and (ii) by extending the US-BIRADS criteria to three dimensions and thereby using data from the entire lesion.

6.5.3   Therapy Monitoring

The mass characterization described earlier was also applied to patients undergoing chemotherapy who were imaged multiple times during their treatment. We tracked more than 20 patients undergoing standard neoadjuvant therapy. Each patient was scanned with the prototype at each cycle of chemotherapy. Volumetric renderings of the acoustic parameter of sound speed were compared with standard ultrasound images for tumor volume and its volume-averaged sound speed. Changes in tumor volume and mean sound speed were successfully tracked for each patient (Figure 6.22). Both tumor volume and its mean sound speed declined for all patients, although the rate of decline was prognostic of complete versus partial tumor response at the time of surgical resection. Each parameter reflected tumor changes at each cycle of therapy more accurately than

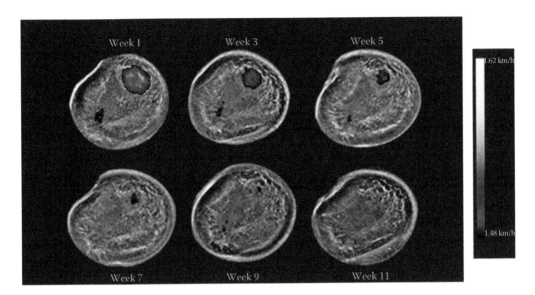

FIGURE 6.22 Thresholded sound speed (color) superimposed on UST reflection image. For approximately the same slice, the repeated scans show a decrease in the size and average sound speed of the cancer over the course of the treatment.

standard ultrasound by quantitating the response of the breast cancer to chemotherapy over time.

UST has the potential to help clinical decision making. The completion of a full course of neoadjuvant therapy versus a quicker surgical resection process for cancers that are more resistant or completely responsive to chemotherapy, respectively, could reduce both morbidity and cost of prolonged courses of neoadjuvant chemotherapy.[103,104]

6.5.4 Predicting Breast Cancer Risk

As noted in the previous section, the ability to fuse images enables multiple parameters to be viewed at the same time. In Figure 6.18, for example, we have fused sound speed and reflection images using a gray-scale representation. The fused images were averaged in stacks of 20 to 25 slices to qualitatively mimic a mammographic-type compression. The four images shown are from four study participants selected to have varying breast density (as characterized by the BIRADS density classification). It is apparent that both fibrous stroma (filamentary structures) and dense glandular tissue (the white "clouds") are recorded by the fused images and that the ratio of stroma to dense tissue changes with breast density in a manner analogous to mammography. It can be further noted that the fatty tissue is represented by the dark regions in the images.

Mammographic density (MD) has been shown to be strongly associated with breast cancer risk. Unlike most risk factors, it has the unique property of being modifiable. The modulation of breast density holds the promise of altering breast cancer risk, thereby reducing the incidence of breast cancer in the general population. Strategies for reducing breast density therefore rely greatly on the ability to accurately measure changes in breast density. Current methods of measuring MD have several major limitations, which include (i) the subjective nature of the measurements, (ii) the absence of an external

Chapter 6

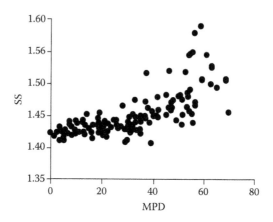

FIGURE 6.23 A scatter plot showing a strong association between mammographic percent density (MPD) and volume averaged sound speed (km/s) for 251 patients.

physical standard, (iii) the characterization of the three dimensions of breast tissue by measuring only the projected area of the breast, and (iv) the subject's exposure to radiation. The characterization of MD by mammography is currently only possible at the age when mammographic screening begins, usually ages 40 to 50 years. However, evidence suggests that prevention and risk assessment based on MD might be introduced at much earlier ages, where a method of assessing MD that uses no radiation would be required. Previous attempts to address these limitations of mammography have not been successful. UST appears capable of overcoming the disadvantages of mammography as a method of characterizing MD for the following reasons.

Sound speed is a quantitative and roughly linear measure of physical tissue density (see Section 6.3). The primary method by which to assess breast density with UST is through the measurement of sound speed. Previous studies have shown results consistent with this relationship.[93–95] For example, in a study involving 251 subjects, both the mammographic percent density and the average volume sound speed were determined. The resulting relationship, shown in Figure 6.23, shows a strong association between the two measurements, suggesting that sound speed may be a biomarker of breast density.

The absence of ionizing radiation, the volumetric aspect of the measurement, and the reproducibility enabled by the absolute measure of sound speed all point to a possible future use of UST to measure breast density. Studies aimed at exploiting this method are being carried out separately.[105]

6.6 Summary and Future Directions

The concept of UST has been around for over half a century. However, its practical medical applications are only now beginning to emerge. Propelled by advances in transducer technology, data acquisition electronics, and computing power, UST is now an active area of clinical research, with multiple groups and companies building breast imaging scanners for both research and commercialization purposes. Recent clinical studies have shown the UST is capable of imaging breast architecture and characterizing lesions. UST images generated in recent years appear at least superficially to be similar to MR images. Ongoing targeted studies are aimed at statistical and quantitative comparisons with MRI to determine if UST's low-cost advantage could be leveraged into clinical practice in situations where UST's performance is comparable with that of MRI.

Other studies are comparing UST's performance against conventional US and mammography to determine if the latter could be supplemented or replaced in a diagnostic setting. Furthermore, multicenter studies are being planned to also assess the potential for UST to become a new, more effective screening modality. Such an outcome would yield great societal impact through the reduction of false negatives, particularly for women with dense breast tissue. By reducing biopsies, UST could also reduce patient anxiety and health care costs. UST could also potentially expand access and frequency of screening by virtue of its speed, comfort, and absence of radiation.

In the arena of personalized medicine, UST also stands to play a major role. Studies have shown that UST can be used to quantify breast density, a major risk factor for developing breast cancer. By identifying women who are at higher risk, UST would offer screening regimens tied to the patient's risk and not limited by radiation concerns. Furthermore, for patients undergoing treatment for breast cancer, UST may offer the data needed to tailor therapies based on an individual's response.

UST is a platform technology that potentially enables applications outside breast imaging. In most current designs, data acquisition systems and computing algorithms can be easily modified to match many transducer configurations. UST development has driven new transducer designs (concave, ring, and 2-D arrays) that open the door for organ-specific arrays. It is therefore possible to think about prostate, liver, brain, and other organ-specific targets in the future.

Finally, as computing power grows further still and as the price of electronics continues to decline, it may be possible to realize an all-body UST scanner. Such a goal is challenging indeed, not only from a data processing perspective but also from the daunting physics required to model wave propagation in soft tissue and also in bone and air. If history is any guide though, patience and time may yield whole-body diagnostics using sound alone. When that day arrives, "listening to your body" will take on a new and literal meaning.

References

1. Berry DA, Cronin KA, Plevritis SK, Fryback DG, Clarke L, Zelen M, Mandelblatt JS, Yakovlev AY, Habbema JD, Feuer EJ. Cancer Intervention and Surveillance Modeling Network (CISNET) Collaborators. Effect of screening and adjuvant therapy on mortality from breast cancer. *New England Journal of Medicince* 2005;1784–1792.
2. Gotzsche PC, Olsen O. Is screening for breast cancer with mammography justifiable? *Lancet* 2000;355:129–134.
3. van den Biggelaar FJ, Nelemans PJ, Flobbe K. Performance of radiographers in mammogram interpretation: A systematic review. *Breast* 2008;17:85–90.
4. Schell MJ, Yankaskas BC, Ballard-Barbash R, Qaqish BF, Barlow WE, Rosenberg RD, Smith-Bindman R. Evidence-based target recall rates for screening mammography. *Radiology* 2007;243:681–689.
5. Armstrong K, Moye E, Williams S, Berlin JA, Reynolds EE. Screening mammography in women 40 to 49 years of age: A systematic review for the American College of Physicians. *Annals of Internal Medicine* 2007;146:516–526.
6. Kuhl CK, Schrading S, Bieling HB, Wardelmann E, Leutner CC, Koenig R, Kuhn W, Schild HH. MRI for diagnosis of pure ductal carcinoma in situ: A prospective observational study. *Lancet* 2007;370:485–492.

7. Saslow D, Boetes C, Burke W, Harms S, Leach MO, Lehman CD, Morris E, Pisano E, Schnall M, Sener S, Smith RA, Warner E, Yaffe M, Andrews KS, Russell CA; American Cancer Society Breast Cancer Advisory Group. American Cancer Society guidelines for breast screening with MRI as an adjunct to mammography. *CA: A Cancer Journal for Clinicians* 2007;57:75–89.

8. Kolb TM, Lichy J, Newhouse JH. Comparison of the performance of screening mammography, physical examination, and breast US and evaluation of factors that influence them: An analysis of 27,825 patient evaluation. *Radiology* 2002;225:165–175.

9. ACRIN website: www.acrin.org.

10. Berg WA, Blume JD, Cormack JB, Mendelson EB, Lehrer D, Böhm-Vélez M, Pisano ED, Jong RA, Evans WP, Morton MJ, Mahoney MC, Hovanessian Larsen L, Barr RG, Farria DM, Marques HS, Boparai K, for the ACRIN 6666 Investigators. Combined screening with ultrasound and mammography vs mammography alone in women at elevated risk of breast cancer. *JAMA* 2008;299(18): 2151–2163.

11. Woodward MJ. Wave-equation tomography. *Geophysics* 1992;5:15–26.

12. Hung SH, Dahlen FA, Nolet G. Frechet kernels for finite-frequency traveltimes—II. Examples. *Geophysics Journal International* 2000;141:175–203.

13. Spetzler J, Snieder R. The effect of small-scale heterogeneity on the arrival time of waves. *Geophysics Journal International* 2001;145:786–796.

14. Monetlli R, Nolet G, Masters G, Dahlen FA, Hung SH. Global P and PP traveltime tomography: Rays versus waves. *Geophysics Journal International* 2004;158:637–654.

15. Spetzler J, Snieder R. Tutorial: The Fresnel volume and transmitted wave. *Geophysics* 2004;69:653–663.

16. Snieder R, Lomax A. Wavefield smoothing and the effect of rough velocity perturbations on arrival times and amplitudes. *Geophysics Journal International* 1996;125:796–812.

17. Schomberg H. An improved approach to reconstructive ultrasound tomography. *Journal of Physics D: Applied Physics* 1978;11:L181–L185.

18. Andersen AH. Ray linking for computed tomography by rebinning of projection data. *Journal of the Acoustical Society of America* 1987;81:1190–1192.

19. Norton SJ. Computing ray trajectories between two points: A solution to the ray-linking problem. *Optical Society of America* 1987;4:1919–1922.

20. Andersen AH. A ray tracing approach to restoration and resolution enhancement in experimental ultrasound tomography. *Ultrasound Imaging* 1990;12:268–291.

21. Littrup PJ, Duric N, Azevedo S, Chambers DH, Candy JV, Johnson S, Auner G, Rather J, Holsapple ET. Computerized ultrasound risk evaluation (CURE) system: Development of combined transmission and reflection ultrasound with new reconstruction algorithms for breast imaging. *Acoustical Imaging* 2001;28:175–182.

22. Littrup PJ, Duric N, Leach Jr. RR, Azevedo SG, Candy JV, Moore T, Chambers DH, Mast JE, Holsapple ET. Characterizing tissue with acoustic parameters derived from ultrasound data. In *Proceedings of SPIE, Medical Imaging 2002: Ultrasonic Imaging and Signal Processing*, 4687–4643. San Diego, California, February 23–28, 2002.

23. Duric N, Littrup PJ, Babkin A, Chambers D, Azevedo S, Kalinin A, Pevzner R, Tokarev M, Holsapple E, Rama O, Duncan R. Development of ultrasound tomography for breast imaging: Technical assessment. *Medical Physics* 2005;32(5):1375–1386.

24. Duric N, Littrup P, Poulo L, Babkin A, Pevzner R, Holsapple E, Rama O, Glide C. Detection of breast cancer with ultrasound tomography: First results with the Computed Ultrasound Risk Evaluation (CURE) prototype. *Medical Physics* 2007;34:773–785.

25. Li C, Duric N. In vivo breast sound-speed imaging with ultrasound tomography. *Medicine and Biology* 2009;35:1615–1628.

26. Hormati A, Jovanovic I, Roy O, Vetterli M. Robust ultrasound travel-time tomography using the bent ray model. *SPIE Medical Imaging* March 2010;7629.

27. Maeda N. A method for reading and checking phase times in autoprocessing system of seismic wave data. *Zisin* 1985;38:365–379.

28. Baer M, Kradolfer U. An automatic phase picker for local and teleseismic events. *Bulletin of the Seismological Society of America* 1987;77:1437–1445.

29. Ramananantoandro R, Bernitsas N. A computer algorithm for automatic picking of refraction first-arrival-time. *Geoexploration* 1987;24:147–151.

30. Boschetti F, Dentith D, List RD. A fractal-based algorithm for detecting first-arrivals on seismic traces. *Geophysics* 1996;61:1095–1102.

31. Molyneux JB, Schmitt DR. First-break timing: Arrival onset times by direct correlation. *Geophysics* 1999;64:1492–1501.
32. Sleeman R, van Eck T. Robust automatic P-phase picking: An on-line implementation in the analysis of broadband seismogram recordings. *Physics of the Earth and Planetary Interiors* 1999;113:265–275.
33. Zhang H, Thurber C, Rowe C. Automatic P-wave arrival detection and picking with multiscale wavelet analysis for single-component recordings. *Bulletin of the Seismological Society of America* 2003;93:1904–1912.
34. Kurz JH, Grosse CU, Reinhardt HW. Strategies for reliable automatic onset time picking of acoustic emissions and of ultrasound signals in concrete. *Ultrasonics* 2005;43:538–546.
35. Di Stefano R, Aldersons F, Kissling E, Chiarabba C. Automatic seismic phase picking and consistent observation error assessment: Application to the Italian seismicity. *Geophysics Journal International* 2006;165:121–134.
36. Li C, Huang L, Duric N, Zhang H, Rowe C. An improved automatic time-of-flight picker for medical ultrasound tomography. *Ultrasonics* 2009;49:61–72.
37. Kak AC, Dines KA. Signal processing of broadband pulsed ultrasound: Measurement of attenuation of soft biological tissue. *IEEE Transactions on Biomedical Engineering* 1978;25:321–344.
38. Parker KJ, Lerner RM, Waag RC. Attenuation of ultrasound magnitude and frequency dependence for tissue characterization. *Radiology* 1984;153:785–788.
39. Parker KJ, Lerner RM, Waag RC. Comparison of techniques for in vivo attenuation measurements. *IEEE Transactions on Biomedical Engineering* 1988;35:1064–1068.
40. Sams M, Goldberg D. The validity of Q estimates from borehole data using spectral ratios. *Geophysics* 1990;55:97–101.
41. Fujii Y, Itoh K, Shigeta K, Wang Y, Tsao J, Kumasaki K, Itoh R. A new method for attenuation coefficient measurement in the liver. *Journal of Ultrasound Medicine* 2002;21:783–788.
42. Li C, Duric N, Huang L. Comparison of ultrasound attenuation tomography methods for breast imaging. *Proceedings of SPIE* 2008;6920:692015-2.
43. Hansen PC. The L-curve and its use in the numerical treatment of inverse problems. In *Computational Inverse Problems in Electrocardiology* edited by P. Johnston, 119–142. Southampton, UK: WIT, 2001.
44. Farquharson CG, Oldenburg DW. Automatic estimation of the trade-off parameter in nonlinear inverse problems using the GCV and L-curve criteria. http://www.esd.mun.ca/~farq/PDFs/papr0502.pdf.
45. Paige CC, Saunders MA. LSQR: Sparse linear equations and least-squares problems. *ACM Transactions on Mathematical Software* 1982;82:43–71.
46. Bissantz N, Hohage T, Munk A, Ruymgaart F. Convergence rates of general regularization methods for statistical inverse problems and applications. *SIAM Journal of Numerical Analysis* 2007;45:2610–2636 (electronic).
47. Loubes JM, Rivoirard V. Review of rates of convergence and regularity conditions for inverse problems. http://www.ceremade.dauphine.fr/~rivoirar/IJAMAS_example.pdf.
48. Chemov LA. *Wave Propagation in a Random Medium.* New York: McGraw-Hill, 1960.
49. Keller JB. Accuracy and validity of the Born and Rytov approximations. *Journal of the Optical Society of America* 1969;59:1003–1004.
50. Devaney A. Inverse scattering theory within the Rytov approximation. *Optics Letters* 1981;6:374–376.
51. Devaney A. Inverse formula for inverse scattering within the Born approximation. *Optics Letters* 1982;7:111–112.
52. Slaney M, Kak AC, Larsen L. Limitations of imaging with first-order diffraction tomography. *IEEE Transactions on Microwave Theory and Techniques* 1984;MTT-32:860–873.
53. Wu RS, Toksoz MN. Diffraction tomography and multisource holography applied to seismic imaging. *Geophysics* 1987;52:11–25.
54. Kak AC, Slaney M. *Principles of Computerized Tomographic Imaging.* IEEE Press, 1988.
55. Schmidt S, Duric N, Li C, Roy O, Huang Z.-F. Modification of Kirchhoff migration with variable sound speed and attenuation for acoustic imaging of media and application to tomographic imaging of the breast. *Medical Physics* 2011;38:998.
56. Pratt RG, Huang L, Duric N, Littrup P. Sound-speed and attenuation imaging of breast tissue using waveform tomography of transmission ultrasound data. *Proceedings of SPIE, Medical Imaging* March 2007;6510.
57. Tarantola A. *Inverse Problem Theory and Methods for Model Parameter Estimation.* SIAM, 2005.
58. Natterer F, Wubbeling F. *Mathematical Methods in Image Reconstruction.* Society for Industrial and Applied Mathematics, Philadelphia, PA, USA, 2001.

Chapter 6

59. Pratt RG. Seismic waveform inversion in the frequency domain, part 1: Theory, and verification in a physical scale model. *Geophysics* 1999;64(3):888–901.
60. Yang DH, Lu M, Wu RS, Peng JM. An optimal nearly analytic discrete method for 2D acoustic and elastic wave equations. *Bulletin of the Seismological Society of America* 2004;94(5):1982–1992.
61. Engquist B, Majda A. Absorbing boundary conditions for the numerical simulation of waves. *Mathematics of Computation* 1977;31(139):629–651.
62. Roy O, Jovanovic I, Hormati A, Parhizkar R, Vetterli M. Sound speed estimation using wave-based ultrasound tomography: Theory and GPU implementation. *Proceedings of SPIE, Medical Imaging* March 2010;7629.
63. Micikevicius P. 3D finite difference computation on GPUs using CUDA. In *2nd Workshop on General Purpose Processing on Graphics Processing Units* 2009:79–84.
64. Wild JJ, Reid JM. Application of echo-ranging techniques to the determination of structure of biological tissues. *Science* 1952;115(2983):226–230.
65. Howry DH, Bliss WR. Ultrasonic visualization of soft tissue structures of the body. *Journal of Laboratory and Clinical Medicine* 1952;40:579–592.
66. Stavros AT, Thickman D, Rapp CL, Dennis MA, Parker SH, Sisney GA. Solid breast nodules: Use of sonography to distinguish between benign and malignant lesions. *Radiology* 1995;196:123–134.
67. Norton SJ, Linzer M. Ultrasonic reflectivity tomography: Reconstruction with circular transducer arrays. *Ultrasonic Imaging* April 1979;1(2):154–184.
68. http://www.u-sys.com/.
69. Greenleaf JF, Johnson SA, Bahn RC, Rajagopalan B. Quantitative cross-sectional imaging of ultrasound parameters. In *Proceedings of the IEE Ultrasonics Symposium*, IEEE Cat. No. 77CH1264-1SU, 1977, 989–995.
70. Greenleaf JF, Johnson SA, Lent AH. *Ultrasound in Medicine and Biology* 1978;3:327–339.
71. Carson PL, Meyer CR, Scherzinger AL, Oughton TV. Breast imaging in coronal planes with simultaneous pulse echo and transmission ultrasound. *Science* 1981;214(4525):1141–1143.
72. Andre MP, Janee HS, Martin PJ, Otto GP, Spivey BA, Palmer DA. High-speed data acquisition in a diffraction tomography system employing large-scale toroidal arrays. *International Journal of Imaging Systems and Technology* 1997;8(1):137–147.
73. Johnson SA, Borup DT, Wiskin JW, Natterer F, Wuebbling F, Zhang Y, Olsen C. Apparatus and method for imaging with wavefields using inverse scattering techniques. United States Patent 6005916, 1999.
74. Marmarelis VZ, Kim T, Shehada RE. *Proceedings of SPIE, Medical Imaging 2003: Ultrasonic Imaging and Signal Processing* (paper 5035–5036). San Diego, California, February 23–28, 2002.
75. Liu D-L, Waag RC. Propagation and backpropagation for ultrasonic wavefront design. *IEEE Transactions on Ultrasonics, Ferroelectrics and Frequency Control* 1997;44(1):1–13.
76. Ruiter NV, Göbel G, Berger L, Zapf M, Gemmeke H. Realization of an optimized 3D USCT. *Proceedings of SPIE* 2011;7968:796805.
77. Glover GH, Sharp JC. Reconstruction of ultrasound propagation speed distributions in soft tissue: Time-of-flight tomography. *IEEE Transactions on Sonics and Ultrasonics* July 1977;24(4):229–234.
78. Greenleaf JF, Bahn RC. Clinical imaging with transmissive ultrasonic computerized tomography. *IEEE Transactions on Biomedical Engineering* 1981;BME-28(2):177–185.
79. SEER website. http://seer.cancer.gov/.
80. American Cancer Society. *Cancer Prevention and Early Detection Facts & Figures 2009*. Atlanta, GA: American Cancer Society, 2009, 34–37.
81. Chen J, Pee D, Ayyagari R, Graubard B, Schairer C, Byrne C, Benichou J, Gail MH. Projecting absolute invasive breast cancer risk in white women with a model that includes mammographic density. *Journal of the National Cancer Institute* 2006;98:1215–1226.
82. Ursin G, Hovanessian-Larsen L, Parisky YR, Pike MC, Wu AH. Greatly increased occurrence of breast cancers in areas of mammographically dense tissue. *Breast Cancer Research* 2005;7:R605–R608.
83. Martin LJ, Boyd N. Potential mechanisms of breast cancer risk associated with mammographic density: Hypotheses based on epidemiological evidence. *Breast Cancer Research* 2008;10:1–14.
84. Turnbull, LW. Dynamic contrast-enhanced MRI in the diagnosis and management of breast cancer. *NMR in Biomedicine* 2009;22(1):28–39.
85. Jansen SA, Fan X, Karczmar GS, Abe H, Schmidt RA, Newstead GM. Differentiation between benign and malignant breast lesions detected by bilateral dynamic contrast-enhanced MRI: A sensitivity and specificity study. *Magnetic Resonance in Medicine* 2008;59(4):747.
86. Chen JH, Feig B, Agrawal G, Yu H, Carpenter PM, Mehta RS, Nalcioglu O, Su MY. MRI evaluation of pathologically complete response and residual tumors in breast cancer after neoadjuvant chemotherapy. *Cancer* 2008;112(1):17–26.

87. Sharma U, Danishad KK, Seenu V, Jagannathan NR. Longitudinal study of the assessment by MRI and diffusion-weighted imaging of tumor response in patients with locally advanced breast cancer undergoing neoadjuvant chemotherapy. *NMR in Biomedicine* 2009;22(1):104–113.

88. Bando H, Tohno E, Katayama H, Hara H, Yashiro T, Noguchi M, Ueno E. Imaging evaluation of pathological response in breast cancer after neoadjuvant chemotherapy by real-time sonoelastography and MRI. *European Journal of Cancer Supplements* 2008;6(7):66–67.

89. Bhattacharyya M, Ryan D, Carpenter R, Vinnicombe S, Gallagher CJ. Using MRI to plan breast-conserving surgery following neoadjuvant chemotherapy for early breast cancer. *British Journal of Cancer* 2008;98(2):289–293.

90. Partridge S. Recurrence rates after DCE-MRI image guided planning for breast-conserving surgery following neoadjuvant chemotherapy for locally advanced breast cancer patients. *Breast Diseases: A Year Book Quarterly* 2008;19(1):91.

91. Tozaki M. Diagnosis of breast cancer: MDCT versus MRI. *Breast Cancer* 2008;15(3):205–211.

92. Partridge S, Gibbs JE, Lu Y, Esserman LJ, Sudilovsky D, Hylton NM. Accuracy of MR imaging for revealing residual breast cancer in patients who have undergone neoadjuvant chemotherapy. *American Journal of Roentgenology* 2002:1193–1199.

93. Mast TD. Empirical relationships between acoustic parameters in human soft tissues. *Acoustics Research Letters Online* 2000;1(2):37–42.

94. Masugata H, Mizushige K, Senda S, Kinoshita A, Sakamoto H, Sakamoto S, Matsuo H. Relationship between myocardial tissue density measured by microgravimetry and sound speed measured by acoustic microscopy. *Ultrasound in Medicine and Biology* 1999;25(9):1459–1463.

95. Weiwad W, Heinig A, Goetz L, Hartmann H, Lampe D, Buchmann J, Millner R, Spielmann RP, Heywang-Koebrunner SH. Direct measurement of sound velocity in various specimens of breast tissue. *Investigative Radiology* 2000;35(12):721–726.

96. Boyd NF, Martin LJ, Chavez S, Gunasekara A, Salleh A, Melnichouk O, Yaffe M, Minkin S, Bronskill MJ. Breast-tissue composition and other risk factors for breast cancer in young women: A cross-sectional study. *Lancet Oncology* 2009;10:569–580.

97. Glide C, Duric N, Littrup P. Novel approach to evaluating breast density utilizing ultrasound tomography. *Medical Physics* 2007;34(2):744–753.

98. Glide-Hurst CK, Duric N, Littrup P. Volumetric breast density evaluation from ultrasound tomography images. *Medical Physics* 2008;35(9):3988–3997.

99. Duric N, Littrup P, Li C, Rama O, Bey-Knight L, Schmidt S, Lupinacci J. Detection and characterization of breast masses with ultrasound tomography: Clinical results. *Proceedings of SPIE, Medical Imaging* 2009;7265:72651G-1-8.

100. Duric N, Littrup P, Chandiwala-Mody P, Li C, Schmidt S, Myc L, Rama O, Bey-Knight L, Lupinacci J, Ranger B, Szczepanski A, West E. In-vivo imaging results with ultrasound tomography: Report on an ongoing study at the Karmanos Cancer Institute. *Proceedings of SPIE* 2010;7629:76290M.

101. Ranger B, Littrup P, Duric N, Li C, Lupinacci J, Myc L, Rama O, Bey-Knight L. Breast imaging with acoustic tomography: A comparative study with MRI. *Proceedings of SPIE, Medical Imaging* 2009;7265:726510-1-8.

102. Ranger B, Littrup P, Duric N, Li C, Schmidt S, Lupinacci J, Myc L, Szczepanski A, Rama O, Bey-Knight L. Breast imaging with ultrasound tomography: A comparative study with MRI. *Proceedings of SPIE* 2010;7629:76291C.

103. Lupinacci J, Duric N, Littrup P, Wang D, Li C, Schmidt S, Rama O, Myc L, Bey-Knight L. Monitoring breast masses with ultrasound tomography in patients undergoing neoadjuvant chemotherapy. *Proceedings of SPIE, Medical Imaging* 2009;7265:726517-1-9.

104. Lupinacci J, Duric N, Littrup P, Wang D, Li C, Schmidt S, Ranger B, West E, Szczepanski A, Rama O, Bey-Knight L, Myc L. Monitoring breast masses with ultrasound tomography in patients undergoing neoadjuvant chemotherapy. *Proceedings of SPIE* 2010;7629.

105. Myc L, Duric N, Littrup P, Li C, Ranger B, Lupinacci J, Schmidt S, Rama O, Bey-Knight L. Volumetric breast density evaluation by ultrasound tomography and magnetic resonance imaging: A preliminary comparative study. *Proceedings of SPIE* 2010;7629:76290N.

7. Task-Based Design and Evaluation of Ultrasonic Imaging Systems

Nghia Q. Nguyen, Craig K. Abbey, and Michael F. Insana

Chapter 7

Ultrasound Imaging and Therapy. Edited by Aaron Fenster and James C. Lacefield © 2015 CRC Press/Taylor & Francis Group, LLC. ISBN: 978-1-4398-6628-3

7.1 Introduction

Ultrasonic imaging systems have traditionally been designed by balancing a host of engineering trade-offs, including echo signal-to-noise ratio (eSNR), contrast, spatial and temporal resolutions, safety, and cost. Diagnostic performance is often evaluated subjectively by experts using the systems in clinical practice. Our goal has been to develop objective task-based assessments where the reason for obtaining the image—the task—is related mathematically to the engineering properties listed earlier and in a manner that generates evaluation metrics equivalent to receiver operating characteristic (ROC) analysis, the industry's gold standard [1]. A task can be to detect a pathology-verified disease condition or to simply discriminate among visual features that characterize a particular condition. Task-based assessments can show us that systems tuned for abdominal or breast applications are often not optimized to give the best lesion contrast resolution for other applications, such as an echocardiography that requires superior spatial and temporal resolutions. Task-based evaluations become particularly important as instruments gain greater computational capabilities so they can quickly reconfigure and optimize the conditions of a specific patient exam. During the past few decades, system adaptability has already begun to increase the performance-to-cost ratio for general sonography. Adaptive reconfiguration takes the form of transmit- and receive-channel echo-signal processing to maximize the transfer of information from the patient to the image for observers.

Imaging systems are evaluated from objective assessments that relate ultimately to observer performance. The observer can be an expert human or an algorithm evaluating criteria based on decision theory [2]. Prominent among the latter is the Bayesian ideal observer—often referred to simply as the ideal observer—that combines all available information to make the decision, and thus it achieves optimal task performance [1]. If the performance measured by a panel of expert radiologists is significantly less than the ideal, the system should be redesigned, but only if it is determined that the acquisition stage of image formation (including output power, noise, transducer properties, and beamforming aspects) is limiting performance. Sometimes task information is present in the image but difficult to observe; for example, flowing-blood echoes are found in recorded echo signal but are difficult to see without Doppler processing and color overlays. When the display stage limits performance, image processing is often very helpful.

To compute the response of the ideal observer, we must first obtain complete statistical knowledge of the data under consideration, which often limits the analysis to tasks far simpler than clinical diagnosis. The analysis, however, is well suited to component tasks involving specific signals that are known exactly in backgrounds known exactly (SKE/BKE tasks) or to signals known exactly where the background is known statistically (SKE/BKS tasks). Examples of five SKE/BKS component tasks related to breast lesion diagnosis are given in the subsequent section.

Task-based assessment was an application of statistical detection theory first used in radar imaging during World War II, with scientific background provided decades earlier by Hotelling, Thurstone, and Neyman and Pearson [3]. The approach was subsequently applied to the evaluation of medical imaging systems by Swets and Pickett [4,5], Wagner and Brown [6], Barrett and Myers [1], and many others. Throughout the 1990s, Barrett et al. [2,7–9] used this approach to develop a rigorous formalism for evaluating photon-based imaging systems.

The approach was first applied to sonography by Wagner and Smith [10,11] in their award-winning papers of 1983. They derived an ideal observer viewing B-mode images for the binary task of low-contrast lesion detection. Analyzing sonograms, they discovered how properties of the transducer combine with object features to influence speckle properties and thus lesion detectability. They realized that diagnostic information in a sonogram was contained in speckles, similar to how recognizable images emerged in paintings from the 19th century impressionistic pointillists. However, image data was formed through the nonlinear processes of demodulation and amplitude compression, and so a rigorous analysis was very challenging. To apply ideal observer analysis to sonograms, Wagner and Smith made numerous limiting assumptions. Among them, they assumed that measurement noise does not exist, only large-area low-contrast lesions may be present, the system provided shift-invariant focal-zone impulse responses, and speckle spots rather than pixels determine statistical properties of image data. Despite these limitations, their analysis provided design criteria that guided subsequent work in speckle reduction [12], beamformation [13], postprocessing [14], and transducer selection [15].

Today, many ultrasound systems provide an option for recording digitized radio frequency (RF) or in-phase–quadrature signals [16]. These signals are generated by known and linear processes that we can model for an ideal observer analysis. The results measure information transfer at the acquisition stage of recording, that is, up to the point of demodulating beamformed RF signals before scan conversion. We avoid these limiting assumptions by analyzing task information recorded by the RF echo signals. However, we will show that from measurements using the Smith–Wagner (SW) computational observer and human observer responses, it is possible to follow the flow of task information from the patient through each stage of image formation to ultimately the diagnostician.

This chapter reviews our laboratory's research conducted during the last decade for improving lesion discrimination in breast sonography [17–19]. Relationships among data at various stages and by various observers are illustrated in Figure 7.1. In the approach, the imaging system is described as a linear device transferring task information from an object on the left to one of three observers who make decisions on the right. Image formation is divided into acquisition and display stages [6]. The acquisition stage produces discrete-time RF echo signals given by vector $\mathbf{g}$ from continuous objects $f(\mathbf{x}, t)$ via the continuous-to-discrete linear imaging operator $\mathcal{H}$ that describes all aspects of pulse transmission, echo reception, and beamforming. The displaying stage is where RF echo data are mapped into B-mode image vectors $\mathbf{b}$ through the discrete-to-discrete nonlinear display operator $\mathcal{O}$. This operator includes all processing applied after demodulation, including postprocessing and scan conversion. We use common lexicographical

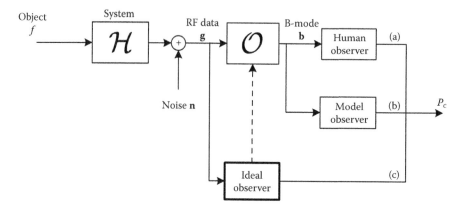

FIGURE 7.1   Ultrasonic image formation is diagrammed to explain task-based performance analysis. Path (a) describes image formation leading to the human observer. Path (b) indicates the same images can be viewed by model observers such as the SW observer. Path (c) indicates the ideal observer observes RF echo data, but this performance can be related to the performance of observers of the image. All observer responses are from 2AFC observer experiments and are given as the proportion of correct responses (P_c).

reordering [20] to represent $M \times M$ data mathematically as $M^2 \times 1$ vectors, but we always present images to observers in conventional formats.

It is convention to use ROC studies to evaluate system performance using path [a]. Patients are scanned and images are formed and read by diagnosticians using 2AFC or graded response observer experiments [3,5]. Controlling for case and reader correlations and variability to implement hypothesis testing, evaluators compute the area under the ROC curve (AUC) to find system parameters and clinical practices that yield large AUC values. The largest AUC value is 1 for perfect discrimination performance, and the smallest is 0.5, resulting from use of a worthless diagnostic detector/test. The same images can be read by a computational observer (also called model observer, path [b]) to eliminate reader variability and speed the reading process, if the model is first shown to mimic human observer responses.

We extend the ideal observer analysis for being applied to RF data in path [c]. It tells us how well the task information is transferred from the patient to recorded data by the acquisition stage. Combining these results with those from human and model observers, we can evaluate the efficiency of the display stage of image formation to generate tasks information in the image that is accessible by the observer. We will show that demodulation is a lossy step needed for human viewing, but there are filtering processes that may be applied to the beamformed RF signals to improve the human observer performance. These methods are able to provide a rational procedure for system design based on task performance.

In subsequent sections, we detail concepts underlying the task-based approach to system design and evaluation. We consider various strategies interpreted from the ideal observer equations to improve discrimination efficiency on several key features of breast cancer sonography. The effectiveness of those strategies is evaluated through a combination of ideal and human observer studies. The chapter concludes with a summary and discussions for future research directions. Our goal is to provide an analytical framework for system design.

7.2 Tasks and Signal Modeling for Breast Cancer Sonography

7.2.1 Tasks

Medical imaging tasks have been categorized as classification or estimation [1], the difference being that the former has a finite number of possible solutions whereas the latter has an infinite number. We focus on two-class classification tasks, in which the observer must classify the object as either benign (class 1) or malignant (class 2). We establish five visual tasks representing for five typical features in the BIRADS atlas [21]. Those features are often considered by radiologists when discriminating malignant from benign breast lesions [17]. Features became visual discrimination tasks by defining a malignant S_2 and a benign S_1 matrix pair for each as in Figure 7.1. Task 1 involves detecting a low-contrast hypoechoic lesion versus a no-lesion background; task 2 requires discrimination of an elongated eccentric lesion from a circular lesion; task 3 is to discriminate a soft, poorly defined boundary from a well-circumscribed boundary; task 4 requires discrimination of spiculated boundary irregularities from a smooth circular boundary; and task 5 involves discriminating a very weakly scattering hypoechoic interior from an anechoic (cyst-like) lesion interior. Although tasks 2 to 4 define lesion boundary features, tasks 1 and 5 challenge the system to image large-area diagnostic features. Task 5 was distinct in that it was the only lager-area, high-contrast lesion involved in a discrimination task. Many tasks with higher complexity features can be synthesized from these five element features. Differences between the two classes (ΔS) are also displayed in the bottom row of Figure 7.2. They illustrate the task information we hope our imaging system will deliver [6].

7.2.2 Signal Modeling

The image formation process diagrammed in Figure 7.1 is graphically illustrated in Figure 7.3. It begins with a feature template S_i representing object classification $i = 1, 2$, indicating benign or malignant, respectively. Note that because of the reordering of data into vectors, matrix S_i represents the 2-D objects in Figure 7.2 as a diagonal matrix. Multiplying the template by a zero-mean white Gaussian random field of variance σ_{obj}^2, we form scattering object $f(\mathbf{x})$ that we assume is stationary in time but spatially variable. The template and the random field are multiplied to generate a scattering field representing amplitude-modulated, incoherent Rayleigh scattering. Acoustic physics indicates that the incoherent backscattered signals encode spatial features in their spatial fluctuations (the covariance matrix) rather than the mean [22]. Scatterers are spatially random in an ensemble sense; however, multiplication by the feature template makes object scattering spatially nonstationary.

The interaction of pulse-echo ultrasound with the scattering media is represented by the system operator $\mathcal{H}$. It is well approximated by a linear transform under the first Born approximation [20]. At this moment, we assume that the RF data is recorded using a delay-and-sum (DS) beamforming strategy. RF echo data may be subjected to further processing, denoted as operator $\mathcal{W}$, before envelope detection. The entire process is summarized by the equations for RF and B-mode data,

$$\mathbf{g} = \mathcal{H}f(x) + \mathbf{n} \text{ and either } \mathbf{b} = \mathcal{O}\mathbf{g} \text{ or } \mathbf{b} = \mathcal{O}\mathcal{W}\mathbf{g}.$$

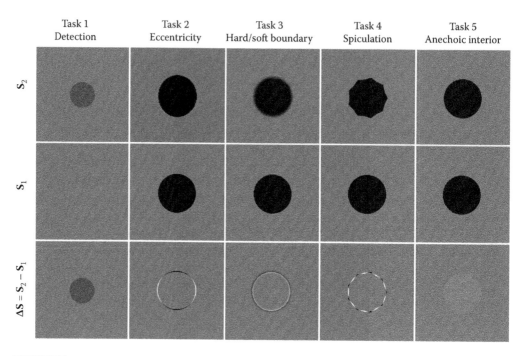

FIGURE 7.2   Geometries of "malignant" and "benign" features defining five component tasks related to breast lesion diagnosis. The lesion diameter is 3 mm. (From Abbey, C. K. et al., *IEEE Trans Med Imaging*, 25, 198–209, 2006. With permission.)

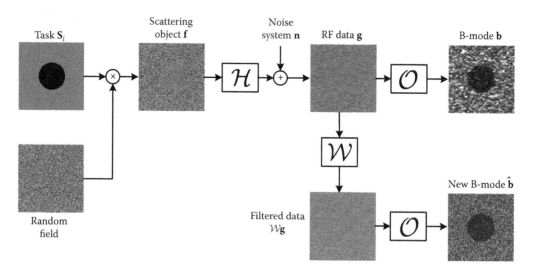

FIGURE 7.3   A graphical model of the sonographic process to generate B-mode images. The top row is conventional processing, and the bottom row includes application of an additive filter before taking the envelope. (Figure 1 from Nguyen, N. Q. et al., *IEEE Trans Med Imaging*, 30, 28–37, 2011. With permission.)

Measurement noise is represented by the additive vector $\mathbf{n}$. It is modeled by an independent white Gaussian noise process $\mathcal{N}(\mathbf{0}, \sigma_n^2 \mathbf{I})$. Because $\mathcal{H}$ is approximated by a linear operator, the first equation can be written in terms of a matrix multiplication, given by:

$$\mathbf{g} = \mathbf{Hf} + \mathbf{n}. \tag{7.1}$$

The spatiotemporal impulse response of the system is spread over the mth row of $\mathbf{H}$ to yield the RF echo sample $g[m]$. Under the assumption of shift invariance, $\mathbf{H}$ is a block-Toeplitz matrix but approximated by the corresponding circulant one, which provides advantages for computation [1,17]. The spatiotemporal impulse responses used to construct matrix $\mathbf{H}$ is generated by the Field II program [23,24], with parameters extracted from a commercial system (SONOLINE Antares™—Siemens Medical Solutions, Mountain View, CA) [19].

An advantage of working in the sampled RF data domain is that the signal is well modeled by a noisy linear transformation, specified by system matrix $\mathbf{H}$ and noise power σ_n^2. The ideal observer operating on RF data is derived in the next section.

7.3   Ideal Observer

7.3.1   Likelihood Functions and the Ideal Observer Test Statistic

With the generation of the scattering function described in Section 7.2.2, the vector object $\mathbf{f}$ has a zero-mean multivariate normal distribution (MVN) with a nonstationary, diagonal covariance matrix $\sigma_{\text{obj}}^2 (\mathbf{I} + \mathbf{S}_i)$, where σ_{obj}^2 is the background region variance. Multivariate normal processes remain multivariate normal after linear transformations. Thus, passing the object through the noisy linear transformation in Equation 7.1 results in another zero-mean Gaussian process for each class. The covariance matrix for the RF echo data $\mathbf{g}$ becomes

$$\mathbf{\Sigma}_i = \sigma_{\text{obj}}^2 \mathbf{H}(\mathbf{I} + \mathbf{S}_i)\mathbf{H}^t + \sigma_n^2 \mathbf{I}. \tag{7.2}$$

The covariance matrices for both classes still capture all the relevant statistics of the task, but they are no longer diagonal because of blurring by the imaging system via $\mathbf{H}$. The likelihood function of the data $\mathbf{g}$ under each class is a zero-mean MVN given by

$$\mathbf{g}_i \sim \text{MVN}(\mathbf{0}, \mathbf{\Sigma}_i). \tag{7.3}$$

This SKE task has the diagnostic feature (signal) encoded in $\mathbf{\Sigma}_i$. The scalar test statistic of the ideal observer response to this discrimination task is derived from the following log-likelihood ratio [7,17]:

$$\lambda(\mathbf{g}) = \log \frac{\text{pdf}(\mathbf{g} \,|\, 2)}{\text{pdf}(\mathbf{g} \,|\, 1)} \cong \mathbf{g}^t \left(\mathbf{\Sigma}_1^{-1} - \mathbf{\Sigma}_2^{-1} \right) \mathbf{g}, \tag{7.4}$$

where pdf $(\mathbf{g}|i)$ is the probability density function for the data given the i-th condition. A distinguishing feature of this test statistic is its quadratic dependence on $\mathbf{g}$, whereas the equivalent radiographic task is linear in the data [1]. The second form of Equation 7.4 has terms unrelated to $\mathbf{g}$ removed from the decision variable because they do not affect performance. Decisions are made by comparing $\lambda(\mathbf{g})$ with a threshold t, and performance is accessed from the proportion of correct responses P_c, which is related to the AUC.

Although the test statistic of the ideal observer is well defined, calculating it is very challenging because of the high dimensionality of the covariance matrices. For example, if the scattering object can be represented by a 128×128 matrix, the corresponding covariance matrix has a size of $16,384 \times 16,384$. Therefore, inverses cannot be computed in a straightforward manner. To resolve this problem, we proposed a power series expansion of the covariance matrices to compute matrix products involving inverse covariance matrices [17].

7.3.2   Power Series Inversion

The power series relies on the decomposition of the image covariance matrices into background and task-specified components, given as follows:

$$\boldsymbol{\Sigma}_i = \sigma_{\text{obj}}^2 \mathbf{H}(\mathbf{I} + \mathbf{S}_i)\mathbf{H}^t + \sigma_n^2 \mathbf{I}$$
$$= \boldsymbol{\Sigma}_0 + \Delta\boldsymbol{\Sigma}_i, \tag{7.5}$$

where $\boldsymbol{\Sigma}_i = \sigma_{\text{obj}}^2 \mathbf{H}\mathbf{H}^t + \sigma_n^2 \mathbf{I}$ is the stationary background term and $\Delta\boldsymbol{\Sigma}_i = \sigma_{\text{obj}}^2 \mathbf{H}\mathbf{S}_i\mathbf{H}^t$ is the nonstationary task feature term. From Golub and Van Loan [25], we have

$$(\mathbf{I} - \mathbf{A})^{-1} = \sum_{k=0}^{\infty} \mathbf{A}^k, \tag{7.6}$$

which holds if the eigenvalues of $\mathbf{A}$ are between -1 and 1. To apply Equation 7.6, we first write Equation 7.5 in the following form:

$$\boldsymbol{\Sigma}_i = \boldsymbol{\Sigma}_0^{1/2}\left(\mathbf{I} + \boldsymbol{\Sigma}_0^{-1/2}\Delta\boldsymbol{\Sigma}\boldsymbol{\Sigma}_0^{-1/2}\right)\boldsymbol{\Sigma}_0^{1/2}, \tag{7.7}$$

to find the following inverse covariance matrix expansion:

$$\boldsymbol{\Sigma}_i^{-1} = \boldsymbol{\Sigma}_0^{-1/2}\left(\sum_{k=0}^{\infty}\left(-\boldsymbol{\Sigma}_0^{-1/2}\Delta\boldsymbol{\Sigma}\boldsymbol{\Sigma}_0^{-1/2}\right)^k\right)\boldsymbol{\Sigma}_0^{-1/2}. \tag{7.8}$$

If we can assume a circulant matrix form for $\mathbf{H}$, we may diagonalize $\mathbf{H}$ by a Fourier transform, as follows:

$$\mathbf{H} = \mathbf{F}^{-1}\mathbf{T}\mathbf{F} \tag{7.9}$$

where $\mathbf{F}$ is the 2-D forward discrete Fourier transform matrix and $\mathbf{T}$ is a diagonal matrix whose elements are the eigenvalues of $\mathbf{H}$ [1]. Consequently, $\boldsymbol{\Sigma}_0$ can be decomposed as

$$\boldsymbol{\Sigma}_0 = \mathbf{F}^{-1}\mathbf{N}_0\mathbf{F}, \tag{7.10}$$

where $\mathbf{N}_0$ is also diagonal with the following elements:

$$[\mathbf{N}_0]_{ii} = \sigma^2_{\text{obj}}\left|[\mathbf{T}]_{ii}\right|^2 + \sigma^2_n. \tag{7.11}$$

Thus, the only inverse required is of the stationary component $\boldsymbol{\Sigma}_0$, which is quickly computed. Terms from Equation 7.4, that is, $\mathbf{g}^t\boldsymbol{\Sigma}_i^{-1}\mathbf{g}$, are efficiently calculated numerically using the Fourier transform. The computation is implemented through an iterative process, with details provided by Abbey et al. [17]. Once test statistics are found, the ideal observer's proportion of correct responses is measured from hundreds of 2AFC data pairs and converted to AUC performance measures.

7.3.3   Performance through ROC Analysis

ROC analysis is the standard method for assessing observer performance in binary classification problems [4]. The ROC curve depicts the probability of detection P_D as a function of the false alarm rate P_F. P_D is also called the sensitivity of the test for detecting malignant features that are present. Curves such as those in Figure 7.4a are generated from overlapping histograms of the test statistic responses for each of the two classes of data. These approximate probability density functions, $\text{pdf}(\lambda(\mathbf{g})|i)$. Selecting threshold t and integrating, we find the cumulative distributions P_D and P_F as we sweep through the range of t ([1], Chapter 13),

$$P_D(t) = \Pr\left(\lambda(\mathbf{g})|2\right) = \int_t^\infty d\lambda(\mathbf{g})\text{pdf}\left(\lambda(\mathbf{g})|2\right),$$

$$\tag{7.12}$$

$$P_F(t) = \Pr\left(\lambda(\mathbf{g})|1\right) = \int_t^\infty d\lambda(\mathbf{g})\text{pdf}\left(\lambda(\mathbf{g})|1\right).$$

With $-\infty \leq t \leq \infty$, P_D and P_F range from 0 to 1. Plotting P_D against P_F, the ROC curve is generated as in Figure 7.4b. The three points labeled A, B, and C represent three pairs (P_D, P_F) calculated at different thresholds t. Often, $\lambda(\mathbf{g}|1)$ and $\lambda(\mathbf{g}|2)$ are normally distributed with equal variance [7]. In that case, the AUC uniquely defines observer performance as the scalar metric

$$\text{AUC} = \int_0^1 dP_F P_D(P_F). \tag{7.13}$$

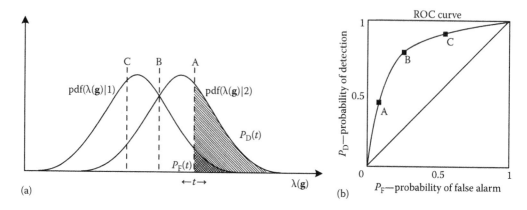

FIGURE 7.4   (a) The pdfs of the test statistic λ under two hypotheses and (P_D, P_F) at a specific threshold t. (b) An ROC curve with three threshold levels. (Modified from Barrett, H. H., and K. J. Myers: *Foundations of Image Science*. 2004. Copyright Wiley-VCH Verlag GmbH & Co. KGaA.)

By adopting $\lambda(\mathbf{g})$ as its test statistic, the ideal observer maximizes P_D for each P_F, which is the *Neyman–Pearson theorem* [26]. Consequently, the ideal observer is guaranteed to maximize the AUC.

7.3.4   2AFC Interpretations

Ideal observer performance can be calculated without first determining the shape of the ROC curve. We see this by combining Equations 7.12 and 7.13 to show (see [7] for the proof)

$$\text{AUC} = \int d\mathbf{g} \int d\mathbf{g}' \text{pdf}\left(\lambda(\mathbf{g})|2\right)\text{pdf}\left(\lambda(\mathbf{g}')|1\right)\text{step}\left[\lambda(\mathbf{g})-\lambda(\mathbf{g}')\right]$$

$$= \Pr\left[\lambda(\mathbf{g}) > \lambda(\mathbf{g}')\right],$$

(7.14)

where the step function equals 1 for positive arguments and 0 for negative arguments. Equation 7.14 is the basis of our procedure for estimating AUC from 2AFC samples that is often used in human observer studies [17].

Assume we have N pairs of RF data generated under the two classes i where N is very large. And assume we use these data in an observer study to compute $\lambda(\mathbf{g}_{i,j})$. We define the score for trial j $(1 \leq j \leq N)$ is

$$o_j = \text{step}\left(\lambda\left(\mathbf{g}_{2,j}\right) - \lambda\left(\mathbf{g}_{1,j}\right)\right).$$

(7.15)

Because $o_j = 0$ or 1 only, the net scores yield P_c and therefore AUC for the observer. To compare with other observer performance, AUC of the ideal observer is converted to the detectability index through the following equation:

$$d_A = 2\text{erf}^{-1}(2\text{AUC} - 1).$$

(7.16)

where d_A is the metric by which observer efficiency is measured.

7.3.5   Detection of Signal-to-Noise Ratio

Another figure of merit for the ideal observer is the detection of signal-to-noise ratio, calculated through the moments of the test statistic $\lambda(\mathbf{g})$, given by

$$\mathrm{SNR}_I = \frac{E_2\{\lambda\} - E_1\{\lambda\}}{\sqrt{\left(\mathrm{var}_2\{\lambda\} + \mathrm{var}_1\{\lambda\}\right)/2}},$$ (7.17)

where $E_i\{\lambda\}$ and $\mathrm{var}_i\{\lambda\}$ are means and variances conditioned on hypothesis i being true. It measures the separation between the two normal pdfs for λ, as illustrated in Figure 7.4a in units of their common standard deviation. It is related to AUC through the following equation [7]:

$$\mathrm{AUC} = \frac{1}{2} + \frac{1}{2}\mathrm{erf}\left(\frac{\mathrm{SNR}_I}{2}\right).$$ (7.18)

The SNR_I played a very important role in the task-based analysis for photon-based imaging. It allows us to factorize the ideal performance into the product of task information and the system's contribution over the spatial frequency. This factorization was known as Wagner's theory for medical image quality [27] and is based on the unification between AUC and SNR_I. The theory, for the first time, allowed us to state explicitly the dependence of the observer performance on the task at the input. In Ref. [7], Barrett et al. showed rigorously the relation between AUC and SNR_I in Equation 7.18 under the normal distribution for the test statistic. This generalization was found for photon-based imaging applications, where the test statistic is a linear function of the data and normally distributed. In sonography, however, the test statistic is a quadratic function of imaging $\mathbf{g}$ (see Equation 7.4) and has a χ^2 distribution. Thus, the normal condition for $\lambda(\mathbf{g})$ is not guaranteed; even the RF data statistics follow an MVN distribution [28]. Without the normality condition, the unification between the AUC and the SNR_I breaks down, and the relationship to the ROC curve described earlier is lost. In the next section, we will investigate the robustness of these figures of merit for sonography by introducing a new metric from information theory.

7.3.6   Task Information

Information-theoretic approaches to image formation involve optimization of performance metrics rooted in information theory. These include likelihood, divergence, discrimination, and entropy [29]. Describing an imaging system as a device that transfers task information from object to observer, channel capacity based on Shannon entropy is a figure of merit that has been used [30]. By incorporating the task in the assessment of image quality, however, we find that the Kullback–Leibler entropy is the more appropriate matrix.

It measures the divergence between test statistic distributions for two classes [31]. We have adapted the Kullback–Leibler divergence to the sonographic image quality framework as the primary metric to assess the robustness of AUC and SNR_I for the ideal observer [32]. Because AUC could not be computed in closed form without the normality

Chapter 7

assumption for λ, we developed a numerical technique to measure the Kullback–Leibler divergence, denoted by J, for sonography. J is defined as follows:

$$J = \int d\mathbf{g} \left(\mathrm{pdf}(\mathbf{g}|2) - \mathrm{pdf}(\mathbf{g}|1) \right) \log \frac{\mathrm{pdf}(\mathbf{g}|2)}{\mathrm{pdf}(\mathbf{g}|1)}, \tag{7.19}$$

to be a unitless scalar value that quantifies task information as the distance between $\mathrm{pdf}(\mathbf{g}|2)$ and $\mathrm{pdf}(\mathbf{g}|1)$. The main difficulty in calculating J is computing inverses and determinants for large covariance matrices. To overcome this computational challenge, we relate J to the integration of the difference between probabilities of detection P_D and false alarm P_F given in Equation 7.12 over decision threshold t (see Nguyen et al. [32] for the proof),

$$J = \int_{-\infty}^{\infty} dt \left[P_D(t) - P_F(t) \right]. \tag{7.20}$$

Figure 7.5 illustrates Equation 7.20 by showing representative plots of P_D and P_F over the range of t. J is the area between the curves.

We computed figures of merit d_A^2, SNR_I^2, and J by applying Equations 7.16, 7.17, and 7.20, respectively, to the decision measurements. They all are functions of the object contrast factor, controlling the difficulty of the task, given as follows [17]:

$$C = A \sum_i \left| \left[\mathbf{S}_2 - \mathbf{S}_1 \right]_{ii} \right|, \tag{7.21}$$

where $A = \Delta x \Delta y$ is the sampling interval area for RF data (equivalent to pixel area in an image). The numerical calculations began with simulating a large number of RF echo data pairs of $\mathbf{g}$ for each visual task. The AUC was measured from 2AFC studies, and SNR_I was calculated from the moments of $\lambda(\mathbf{g})$. The metric J was computed from $P_D(t)$

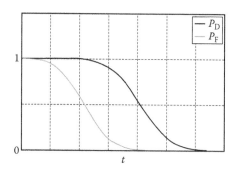

FIGURE 7.5 Plots of P_D and P_F as functions of the decision threshold t. The area between the two curves equals the Kullback–Leibler divergence, J. (From Nguyen, N. Q. et al., Objective assessment of sonographic quality I: Task information, *IEEE Transactions on Medical Imaging*, 32: 683–690, 2013. With permission.)

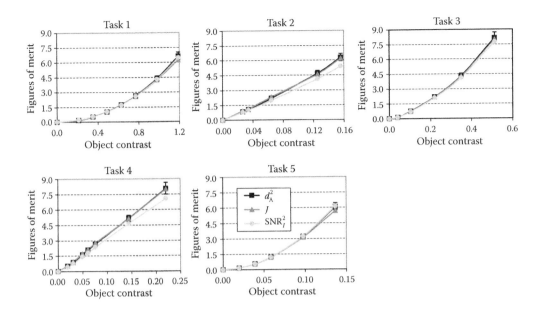

FIGURE 7.6 Comparison of detectability index d_A^2, SNR_I^2, and J for the five visual tasks. The three curves in a graph is plotted as a function of object contrast. The legend in the plot of Task 5 applies to all plots. The task features are provided in Figure 7.2. (From Nguyen, N. Q. et al., Objective assessment of sonographic quality I: Task information, *IEEE Transactions on Medical Imaging*, 32: 683–690, 2013. With permission.)

and $P_F(t)$, and through Equation 7.20. To control for case variability, we used the same RF echo data pairs to calculate these three metrics at each value of feature contrasts. Plots of the results for the five tasks are shown in Figure 7.6. In each plot, the contrast feature is tuned for the AUC ranging from 0.5 to 0.96, which is equivalent to a range from 0 to 6 for d_A^2. Above this range, d_A^2 increases rapidly with large uncertainties and comparisons become inaccurate.

Figure 7.6 shows that SNR_I^2 is nearly identical to J in tasks 1, 3, and 5, where the test statistic may be considered normally distributed. In tasks 2 and 4, however, SNR_I^2 and J are not equivalent, where SNR_I^2 is lower up to 14.5% for task 2. For the five tasks, the d_A^2 values are statistically equivalent to J. In tasks 1, 3, and 5, we note that the task contrasts are over large area regions. In those tasks, the distribution of the test statistic can be approximated well by a normal distribution. While in the other tasks, the task contrasts are limited around the lesion edges, and therefore, the normal condition was not held. In those tasks, the detectability index d_A^2 is still matched to divergence J. Our finding confirms Wagner's prediction that AUC describes the full information content available in the data from an imaging system [33]. Information in this context is understood to be Kullback–Leibler's entropy, which measures the difficulty of discriminating data from two classes. Estimates from SNR_I^2 can be affected by small deviations from normal.

The relation between d_A^2 and J also allows factorizing the ideal performance to task information and the engineering metrics of the ultrasound imaging system over the spatial frequency, making our framework equivalent to the task-based analysis developed rigorously in photon-based imaging [6]. Details of the work are provided in Ref. [34]. In the next section, we focus on how to use the analysis to derive filtering and beamforming processes applied to sonographic data.

7.4   Ideal Observer Strategies

Wagner argued that the ideal observer performance fully described the task information content of data. If an actual observer does not achieve ideal performance, the information is still in the data [27] and may be accessed through image processing. In this section, we explore ideal observer strategies to search for display stage processing that increases the transfer of information through all parts of the image formation process.

7.4.1   Wiener Filtering

The first-order term in the power series expansion of Equation 7.8 is as follows [17]:

$$\Sigma_1^{-1} - \Sigma_2^{-1} \simeq \Sigma_0^{-1}\left(\Delta\Sigma_2 - \Delta\Sigma_1\right)\Sigma_0^{-1}. \tag{7.22}$$

Consequently, the first-order approximation to the test statistic in Equation 7.4 is as follows:

$$\lambda(\mathbf{g}) \approx \sigma_{\text{obj}}^2 \mathbf{g}^t \Sigma_0^{-1} \mathbf{H}\Delta\mathbf{S}\mathbf{H}^t \Sigma_0^{-1}\mathbf{g}, \tag{7.23}$$

for the task $\Delta\mathbf{S} = (\mathbf{S}_2 - \mathbf{S}_1)$.

The first-order approximation of Equation 7.23 provides insights into ideal strategies for discrimination. The factor $\mathbf{H}^t\Sigma_0^{-1}\mathbf{g} = \left(\mathbf{g}^t\Sigma_0^{-1}\mathbf{H}\right)^t$ is recognized as the spatial Wiener filter acting on RF echo signals. It suggests that the ideal observer does decorrelate RF data before combining it with the task information. That step is necessary, provided the first-order approximation in Equation 7.22 is a good one. Because the ideal observer combines all available information to make decisions, we interpret Equation 7.23 as a generally applicable single strategy for all features short of adding specific task information into the process. In the following section, we will show that Wiener filtering RF signals before envelope detection can increase information transfer through the display stage and into the B-mode image in a way that improves the performance of trained observers [17].

The Wiener filter $\mathbf{H}^t\Sigma_0^{-1}\mathbf{g}$ involves computing the inverse of Σ_0, which is stationary and easily inverted using Fourier techniques. We called it the *stationary Wiener filter* to distinguish it from the *iterative Wiener filter* described in the next section. An envelope generated by first Wiener filtering RF data is referred as a WFB-mode image.

7.4.2   Adaptive Wiener Filtering

The ability of the stationary Wiener filter to improve human observer performance depends on the accuracy of the first-order approximation. The linear approximation to matrix inverses $(\mathbf{I} + \mathbf{S}_i) \simeq \mathbf{I} - \mathbf{S}_i$ is a good one when eigenvalues of $\mathbf{S}_i$ are close to 0, which is true for low-contrast feature. When that is not the case, as in task 5, we no longer separate covariance matrix components into stationary and nonstationary components. Instead, we form *average* and *difference* components, respectively,

$$\Sigma_a = \sigma_{\text{obj}}^2 \mathbf{H}(\mathbf{I} + \mathbf{S}_a)\mathbf{H}^t + \sigma_n^2\mathbf{I} \text{ and } \Delta\Sigma = \sigma_{\text{obj}}^2 \mathbf{H}\Delta\mathbf{S}\mathbf{H}^t, \tag{7.24}$$

where

$$\mathbf{S}_a = 0.5(\mathbf{S}_1 + \mathbf{S}_2) \text{ and } \Delta\mathbf{S} = 0.5(\mathbf{S}_1 - \mathbf{S}_2).$$

As with the stationary filter, we expand the matrices in a power series and truncate after the first term to find

$$\boldsymbol{\Sigma}_1^{-1} - \boldsymbol{\Sigma}_2^{-1} \simeq 2\boldsymbol{\Sigma}_a^{-1} 2\Delta\boldsymbol{\Sigma}\boldsymbol{\Sigma}_a^{-1}. \qquad (7.25)$$

The new first-order approximation of λ becomes

$$\lambda(\mathbf{g}) \approx \sigma_{\text{obj}}^2 \mathbf{g}^t \boldsymbol{\Sigma}_a^{-1} \mathbf{H}\Delta\mathbf{S}\mathbf{H}^t \boldsymbol{\Sigma}_a^{-1} \mathbf{g} \qquad (7.26)$$

Equation 7.26 suggests that the filter $\mathbf{H}^t\boldsymbol{\Sigma}_a^{-1}$ be applied to the echo data, where the average covariance between two states is applied instead of the stationary background covariance. The advantage of this change is to enable the signal strength to vary significantly within any one image provided differences between compared images remain small. The disadvantage of the new filter is that $\boldsymbol{\Sigma}_a$ is nonstationary, so we cannot use Fourier techniques to quickly compute its inverse.

The power series approach can be applied by decomposing $\boldsymbol{\Sigma}_a$ into stationary and nonstationary components, $\boldsymbol{\Sigma}_a = \boldsymbol{\Sigma}_0 + \sigma_{\text{obj}}^2\mathbf{H}\mathbf{S}_a\mathbf{H}^t$, yielding an iterative formula for calculating $\mathbf{H}^t\boldsymbol{\Sigma}_a^{-1}\mathbf{g}$ given by

$$\mathbf{q}_{i+1} = -\sigma_{\text{obj}}^2\mathbf{H}^t\boldsymbol{\Sigma}_0^{-1}\mathbf{H}\mathbf{S}_a\mathbf{q}_i, \text{ and}$$

$$\qquad\qquad\qquad\qquad\qquad\qquad\qquad\qquad\qquad (7.27)$$

$$\mathbf{p}_{i+1} = \mathbf{p}_i + \mathbf{q}_i.$$

The iterative scheme is initialized by $\mathbf{q}_0 = \mathbf{p}_0 = \sigma_{\text{obj}}^2\mathbf{H}^t\boldsymbol{\Sigma}_0^{-1}\mathbf{g}$. Equation 7.27 begins with the stationary Wiener filter and iteratively converges to $\sigma_{\text{obj}}^2\mathbf{H}^t\boldsymbol{\Sigma}_a^{-1}\mathbf{g}$ using the power series inverse approximation for $\boldsymbol{\Sigma}_a$. We refer to the result as the adaptive or *iterative Wiener filter* and the corresponding envelope as the IWFB-mode image [19]. To compare with the stationary Wiener filter, the iterative Wiener filter is combined with the average task information $\mathbf{S}_a$ through the iterations. Thus, IWFB-mode images can adapt to the task as specified by $\mathbf{S}_a$ provided that $\mathbf{S}_a$ is known. To make this filter practical in the clinical environment where $\mathbf{S}_a$ is not known a priori, we propose a method to find it from $\mathbf{S}_1$ and $\mathbf{S}_2$, which are estimated after image segmentation.

7.4.3 Introduce Task Information through Image Segmentation

We adopted a segmentation algorithm that makes use of a Markov random field (MRF) model and for normally distributed image data [35,36]. Ashton et al. [37] successfully applied it to segment sonograms despite the underlying Rayleigh distribution. They decomposed the image data into multiple resolution layers and then invoked the central limit theorem to assume pixel values at the lowest level were distributed by a Gaussian

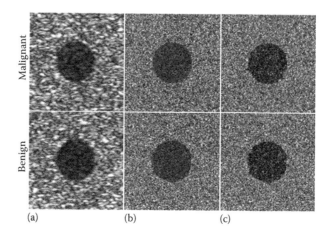

FIGURE 7.7   Examples of (a) standard B-mode, (b) WFB-mode, and (c) IWFB-mode image of benign and malignant pairs of task 5 (linear scale). (From Nguyen, N. Q. et al., *IEEE Transactions on Medical Imaging*, 30: 28–37, 2011.)

function. Segmentation was applied at successively higher resolution layers through a computationally intensive process.

In Nguyen et al. [19], we found that decomposition into multiple layers can be disregarded by directly applying the algorithm to WFB-mode images, which are at least partially decorrelated. The goal of segmentation is to extract information about Si, including the boundary and contrast. Therefore, we adopted a two-level approach that causes errors in task 3 where the malignant features many gray levels at the lesion boundary. Because the new first-order approximation is quite accurate with the average/difference decomposition, the performance of the iterative Wiener filter depends mainly on the quality of the segmentation.

Figure 7.7 shows example images processed three different ways, including the standard B-mode, applying the stationary Wiener filter and iterative Wiener filter with segmentation before taking the envelope. The main effect of the stationary Wiener filter is to clarify lesion edges, whereas that of the iterative Wiener filter is to clarify lesion contrast. Although the effects apparent in these sample images are subtle, the overall performance improvement for train human observers was significant, as shown in the next section. The disadvantage of the iterative Wiener filter is that it is computationally expensive because of segmentation and iterations.

7.4.4   Beamforming Using Precompression Filtering

Thus far, we have only analyzed beamformed RF echo data, that is, signals from receive channels in the array aperture that were appropriately delayed and then summed to form an A-line. The standard DS beamformer can be optimal only when the pulse-echo impulse response from each transducer element is a Dirac delta function [38]. Failing ideal conditions, RF echo signals **g** loses coherence, which lowers eSNR and degrades spatial and contrast resolutions. Coherence would be restored if we could filter each receive-channel signal by the inverse of its impulse response before delaying and summing. However, inverse filtering noisy signals are usually unstable, so instead we turn to the ideal observer analysis for a strategy.

We consider beamforming in the image formation model by expanding the system operator $\mathbf{H}$ to describe individual receive elements of the array (see Figure 7.8). Consequently, Equation 7.1 becomes

$$
\mathbf{g}_T = \begin{bmatrix} \mathbf{g}_1 \\ \mathbf{g}_2 \\ \vdots \\ \mathbf{g}_{N-1} \end{bmatrix} = \begin{bmatrix} \mathbf{H}_1 \\ \mathbf{H}_2 \\ \vdots \\ \mathbf{H}_{N-1} \end{bmatrix} \mathbf{f} + \begin{bmatrix} \mathbf{n}_1 \\ \mathbf{n}_2 \\ \vdots \\ \mathbf{n}_{N-1} \end{bmatrix} = \mathbf{H}_T \mathbf{f} + \mathbf{n}_T,
\tag{7.28}
$$

where $\mathbf{g}_T$ is a vector of RF echo signals from all receive channels before delay and summation. $\mathbf{H}_i$ is the pulse-echo system matrices for each element, and $\mathbf{H}_T$ is the system matrix of prebeamformed RF data. $\mathbf{n}_T$ is the acquisition noise. Assuming noise on the ith channel is a WGN process with variance $\sigma_{n,T}^2$, the variance of the beamformed noise signal is $\sigma_n^2 = N\sigma_{n,T}^2$, where N is the number of elements in the receive aperture. Denoting $\mathcal{B}$ as a beamforming operator, we can write

$$\mathbf{H} = \mathcal{B}\mathbf{H}_T \text{ or } \mathbf{H} = \mathbf{B}\mathbf{H}_T \text{ and } \mathbf{n} = \mathbf{B}\mathbf{n}_T.$$

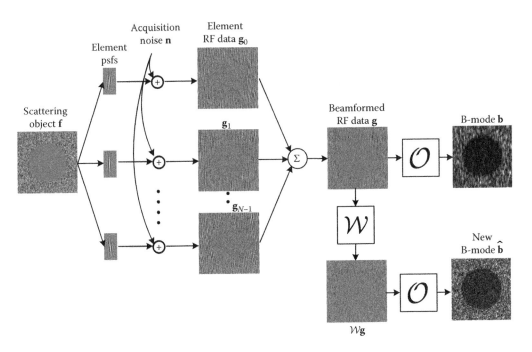

FIGURE 7.8   A graphical model is shown for image formation that includes beamforming. The acquisition stage is extended to each element of the transducer, where the data are initially acquired.

Where the matrix forms hold if B is a linear operator.

To discover optimal combinations of element data resulting in an A-line that yields maximum task performance, we write the pdf of $\mathbf{g}_T$ for SKE data conditioned on the *i*th hypothesis as

$$\mathrm{pdf}\left(\mathbf{g}_T \mid i\right)=\frac{1}{\left(\sqrt{2\pi}\sigma_{n,T}\right)^M}\int \mathrm{d}\mathbf{f}\exp\left(-\frac{1}{2\sigma_{n,T}^2}(\mathbf{g}_T-\mathbf{H}_T\mathbf{f})^t(\mathbf{g}_T-\mathbf{H}_T\mathbf{f})\right)\mathrm{pdf}\left(\mathbf{f}\mid i\right), \quad (7.29)$$

where the multidimensional integral is over the ensemble of $\mathbf{f}$ and M is the length of vector $\mathbf{g}_T$. The ideal observer test statistic, now expressed as the likelihood ratio, is

$$\Lambda(\mathbf{g}_T)=\frac{\mathrm{pdf}\left(\mathbf{g}_T \mid 2\right)}{\mathrm{pdf}\left(\mathbf{g}_T \mid 1\right)}=\frac{\displaystyle\int \mathrm{d}\mathbf{f}\exp\left(-\frac{1}{2\sigma_{n,T}^2}(\mathbf{g}_T-\mathbf{H}_T\mathbf{f})^t(\mathbf{g}_T-\mathbf{H}_T\mathbf{f})\right)\mathrm{pdf}\left(\mathbf{f}\mid 2\right)}{\displaystyle\int \mathrm{d}\mathbf{f}\exp\left(-\frac{1}{2\sigma_{n,T}^2}(\mathbf{g}_T-\mathbf{H}_T\mathbf{f})^t(\mathbf{g}_T-\mathbf{H}_T\mathbf{f})\right)\mathrm{pdf}\left(\mathbf{f}\mid 1\right)}. \quad (7.30)$$

Canceling common terms in the numerator and denominator yields the following:

$$\Lambda(\mathbf{g}_T)=\frac{\displaystyle\int \mathrm{d}\mathbf{f}\exp\left(\frac{1}{2\sigma_{n,T}^2}\left(2\mathbf{f}^t\mathbf{H}_T^t\mathbf{g}_T-\mathbf{f}^t\mathbf{H}_T^t\mathbf{H}_T\mathbf{f}\right)\right)\mathrm{pdf}\left(\mathbf{f}\mid 2\right)}{\displaystyle\int \mathrm{d}\mathbf{f}\exp\left(\frac{1}{2\sigma_{n,T}^2}\left(2\mathbf{f}^t\mathbf{H}_T^t\mathbf{g}_T-\mathbf{f}^t\mathbf{H}_T^t\mathbf{H}_T\mathbf{f}\right)\right)\mathrm{pdf}\left(\mathbf{f}\mid 1\right)}. \quad (7.31)$$

We stop here because the point of the derivation is to show that $\mathbf{g}_T$ is always matched filtered (multiplied by $\mathbf{H}_T^t$) when it appears. Consequently, optimal performance is achieved by matched filtering individual channel signals before summation. Observer performance cannot be increased by beamforming, but matched filtering is an ideal strategy for preserving information before summation. Note that matched filtering naturally time delays channels signals to be summed, and it is robust to measurement noise.

The beamformed RF data must still be envelope detected and scan converted to form images for human observers and diagnostic information may be lost through this nonlinear processing. In the next section, we use ideal observer analysis to assess the efficiency of the matched filter beamformer and compare it with the conventional DS beamformer. To explore the value of signal envelope and phase in matched filtering, we also consider amplitude matched filter (AMF) and phase matched filter (PMF) beamformers. In addition, we consider a 1-D matched filter, applied along the direction of the beam axis only, to separate axial and lateral beam effects on observer performance. In Section 7.5.4, we investigate the loss of observer efficiency due to the display stage of image formation for each of these beamformers.

7.5 Assessments of Image Quality

7.5.1 Human Observer Studies

In this section, we compare those results with human observer performance to study visual discrimination efficiency. Human observer measurements are time consuming to obtain and fraught with many sources of potential uncertainty, and yet they are the state of the art for medical diagnosis. Comparing human and ideal observer performances points to inefficiencies in the image formation process and sometimes the analysis will suggest processing that improves the transfer of task information. Ideal observer performance, as we define it, depends only on the acquisition stage. Human performance, however, depends on the acquisition and display stages; the latter includes the effects of postbeamforming data processing, for example, Wiener filtering, as well as the effects of envelope detection, amplitude compression, and scan conversion. It also includes any intrinsic observer limitations such as training and internal noise of the human eye–brain system.

Humans are shown pairs of standard B-mode images in one study and pairs of Wiener filtered (WFB-mode) or iterative Wiener filtered (IWFB-mode) images in other studies. They are asked to identify the one with malignant features using the 2AFC testing paradigm [3,17]. Examples of these images are shown in Figure 7.7. Observers also view the signal template in a separate image showing them the malignant feature they are asked to identify.

The goal of this first study is to compare different imaging methods to evaluate the effectiveness of postbeamforming spatial filtering of the RF echo signals that is applied before envelope detection. Observers were informed of all feature parameters such as target amplitude and location. After a training period, each observer viewed 400 randomized image pairs per study, and the proportion of correct responses (P_c) was measured. From P_c, we can compute d_A and AUC metrics. The correctness of each response was immediately indicated. The background region echo SNR was 32 dB for all simulated images, which was measured experimentally using tissuelike phantoms. Five observers each participated in 15 studies, involving five tasks under three imaging conditions labeled B-mode, WFB-mode, and IWFB-mode. Although all images in a study are statistically independent, we controlled for case variability by applying different filters to the same RF data.

Human observer results are summarized in Figure 7.9. We find humans viewing B-mode images for tasks 1 to 4 yield the lowest performance compared with images where filtering was applied. Filtering before envelope detection preserves more of the task information that is normally lost at the display stage. Both filters increased human performance about the same amount except for task 5, where Wiener filtering reduced P_c substantially from 79% for B-mode images to 63% for WFB-mode images. Task 5 results are examples of what occurs when the linear approximation to the covariance matrix inverse fails to hold. The Wiener filter produced in this way was well matched to the background but not to the interior of the lesion area where the discriminating signal was located. The Wiener filter inappropriately amplified lesion noise, where eSNR < 0 dB (signal is dominated by noise). Because this Wiener filter does not take the task into consideration, it can be expected to enhance human performance only when specific properties of the task are not very important. Meanwhile, the iterative WF significantly improved performance for task 5 because it includes task information estimated

Chapter 7

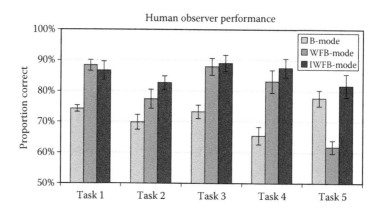

FIGURE 7.9 Average proportion correct for five human observer studies involving three forms of signal processing as illustrated in Figure 7.7 and the five diagnostic tasks described in Section 7.2.1. Error bars indicate standard errors. (From Nguyen, N. Q. et al., *IEEE Transactions on Medical Imaging*, 30: 28–37, 2011.)

from the images in the filter. The IWF performance improvement comes at the cost of an approximately fivefold increase in computational load. Variations of the degree of improvements over the five tasks just underscore the task-dependent nature of performance. As designers, we need to understand which clinical examination conditions warrant the extra effort.

7.5.2 Observer Efficiency as System Design Criteria

Ideal observer performance quantifies the amount of task information available in the RF echo signals. Deficiencies in the acquisition stage must be addressed with improved hardware, primarily transducer designs. Human observer performance quantifies the total information available in the image that is accessible by decision makers. Differences between the two are display stage issues that suggest postprocessing opportunities to enhance diagnostic performance. Differences between human and ideal observer performance are quantified by the observer efficiency [39],

$$\eta_H = \frac{d_H^2}{d_A^2},\tag{7.32}$$

where d_H and d_A are the detectability indices of the human and ideal observer at the same contrast level. In photon imaging, d_A is identical to SNR_I. In that case, η_H is analogous to the detective quantum efficiency (DQE) of the imaging system [1]. In sonography, however, we find that d_A is better described by the Kullback–Leibler divergence, so η_H is equivalent to discrimination efficiency [31]. When efficiency is low, d_A can be so large that it is difficult to calculate accurately in the 2AFC procedure. There are advantages to defining η_H as the ratio of squared feature contrasts for the two observers that give the same detectability index; that is [17],

$$\eta = \left(\frac{C_I}{C_H}\right)^2,\tag{7.33}$$

where C_H and C_I are the contrast factors for human and ideal observers that generate under equivalent performance, or $d_A = d_H$. For statistical reasons, we adjust task difficulty in simulated image to achieve $P_c \sim 0.7$–0.8 for human observers.

For reasons described in Section 7.1, we cannot compute the exact ideal observer for any B-mode imaging data. The SW analysis [10,11] does provide the ideal observer's test statistic for our task 1, if readers are willing to enable us to disregard violations of their assumptions. Acknowledging these violations, we refer to their test statistic as the SW observer. We bravely apply it to all five tasks while acknowledging it is an approximation, albeit the best we have today. In our notation, the test statistic of the SW observer is [17]

$$\lambda_{SW}(\mathbf{b}) = \mathbf{b}^t(\mathbf{S}_2 - \mathbf{S}_1)\mathbf{b}. \tag{7.34}$$

This is similar to the linear approximation of the RF-signal result in Equation 7.23, except the image data vector $\mathbf{b}$ is not decorrelated. Assuming the utility of the SW observer, η_H can be decomposed into

$$\eta_H = \eta_{SW}\eta_{H|SW}, \tag{7.35}$$

where η_{SW} is the efficiency of the SW observer acting on envelope images with respect to the ideal observer acting on RF data, and $\eta_{H|SW}$ is the efficiency of human observer with respect to the SW observer both acting on B-mode images. Separating efficiency in this manner enables us to identify sources of information loss. η_{SW} is a measure of information lost by the demodulation process, and $\eta_{H|SW}$ is a measurement of information lost by the human-observer system.

7.5.3 Postprocessing Filtering

The three efficiencies obtained from the previous human and ideal observer performance measurements and that from applying the SW observer to the same data are plotted in Figure 7.10. Figure 7.10a shows the efficiency of human observers relative to the ideal observer. It is the product of the results of Figure 7.10b and c, which describes the component efficiencies on the right-hand side of Equation 7.35. Figure 7.10a shows the human observer efficiency ranges from 0.2% to 40% over our five tasks. Discrimination performance is better for the large-area contrast features (tasks 1 and 5) than for the edge detection tasks (2–4), but there seems to be room for improvement overall. Except for task 4, where humans struggle to see the spiculated boundary (Figure 7.10c), it appears that most of the information is lost in B-mode images during the envelope detection process (Figure 7.10b). RF echo filtering greatly improves the passage of task information through the envelope detection process. Surprisingly, filtering seems to hamper accessibility of task information by humans, but the net effect is that filtering improves human performance for all five tasks. We find these results very enlightening from a designers perspective because it points to the stage where information is lost and therefore possible solutions.

Chapter 7

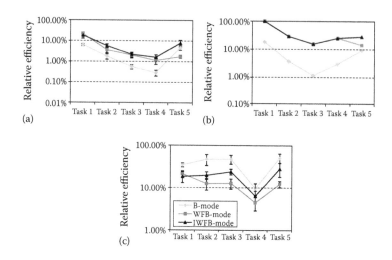

FIGURE 7.10   (a) Human observer efficiencies measured on B-mode, WFB-mode, and IWFB-mode images, η_H, from Equation 7.33. (b) SW observer efficiency, η_{SW}. (c) Human efficiency relative to the SW observer, $\eta_{H|SW}$. Values in (a) equal the product of corresponding values from (b) and (c). Note that the ordinate scaling of relative efficiency axis is changed among the figures. (From Nguyen, N. Q. et al., *IEEE Transactions on Medical Imaging*, 30: 28–37, 2011.)

7.5.4   Preprocessing Filtering

The five beamformers described in Section 7.4.4 are considered in this section. Before summation, each includes individual receive-channel filtering that we refer to as 2-D matched filter (2DMF), AMF, PMF, 1-D (axial only) matched filter (1DMF), and DS beamformers. We used the Field II program to simulate the impulse response for each receive channel [23,24] to specify $\mathbf{H}_T$. The entire aperture was used to transmit a focused pulse, and echoes received on each receive channel were filtered and summed to give the pulses shown in Figure 7.11. Also shown are the system transfer spectrum, noise power spectrum, and eSNR maps in spatial frequency. We see how bandwidth depends on the beamforming approach. The 2DMF and AMF beamformers appear to be somewhat more narrow-band than either the DS or the PMF beamformers. The 1DMF agrees with the 2DMF in the axial direction but has slightly broader bandwidth in the lateral direction. Noise power spectrum in the 2DMF and AMF beamformers is also concentrated around the transfer function bandwidth. By contrast, the phase-only DS and PMF beamformers leave the noise white with a flat power spectrum over the Nyquist band. Noise power in the 1DMF is limited to the band of the 2DMF on the axial direction and is white to the Nyquist frequency in the lateral direction [18].

To investigate the efficiency of transferring information from prebeamformed RF data to final B-mode images, we again approximate the SW observer as the ideal observer for B-mode images. The stationary Wiener filter is also applied after summation to investigate the role of further processing before demodulation. The results are plotted in Figure 7.12.

From the analysis in Section 7.4.4, we know that using the 2DMF beamformer to compress the data volume preserves ideal performance. It is illustrated in Figure 7.12

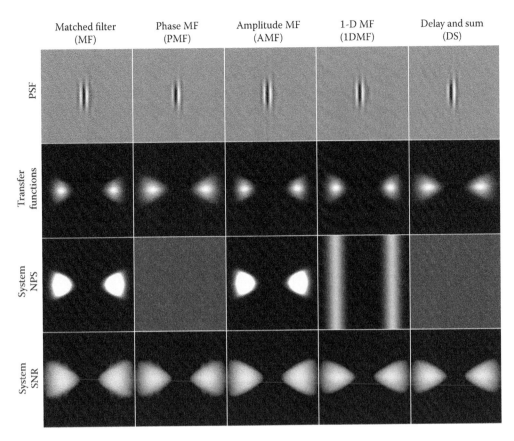

FIGURE 7.11 The five beamforming approaches used in this study are analyzed here for their system transfer and noise power properties. The top row gives the beamformed pulse profile (pulses propagate from left to right), which is effectively the PSF of the beamformed RF data. The second row of the panel is the amplitude of the transfer function of each pulse (zero-frequency origin is at the center). The third row of the panel shows the noise power spectrum of the composite acquisition noise term. The fourth row of the panel displays eSNR in spatial frequency with intensity log-compressed to a decibel scale. (From Abbey, C. K. et al., *IEEE Transactions on Ultrasonics, Ferroelectrics and Frequency Control*, 57: 1782–1796, 2010.)

by showing that the 2DMF acting on RF data has an efficiency of 100% in all tasks. The efficiency drops consistently across the different beamformers arranged from the 2DMF to DS beamformers. For standard DS beamforming, efficiency values range from 24% to 75% with an average across tasks of 45%. Therefore, interpreting efficiency as a measure of diagnostic information suggests that DS beamforming loses roughly half the diagnostic information in the RF echo data. The PMF has an efficiency ranging from 71% to 90% with an average value across tasks of 83%. Thus, incorporating only the phase component of the matched filter makes up roughly 60% of the efficiency drop between the 2DMF and the DS beamformers. The AMF efficiency ranges from 34% to 87% and makes up an average of 20% in the difference between the 2DMF and the DS beamformers. This shows that amplitude has a significant but substantially smaller role than phase in transferring information to the compressed RF signal. A surprising finding was that the 1DMF showed no improvement over the DS beamformer. Beamforming is often viewed as an operation that combines signals from within an aperture to produce

Chapter 7

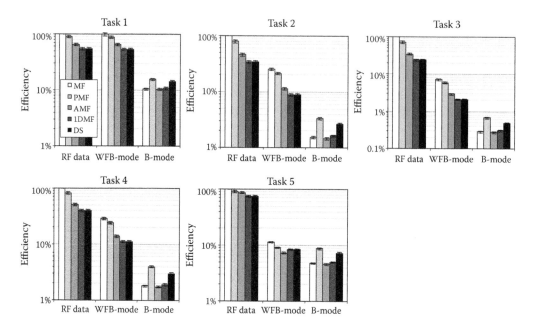

FIGURE 7.12   Efficiencies of the ideal observer acting on pre-envelope RF data and SW approximations to the ideal observer acting on B-mode and WFB-mode images are shown across the beamforming system for each task. The legend in the plot of task 1 is applied to the plots of the other tasks. Tasks information is illustrated in Figure 7.2. (From Abbey, C. K. et al., *IEEE Transactions on Ultrasonics, Ferroelectrics and Frequency Control*, 57: 1782–1796, 2010.)

a single A-scan line. Findings from ideal observer analysis indicate that element correlations from different apertures play a crucial role in the transfer of diagnostic information in beamforming.

Efficiency of the stationary Wiener filtered envelope images in Figure 7.12 is very close to the ideal observer performance in task 1, suggesting that the filter helps recover most of information on the B-mode images. In tasks 2 to 4, efficiency is substantially less than the ideal observer acting on the RF data, but the relative ordering of the different beamformers is quite similar. Task 5 is the exception, with much less of a drop in efficiency going from the 2DMF to DS beamformers. There is also less of a relative advantage with the PMF and an observed disadvantage with the AMF beamformers. This is consistent with findings for the stationary Wiener filter, which provides less benefit in task 5 because the filter is not tuned for regions containing signal. Nonetheless, the 2DMF beamformer maintains roughly a twofold advantage over DS for Wiener filtered envelope images.

Somewhat surprisingly, the 2DMF beamformer loses its advantage in standard B-mode envelope images. The 2DMF and the AMF beamformers have the lowest efficiency of the five beamformers tested, roughly two thirds the efficiency of DS, and even less when compared with the PMF, which performed the best of the methods applied to B-mode envelope images. The 1DMF also underperforms DS beamforming in this case. These results show that a substantial advantage in diagnostic information in the compressed RF data can subsequently be lost in the transformation to an envelope image.

The advantage of the 2DMF beamformer is only realized with a postprocessing step before the envelope image is computed.

7.6   Measurements of Spatiotemporal Impulse Response Functions

In this analytical framework, the system matrix **H** plays a central role in modeling the RF signals and developing pre- and postsummation filters that are suggested by ideal observer strategies. Therefore, it is crucial to accurately estimate **H** experimentally. As described in Section 7.2.2, **H** is constructed from the spatiotemporal impulse responses of the system, which can vary in space and time. Accurate measurements of the impulse response function are difficult to obtain. The impulse response is also difficult to predict numerically because even small, unknown perturbations in the linear array geometry can make significant changes in the pulse-echo field patterns, especially in the near field. In this section, we propose two methods to measure the impulse response and list advantages and disadvantages for measuring each.

7.6.1   Scattering Spheres

Following the linear pulse-echo model describing RF data developed by Zemp et al. [20], we can measure the system impulse response by scanning a single scatterer that approximates a Dirac delta function applied to the system input. We used a gelatin gel volume into which 0.04 mm glass spheres are randomly suspended. The density of the glass sphere is 2.38 g/cc, and the speed of sound in glass is 5570 m/s. The density and sound speed in gelatin is 1.06 g/cc and 1500 m/s. With a pulse center frequency at 10 MHz, the sphere diameter is less than one third of the wavelength, and thus the high-scattering point targets are reasonable approximations to delta functions.

Figure 7.13 shows a B-mode image through a cross section of the gel in the axial–lateral plane. Notice how shift-varying the impulse response appears with depth. Because the spheres are so small, scattering from each sphere is weak, so we improve the eSNR by averaging 1000 RF echo without moving the transducer. A Siemens Antares Sonoline system was used with a VF10-5 linear array. A fixed focus of 40 mm was on both transmit and receive. The lateral array pitch is 0.2 mm. The point-spread functions are most compact near the focal length.

By elevating the transducer at 0.5 mm increments, we acquired many planes to synthesize 3-D impulse responses for the system. Figure 7.14 shows impulse responses recorded at three depths and in two orthogonal planes. When interpreting these pictures of impulse responses, remember that the sections are through planes of generally curved functions that have not been demodulated.

This method is fast and convenient; however, it has two disadvantages. First, the scattering signals from tiny glass spheres are weak, so measurements may be affected by scattering from gel impurities and surface reflections. These are apparent in Figures 7.13 and 7.14. Second, the shift-varying impulse response requires scatterers to be placed far enough apart to not interfere and yet dense enough to capture the spatial variations in the impulse response. These problems can be avoided by reconstructing impulse responses from line scatterers, as described in the next section.

Chapter 7

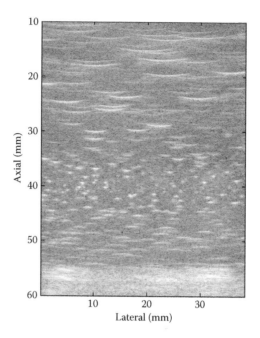

FIGURE 7.13 A log-compressed B-mode image of a cross section of a gelatin gel block with a random distribution of 0.04 mm diameter spheres. eSNR is improved by averaging 1000 RF data frames.

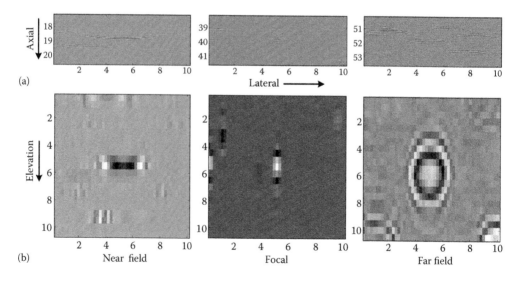

FIGURE 7.14 The 3-D spatiotemporal impulse response at near-field, focal-length, and far-field distances. (a) A slice near the center of the 3-D impulse response function in the axial–lateral plane. (b) A slice near the center of the 3-D impulse response function in the lateral elevation plane. Distances are in millimeters. When more than one sphere is in the field, multiple copies of impulse response functions are found.

7.6.2   Reconstruction from Projections of Line Scatterer Echoes

Reconstructing impulse response functions from line scatterer echoes is also based on a linear pulse-echo model of RF formation. We placed a thin line scatterer in water within a plane normal to the beam axis (Figure 7.15). The plane was a known distance from the aperture, and the line was at a known angle relative to the 1-D transducer array axis. We then record an RF echo scan plane. The A-lines in this recording are equivalent to translating the line across the aperture. Rotating the line and rescanning, we built up a set of data equivalent to a sinogram in tomographic projection imaging. Line scattering echo data are projection data that can be reconstructed to estimate the impulse response in three spatial and one temporal axis. There are a few important differences between this problem and the CT reconstructions that we describe in a previous study [40]. Essentially, we perform reconstructions using a pseudoinverse approach implemented through application of singular value decomposition [1]. Although there are many similarities to basic CT reconstructions using filtered backprojection [41], the main difference is the presence of a large void in the data caused by use of 1-D linear arrays. Readers are referred to our previous study [40] for mathematical details.

Line scatterers are easier to position at known depths than glass spheres, and the line provides much stronger scattering signals. We provide examples of reconstructed

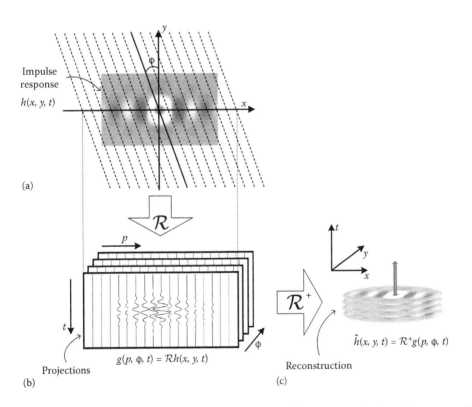

FIGURE 7.15   Geometry used to acquire echo projections and reconstruct the impulse response [39]. (a) View from the aperture looking down at the line scatterer with the beam profile superimposed. Parallel lines show the scatter as it is translated across the aperture at angle ϕ. The 3-D volume of projection data (b) is reconstructed via operator $\mathcal{R}^+$ to form impulse response estimates (c).

Chapter 7

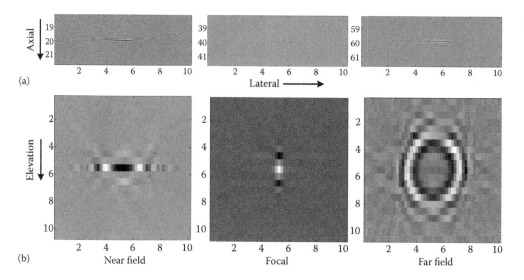

FIGURE 7.16 The reconstructed 3-D spatiotemporal impulse response at near field, focal field, and far field by scanning a rotated line. (a) A slice near the center of the 3-D impulse response function in the axial–lateral plane. (b) A slice near the center of the 3-D impulse response function in the lateral elevation plane. Distances are in millimeters.

impulse responses measured using the Antares system at near-field (20 mm), focal-length (40 mm), and far-field (60 mm) distances in Figure 7.16. Comparing these results to those from the first method, the high eSNR of the line scattering signal allows us to more clearly visualize phase changes in the lateral elevation plane. However, in the far field, where the function is less compact, we could not reconstruct the impulse well along the elevated direction. Reconstruction errors are caused by missing data in the sinogram. The missing information can be recovered using 2-D array transducers to scan the line or by moving the 1-D array along the elevation, but these acquisitions are very time consuming. Long acquisitions add additional error if the water temperature changes. This method requires significant automation for efficient implementation. Nevertheless, task-based design and assessments of performance require accurate estimates of the impulse response.

7.7 Conclusions

This chapter is a review of the concepts behind task-based design and an evaluation of sonographic systems for clinical diagnosis. The applications have been to optimize breast cancer diagnosis, but the concepts are generally applicable. Although much is known on this topic in radiography and other photon-based medical imaging methods, the subject is far from mature in sonography. There are several reasons for this. One reason is the use of nonlinear processing in the display stage, which complicates system modeling and statistical analysis. Another reason is the low risk of ultrasound imaging modality.

The innovative idea is to extend the ideal observer analysis to the RF domain, where imaging data can be modeled as a (noisy) linear transformation of the scattering object. However, the physics of backscatter place tissue contrast in the covariance matrix of stochastic object functions and not the mean as in radiography. Consequently, the ideal observer's test statistic is a quadratic function of the data rather than a linear function. This major difference creates (i) a potential deviation from a normal distribution for the test statistic and (ii) a challenge in its calculation. To overcome the first issue, we introduce the Kullback–Leibler divergence to the framework and show its robust relation with the ideal observer performance. This connection allows us to link the ideal performance to the clinical features and system engineering metrics, making the framework rigorous for being used to evaluate ultrasound imaging systems. The second issue in test statistic calculation can be accomplished using power series expansion, from which we are able to form filtering and beamforming processes as approximations of the ideal Bayesian strategies. In our study, the filters and beamformers are designed in the context of optimal variables.

Radiography uses ionizing radiation that demands very careful use. In addition to diagnostic errors, the use of x-ray and gamma-ray imaging carries significant patient risks. The risks associated with sonography, fortunately, are much lower. Designers of this low-cost modality have not had to face the same economic and safety design pressures that spawned development a rigorous image quality analysis in radiography. Once such an analysis is also available in sonography, the industry will have better tools to address the value and limitations imposed by the current output power limits. In addition, ever greater computational power means that systems can be quickly reconfigured for different patient body types and examination requirements, which can improve diagnostic performance with little increased cost. Yet this can occur only when a general task-based analysis is available for sonography. We predict that despite its developing difficulties, the need for quantitative task-based assessments of medical sonography is growing.

Acknowledgments

The research has been conducted under award no. CA118294 from the National Institutes of Health. The authors thank Karen Lindfors at the University of California at Davis Medical Center and their colleagues at the University of Illinois and University of California, Santa Barbara, for their advice and support. The authors also acknowledge Robert F. Wagner and Harrison H. Barrett, who inspired and guided them through their seminal contributions.

References

1. Barrett, H.H., and K.J. Myers, *Foundations of Image Science*. Hoboken, NJ: John Wiley & Sons, 2004.
2. Barrett, H.H., Objective assessment of image quality: Effects of quantum noise and object variability, *Journal Optical Society of America A*, 7(1990): 1266–1278.
3. Green, D.M., and J.A. Swets, *Signal Detection Theory and Psychophysics*. New York: Wiley, 1966.
4. Swets, J.A., ROC analysis applied to the evaluation of medical imaging techniques, *Investigative Radiology*, 4(1979): 109–120.
5. Swets, J.A., and R.M. Pickett, *Evaluation of Diagnostic Systems: Methods from Signal Detection Theory*. New York: Academic Press, 1982.
6. Wagner, R.F., and D.G. Brown, Unified SNR analysis of medical imaging systems, *Physics in Medicine and Biology*, 30(1985): 489–518.

Chapter 7

7. Barrett, H.H., C.K. Abbey, and E. Clarkson, Objective assessment of image quality. III. ROC metrics, ideal observers, and likelihood-generating functions, *Journal of the Optical Society of America A*, 15(1998): 1520–1535.
8. Barrett, H.H., J. Yao, J.P. Rolland, and K.J. Myers, Model observers for assessment of image quality, *Proceedings of the National Academy of Science*, 90(1993): 9758–9765.
9. Barrett, H.H., J.L. Denny, R.F. Wagner, and K.J. Myers, Objective assessment of image quality. II. Fisher information, Fourier crosstalk, and figures of merit for task performance, *Journal of the Optical Society of America A*, 12(1995): 834–852.
10. Wagner, R.F., S.W. Smith, J.M. Sandrik, and H. Lopez, Statistics of speckle in ultrasound B-scans, *IEEE Transactions on Sonics and Ultrasonics*, 30(1983): 156–163.
11. Smith, S.W., R.F. Wagner, J.M. Sandrik, and H. Lopez, Low contrast detectability and contrast/detail analysis in medical ultrasound, *IEEE Transactions on Sonics and Ultrasonics*, 30(1983): 164–173.
12. Trahey, G.E., J.W. Allison, S.W. Smith, and O.T. von Ramm, A quantitative approach to speckle reduction via frequency compounding, *Ultrasonic Imaging*, 8(1986): 151–164.
13. Wagner, R.F., M.F. Insana, and S.W. Smith, Fundamental correlation lengths of coherent speckle in medical ultrasonic images, *IEEE Transactions on Ultrasonics, Ferroelectrics and Frequency Control*, 35(1988): 34–44.
14. Thijssen, J.M., B.J. Oosterveld, and R.F. Wagner, Gray level transforms and lesion detectability in echographic images, *Ultrasonic Imaging*, 10(1988): 171–195.
15. Hall, T.J., M.F. Insana, L.A. Harrison, N.M. Soller, and K.J. Schlehr, Ultrasound contrast-detail analysis: A comparison of low-contrast detectability among scanhead designs, *Medical Physics*, 22(1995): 1117–1125.
16. Brunke, S.S., M.F. Insana, J.J. Dahl, C. Hansen, M. Ashfaq, and H. Ermert, An ultrasound research interface for a clinical system, *IEEE Transactions on Ultrasonics, Ferroelectrics and Frequency Control*, 54(2007): 198–210.
17. Abbey, C.K., R.J. Zemp, J. Liu, K.K. Lindfors, and M.F. Insana, Observer efficiency in discrimination tasks simulating malignant and benign breast lesions with ultrasound, *IEEE Transactions on Medical Imaging*, 25(2006): 198–209.
18. Abbey, C.K., N.Q. Nguyen, and M.F. Insana, Optimal beamforming in ultrasound using the ideal observer, *IEEE Transactions on Ultrasonics, Ferroelectrics and Frequency Control*, 57(2010): 1782–1796.
19. Nguyen, N.Q., C.K. Abbey, M.F. Insana, An adaptive filter to approximate the Bayesian strategy for sonographic beamforming, *IEEE Transactions on Medical Imaging*, 30(2011): 28–37.
20. Zemp, R.J., C.K. Abbey, and M.F. Insana, Linear system model for ultrasonic imaging: Application to signal statistics, *IEEE Transactions on Ultrasonics, Ferroelectrics and Frequency Control*, 50(2003): 642–654.
21. American College of Radiology, *Breast Imaging Reporting and Data System Atlas*, Reston, VA, 2003.
22. Insana, M.F., and D.G. Brown, Acoustic scattering theory applied to soft biological tissues, In *Ultrasonic Scattering in Biological Tissues*, edited by K.K. Shung and G.A. Thiemes, Boca Raton, FL: CRC Press, 1993.
23. Jensen, J.A., and N.B. Svendsen, Calculation of pressure fields from arbitrarily shaped, apodized, and excited ultrasound transducers, *IEEE Transactions on Ultrasonics, Ferroelectrics and Frequency Control*, 39(1992): 262–267.
24. Jensen, J.A., Field: A program for simulating ultrasound systems, *Medical and Biological Engineering and Computing*, 34(1996): 351–353.
25. Golub, G.H., and C.F. Van Loan, *Matrix Computations*, 3rd ed. Baltimore, MD: The Johns Hopkins University Press, 1996.
26. Poor, H.V., *An Introduction to Signal Detection and Estimation, Second Edition*. New York: Springer-Verlag, 1994.
27. Myers, K.J., L.S. Kyprianou, and S. Park, Wagner's unified theory of image quality: Three decades later, In *Advances in Medical Physics*, edited by A.B. Wolbarst, A. Karellas, E.A. Krupinski, and W.R. Hendee, Madison, WI: Medical Physics Publishing, 2010.
28. Middleton, D., *An Introduction to Statistical Communication Theory*. Los Altos, CA: Peninsula Publishing, 1987.
29. O'Sullivan, J.A., R.E. Blahut, and D.L. Snyder, Information-theoretic image formation, *IEEE Transactions on Information Theory*, 44(1998): 2094–2122.
30. Shannon, C.E., A mathematical theory of communication, *Bell System Technical Journal*, 27(1948): 379423.

31. Kullback, S., *Information Theory and Statistics*. New York: Dover, 1997.
32. Nguyen, N.Q., C.K. Abbey, and M.F. Insana, Objective assessment of sonographic quality I: Task information, *IEEE Transactions on Medical Imaging*, 32(2013): 683–690.
33. Wagner, R.F., Low contrast sensitivity of radiologic, CT, nuclear medical, and ultrasound medical imaging systems, *IEEE Transactions on Medical Imaging*, MI-2(1983): 105–121.
34. Nguyen, N.Q., C.K. Abbey, and M.F. Insana, Objective assessment of sonographic quality II: Acquisition Information Spectrum, *IEEE Transactions on Medical Imaging*, 32(2013): 691–698.
35. Besag, B.J., On the statistical analysis of dirty pictures, *Journal of the Royal Statistical Society: Series B*, 48(1986): 259–302.
36. Geman, J.S., and D. Geman, Stochastic relaxation, Gibbs distribution and the Bayesian restoration of images, *IEEE Transactions on Pattern Analysis and Machine Intelligence*, 6(1984): 721–741.
37. Ashton, E.A., and K.J. Parker, Multiple resolution Bayesian segmentation of ultrasound images, *Ultrasonic Imaging*, 17(1995): 291–304.
38. Van Trees, H.L., *Detection, Estimation, and Modulation Theory, Part IV, Optimum Array Processing*. New York: Wiley, 2002.
39. Tanner, W.P., and T.G. Birdsall, Definition of d′ and η as psychophysical measures, In *Signal Detection and Recognition by Human Observer: Contemporary Readings*, edited by J.A. Swets, New York: Wiley, 1964.
40. Nguyen, N.Q., C.K. Abbey, R.D. Yapp, and M.F. Insana, Tomographic reconstruction of the pulse-echo spatiotemporal impulse response, In *Proceedings of Society of Photo-Optical Instrumentation Engineers (SPIE)*, 7629(2010): 76290F-1–76290F-11.
41. Kak, A.C., and M. Slaney, *Principles of Computerized Tomographic Imaging*. Philadelphia, PA: SIAM, 1999.

Chapter 7

8. Acoustic Radiation Force–Based Elasticity Imaging

Joshua R. Doherty, Mark L. Palmeri, Gregg E. Trahey, and Kathryn R. Nightingale

Chapter 8

Ultrasound Imaging and Therapy. Edited by Aaron Fenster and James C. Lacefield © 2015 CRC Press/Taylor & Francis Group, LLC. ISBN: 978-1-4398-6628-3

8.1   Introduction

Since ancient Egyptian medicine, manual palpation has been used to detect the size, location, and stiffness of superficial structures within the body [1]. By assessing the elasticity (i.e., stiffness) of structures compared with surrounding tissues, the information gleaned by palpation can help clinicians determine states of disease associated with various pathologic processes. For instance, the presence of a stiff mass within otherwise healthy breast tissue can be an indication of breast cancer [2–5]. Although indispensable for medical diagnosis, manual palpation methods do not allow clinicians to "see" changes deep within the tissue. Furthermore, because these pathologic processes are associated with changes in the mechanical properties of tissue, they can often go unnoticed by conventional imaging techniques, such as diagnostic ultrasound, which distinguishes features based on the acoustic properties of tissue. To that end, ultrasound elasticity imaging methods have been developed as noninvasive tools for probing the elasticity of tissue deep within the body.

Ultrasound elasticity imaging methods consist of both the excitation of soft tissue and the monitoring of the tissue's deformation response. In classifying the methods according to the source of excitation, there exist (1) external methods (e.g., elastography or strain imaging [6], sonoelasticity [7]) that apply forces externally to deform underlying tissue using a mechanical vibration device or by compressing the skin using a transducer, (2) internal physiologic methods that rely on sources such as breathing, cardiac motion, or arterial pulsation to naturally deform surrounding tissues [8–12], and also (3) internal remote (acoustic radiation force [ARF]) methods that use a remotely focused acoustic device to apply an excitation directly within the region of interest (e.g., acoustic radiation force impulse [ARFI] [13], shear wave elasticity imaging [SWEI] [14], and vibroacoustography [15]). Excellent reviews of these methods have been provided by Greenleaf et al. [16], Parker et al. [17], and Wells and Liang [1].

This chapter focuses on ARF-based elasticity imaging methods. As depicted in Figure 8.1, these methods consist of using a focused ultrasound transducer to generate sufficient ARF to cause local deformation that can be monitored with either the

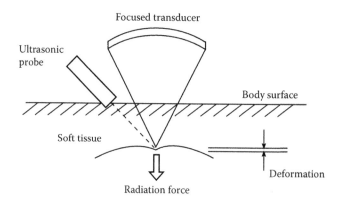

FIGURE 8.1   General concept of ARF-based elasticity imaging methods. Consists of (1) a focused ultrasound transducer to generate sufficient ARF within the focal geometry of the beam to cause localized tissue displacements and (2) measurement of the resulting deformation response using the same or a separate monitoring device.

same ultrasound transducer or a separate monitoring device. Unlike external excitation methods that can be challenged by insufficient coupling of the excitation within the region of interest, ARF methods, which use smaller stresses and are less affected by external boundaries, are advantageous because they apply the force directly to the desired structure. By monitoring the induced deformation, these methods are used to either form qualitative images that depict the relative stiffness of structures compared with surrounding tissues and/or to provide a quantitative estimate of tissue stiffness.

The purpose of this chapter is to familiarize readers of various backgrounds with the phenomenon of ARF and how it can be used to elucidate the mechanical properties of soft tissue. For graduate engineering students or those wishing to better understand the physical principles of ARF, Section 8.2 includes a review of solid mechanics needed to explain the various models that are used to describe tissue deformations along with a derivation of the generated ARF in soft tissue. The mathematical rigor in this section and also in Section 8.3, a description of common algorithms used to estimate the resulting displacements, may be too detailed for some readers. Therefore, clinicians more interested in various ARF-based elasticity imaging methods that have been developed may wish to start with Section 8.4, an overview of several methods including relevant clinical applications and results, referring back to previous sections as needed. For all audiences, important safety considerations for using ARF methods *in vivo* are addressed in Section 8.5.

8.2 ARF-Induced Tissue Deformation

Soft tissues are viscoelastic materials that exhibit properties of both elastic solids and viscous fluids. Soft tissues are often modeled as linear elastic solids at low excitation frequencies (<1 kHz), and they are generally modeled as viscous fluids at ultrasonic frequencies, where soft tissues do not support shear stresses [18,19]. Both types of models are used in the following section to accurately predict both the deformation response, including the propagation of shear waves, and the generation of ARF in soft tissues.

8.2.1 Soft Tissue Mechanics

In response to an applied force, an internal restoring stress ($\vec{\sigma}$) develops within a material that is associated with a resulting deformation described by the strain ($\vec{\varepsilon}$). The strain, which compares the resulting deformation to the initial state of the material, can be expressed in terms of the displacement ($\vec{u}$), where $\nabla \vec{u}$ is a second-order tensor representing the gradient of the displacement vector field as

$$\vec{\varepsilon} = \frac{1}{2}[\nabla \vec{u} + \nabla \vec{u}^T + \nabla \vec{u}^T \nabla \vec{u}]. \tag{8.1}$$

According to Equation 8.1, the strain tensor contains both linear and nonlinear parts. For infinitesimal (small) strains, typically encountered in ARF-based elasticity imaging methods, the nonlinear terms can be neglected, leading to the linearized strain equation

Chapter 8

$$\vec{\varepsilon} \approx \frac{1}{2}[\nabla \vec{u} + \nabla \vec{u}^T]. \tag{8.2}$$

To characterize the mechanical response of soft tissue, a constitutive equation that mathematically relates stress and strain is needed. In general, soft tissues are viscoelastic, inhomogeneous, and anisotropic [20]. Along with the assumption of linearity, the analysis can be further simplified by assuming that soft tissue is purely elastic (i.e., neglecting any viscous contributions), meaning that the deformations do not depend on the rate of applied stress and that the tissue returns back to a nondeformed state. Under these assumptions, a linear (Hookean) elastic model is a good first-order approximation. This constitutive equation can be written in both vector and component form as follows [21]:

$$\vec{\sigma} = \vec{C}\vec{\varepsilon}, \tag{8.3}$$

$$\sigma_{ij} = C_{ijkl}\varepsilon_{kl}. \tag{8.4}$$

Here $\vec{C}$ is the fourth-order Cauchy elasticity (stiffness) tensor, which depends on 81 independent material coefficients (elastic moduli) that describe the stiffness of the material or its ability to resist deformation. If it is also assumed that soft tissue is homogeneous (the mechanical properties of all particles are identical) and isotropic (the material properties are orientation independent), the constitutive model reduces to Equation 8.5, where δ_{ij} is the Kronecker delta function. Solving for strain in terms of stress, this model can be inverted leading to Equation 8.6. These constitutive equations are valid for an isotropic linearly elastic solid, which depend on only two material coefficients known as the Lamé constants (λ and μ):

$$\sigma_{ij} = \lambda \varepsilon_{kk} \delta_{ij} + 2\mu \varepsilon_{ij}, \tag{8.5}$$

$$\varepsilon_{ij} = \frac{1}{2\mu}\left[\sigma_{ij} - \frac{\lambda}{3\lambda + 2\mu}\sigma_{kk}\delta_{ij}\right]. \tag{8.6}$$

Various elastic moduli describe the material's resistance to specific types of deformation. The Young's modulus (E) describes the material's resistance to deformation in uniaxial compression/tension and the shear modulus (μ), one of the Lamé constants, describes the resistance to shear. A material constant called the Poisson ratio (ν) also exists, which describes the ratio of the deformation occurring orthogonal to the deformation occurring in the direction of an applied load. These material constants and elastic moduli are related according to the following equations [21]:

$$E = \frac{\mu(3\lambda + 2\mu)}{\lambda + \mu}, \tag{8.7}$$

$$\nu = \frac{\lambda}{2(\lambda + \mu)}, \tag{8.8}$$

$$\mu = \frac{E}{2(1+\nu)}. \tag{8.9}$$

Considering soft tissues to be incompressible (i.e., $\nu = 0.5$ under these assumptions), the shear modulus can be related to the Young's modulus according to $\mu = E/3$ (Equation 8.9). The stiffness of biological tissues, usually reported in terms of Young's modulus, varies significantly in the literature. Typical values for healthy soft tissue constituents are about 10 kPa for parenchyma, 20 kPa for muscle, and 50 kPa for connective tissue, as reported by Wells and Liang [1]. The differences in stiffness between healthy and diseased tissue can be quite significant, providing a means of contrast for elasticity imaging methods. For instance, Krouskop et al. [22] found that normal fatty breast tissue ($E \approx 20$ kPa) was approximately five times softer than breast tissue with carcinoma ($E \approx 100$ kPa) from mechanical measurements recorded at a static loading of 5% for loading frequencies of 0.1 and 4 Hz.

Two common deviations from the simple linear elastic model used to derive the above relations include modeling the tissue as being a viscoelastic and/or nonlinear solid [20]. In a viscoelastic material, the tissue stiffness is dependent on the frequency of excitation (i.e., $E(f)$ and $\mu(f)$), leading to dispersion, an effect where the propagating wave speed is a function of the excitation frequency. The tissue can also be modeled as being nonlinear, such that the effective stiffness ($E(\varepsilon)$ and $\mu(\varepsilon)$) is dependent on the magnitude of the applied strain.

8.2.2   ARF Derivation

The generation of ARF is a phenomenon associated with a transfer of momentum (via absorption and scattering) from an ultrasonic wave to the propagating medium. Therefore, an equation that can account for ARF generation in soft tissues must include viscous effects describing the appreciable loss of energy from the propagating acoustic wave. At ultrasonic frequencies, where soft tissues do not support shear stresses, a linear viscous fluid model is appropriate. To that end, following the work of Nyborg [23] and Eckart [24], the approach herein considers the case of a linear viscous fluid to derive a first-order approximation of the magnitude and direction of ARF. Beginning with Newton's Second Law ($\vec{F} = m\vec{a}$) and representing the particle acceleration ($\vec{a}$) in a fluid as the sum of the local particle acceleration and the convective acceleration by [25]

$$\vec{a} = \frac{\partial \vec{v}}{\partial t} + \vec{v} \cdot \nabla \vec{v}, \tag{8.10}$$

where $\vec{v}$ is particle velocity and t is time, the force per unit volume ($\vec{f}$) can be described as

$$\vec{f} = \rho \left(\frac{\partial \vec{v}}{\partial t} + \vec{v} \cdot \nabla \vec{v} \right), \tag{8.11}$$

$$\vec{f} = \frac{\partial(\rho \vec{v})}{\partial t} - \vec{v}\frac{\partial \rho}{\partial t} + \rho(\vec{v} \cdot \nabla)\vec{v} \tag{8.12}$$

where ρ represents the density. Note the identity $\vec{v} \cdot \nabla \vec{v} = (\vec{v} \cdot \nabla)\vec{v}$ along with the product rule was used to simplify the equation. From the continuity equation, which describes the conservation of mass and written as [25]

$$\frac{\partial \rho}{\partial t} + \nabla \cdot \rho \vec{v} = 0, \tag{8.13}$$

the force can be expressed as

$$\vec{f} = \frac{\partial(\rho \vec{v})}{\partial t} + \rho(\vec{v} \cdot \nabla)\vec{v} + \vec{v}\nabla \cdot \rho \vec{v}. \tag{8.14}$$

To find an approximate solution for the magnitude and direction of ARF, a perturbation method is used [26]. In this approach, the general equation is nondimensionalized by creating a perturbation parameter whose order is related to the relative effect of the associated terms on the overall solution. For the approximation that follows, the constants V (m/s), T (s), and Z (m) are used along with a dimensionless perturbation parameter N defined such that

$$N = \frac{VT}{Z}. \tag{8.15}$$

The general force equation is nondimensionalized by substituting $\vec{v} = \bar{v}V$, $t = \tau T$, and $\vec{z} = \vec{\gamma}Z$ for velocity, time, and spatial components, respectively, leading to

$$\vec{f} = \frac{\partial(\rho \bar{v}V)}{\partial(\tau T)} + \rho\left(\bar{v}V \cdot \nabla\left(\frac{1}{Z}\right)\right)\bar{v}V + \bar{v}V\nabla\left(\frac{1}{Z}\right) \cdot \rho \bar{v}V, \tag{8.16}$$

$$\vec{f} = \frac{\partial(\rho \vec{v})}{\partial \tau} + N(\rho(\vec{v} \cdot \nabla)\vec{v} + \vec{v}\nabla \cdot \rho \vec{v}). \tag{8.17}$$

The velocity and density terms are expanded as a power series, with terms whose effect decreases with increasing index values as

$$\vec{v} = \vec{v}_0 + N\vec{v}_1 + N^2\vec{v}_2 + \ldots, \tag{8.18}$$

$$\rho = \rho_0 + N\rho_1 + N^2\rho_2 + \ldots \tag{8.19}$$

Substitution of the zeroth-order (N^0) and the first-order (N^1) terms gives the following up to a first-order approximation:

$$\vec{f} = \frac{\partial(\rho_0 \vec{v}_0)}{\partial \tau} + N\left(\frac{\partial(\rho_0 \vec{v}_1 + \rho_1 \vec{v}_0)}{\partial \tau}\right) + N(\vec{v}_0 \nabla \cdot \rho_0 \vec{v}_0 + \rho_0 \vec{v}_0 \cdot \nabla \vec{v}_0). \tag{8.20}$$

A similar analysis can be performed for the continuity equation (Equation 8.13), leading to

$$\frac{\partial \rho_0}{\partial \tau} + N\left(\frac{\partial \rho_1}{\partial \tau} + \nabla \cdot \rho_0 \vec{v}_0\right) = 0, \tag{8.21}$$

where the zeroth-order equation gives a statement of steady-state density $\left(\frac{\partial \rho_0}{\partial \tau} = 0\right)$. Substitution into the force equation and replacing the nondimensionalized parameters with the original parameters leads to

$$\vec{f} = \rho_0 \frac{\partial \vec{v}_0}{\partial t} + N\left(\rho_0 \frac{\partial \vec{v}_1}{\partial t} + \frac{\partial(\rho_1 \vec{v}_0)}{\partial t} + \rho_0(\vec{v}_0 \nabla \cdot \vec{v}_0 + \vec{v}_0 \cdot \nabla \vec{v}_0)\right). \tag{8.22}$$

The zeroth-order equation (Equation 8.23) is a statement of linear momentum. The first-order equation (Equation 8.24) provides an approximate solution to the viscous loss components that are associated with the generation of ARF in terms of density and velocity:

$$\vec{f} = \rho_0 \frac{\partial \vec{v}_0}{\partial t}, \tag{8.23}$$

$$\vec{f} = \rho_0 \frac{\partial \vec{v}_1}{\partial t} + \frac{\partial(\rho_1 \vec{v}_0)}{\partial t} + \rho_0(\vec{v}_0 \nabla \cdot \vec{v}_0 + \vec{v}_0 \cdot \nabla \vec{v}_0). \tag{8.24}$$

The ARF ($\vec{f}_{rad}$) is the time average (indicated by $\langle\rangle$) body force over a suitable number of cycles, which gives

$$\vec{f}_{rad} = \langle\vec{f}\rangle = \frac{1}{T} \int_{-T/2}^{T/2} \vec{f} \, dt, \tag{8.25}$$

$$\vec{f}_{rad} = \langle\rho_0(\vec{v}_0 \nabla \cdot \vec{v}_0 + \vec{v}_0 \cdot \nabla \vec{v}_0)\rangle. \tag{8.26}$$

For the case of an exponentially attenuating plane acoustic wave traveling in the positive z-direction ($\hat{z}$) in an unbounded medium, the velocity can be described as [23]

$$\vec{v}_0 = V_0 e^{-\alpha z} \cos(2\pi f t - kz)\hat{z}, \tag{8.27}$$

where V_0 is the initial velocity magnitude, f is the frequency, α is the absorption coefficient, z is the propagation distance, and k is the wave number. Substitution into Equation 8.26 leads to the following equations:

$$\vec{f}_{rad} = 2\rho_0 \langle\vec{v}_0 \nabla \vec{v}_0\rangle, \tag{8.28}$$

Chapter 8

$$\vec{f}_{rad} = 2\rho_0 V_0^2 e^{-2\alpha z} \left\langle \alpha \cos^2(2\pi ft - kz) - \frac{1}{2}\sin(2\pi ft - kz) \right\rangle \hat{z}, \tag{8.29}$$

$$\vec{f}_{rad} = V_0^2 e^{-2\alpha z} \rho_0 \alpha \hat{z}. \tag{8.30}$$

Expressing this force in terms of the time-average intensity ($\langle I \rangle$), described as

$$\langle I \rangle = \frac{V_0^2 e^{-2\alpha z}}{2} \rho_0 c_0, \tag{8.31}$$

where c_0 is the speed of sound in the medium, the ARF reduces to

$$\vec{f}_{rad} = \frac{2\alpha \langle I \rangle}{c_0} \hat{z}. \tag{8.32}$$

This ARF manifests itself as a body force acting in the direction of wave propagation (i.e., in the direction of the Poynting vector) that applies a low frequency excitation (i.e., the radiation force excitation arises over the time-average of the acoustic pulse, as compared with the ultrasonic frequency of the acoustic pulse) to the interrogated tissue.

In the approach of Nyborg [23] and Eckart [24], who described ARF in terms of a Navier–Stokes equation for a linearly viscous fluid to approximate the fluid streaming velocity, the contribution of scattering in the momentum transfer was neglected, such that all attenuation losses were attributed to absorption alone. This assumption seems reasonably valid in soft tissue where absorption is the dominant component of loss in acoustic wave energy [27]. In practice, the absorption coefficient is typically replaced with an attenuation coefficient, which accounts for losses due to both scattering and absorption. Other derivations, including that of Westervelt [28], account for scattering effects, which are dependent on the angular scattering properties of the target, to describe the phenomena of ARF.

For conventional ultrasound imaging, the magnitude of generated ARF is relatively small such that tissue deformation is negligible (<1 μm). Therefore, ARF methods typically use longer and/or higher power acoustic pulses to achieve the increased intensities needed to generate appreciable ARF, for which the deformation can be measured with ultrasonic techniques. For *in vivo* applications, typical peak acoustic radiation force magnitudes are in the order of dynes, creating tissue deformations typically in the range of 1 to 10 μm.

8.2.3　Shear Wave Generation and Propagation

Recall that the ARF arises over the temporal average of the acoustic pulse, which allows the tissue to respond in an elastic way that can be described by a viscoelastic solid model. Therefore, when ARF is applied, shear stresses develop within the tissue, which generate low frequency shear (transverse) waves that propagate away from the edges of the radiation force field (or region of excitation [ROE]) in a direction orthogonal to the direction of ultrasonic wave propagation. The governing equation for shear wave propagation can be derived using Newton's 2nd Law ($\vec{F} = m\vec{a}$). Written in component form with the acceleration described in terms of displacement ($\vec{u}$), this simplifies to

$$f_i = \frac{\partial(\sigma_{ij})}{\partial x_j} = \rho a_i = \rho \frac{\partial^2 u_i}{\partial t^2}. \tag{8.33}$$

Using the constitutive equation for an isotropic linearly elastic material (Equation 8.5) and making use of the linear strain equation (Equation 8.2), this simplifies in component and vector form to the following equations:

$$(\lambda + \mu)\left\{\frac{\partial u_i}{\partial x_j \partial x_i}\right\} + \mu\left\{\frac{\partial u_i}{\partial x_j \partial x_j}\right\} = \rho \frac{\partial^2 u_i}{\partial t^2}, \tag{8.34}$$

$$(\lambda + \mu)\{\nabla(\nabla \cdot \vec{u})\} + \mu\{\nabla^2 \vec{u}\} = \rho \frac{\partial^2 \vec{u}}{\partial t^2}. \tag{8.35}$$

Using the vector identity that $\nabla \times \nabla \times \vec{u} = \nabla(\nabla \cdot \vec{u}) - \nabla^2 \vec{u}$, this becomes

$$(\lambda + \mu)\{\nabla(\nabla \cdot \vec{u})\} + \mu\{\nabla(\nabla \cdot \vec{u}) - \nabla \times \nabla \times \vec{u}\} = \rho \frac{\partial^2 \vec{u}}{\partial t^2}, \tag{8.36}$$

$$(\lambda + 2\mu)\{\nabla(\nabla \cdot \vec{u})\} - \mu\{\nabla \times \nabla \times \vec{u}\} = \rho \frac{\partial^2 \vec{u}}{\partial t^2}. \tag{8.37}$$

Using the Helmholtz theorem, which states that any vector field can be decomposed into the superposition of a curl-free ($\vec{\psi}$) vector field and a divergence-free ($\vec{\vartheta}$) vector field, the displacement vector field ($\vec{u}$) can be resolved as [29]

$$\vec{u} = \nabla \vec{\psi} + \nabla \times \vec{\vartheta}. \tag{8.38}$$

Because $\nabla \times (\nabla \vec{\psi}) = 0$, taking the curl of Equation 8.37 will remove the curl-free component of the displacement, thereby isolating the divergence-free component of the displacement vector field that represents the equivoluminal components, which occur in the transverse direction of wave propagation (i.e., shear components), leading to the following equations:

$$-\mu\{\nabla \times \nabla \times (\nabla \times \vec{u})\} = \rho \frac{\partial^2 (\nabla \times \vec{u})}{\partial t^2}, \tag{8.39}$$

$$\mu\{\nabla^2 (\nabla \times \vec{u})\} - \rho \frac{\partial^2 (\nabla \times \vec{u})}{\partial t^2} = 0. \tag{8.40}$$

The use of the identity $\nabla \times \vec{u} = \vec{0}$ leads to the wave equation governing shear wave propagation, often referred to as the Helmholtz equation, described using the Laplacian operator (∇^2) as [29] follows:

$$\mu \nabla^2 \vec{\vartheta} - \rho \frac{\partial^2 \vec{\vartheta}}{\partial t^2} = 0. \tag{8.41}$$

The speed of these propagating shear waves is related to the shear modulus by

$$c_T = \sqrt{\frac{\mu}{\rho}}. \tag{8.42}$$

In a similar manner, the dilatational component of the displacement vector field, which occurs in the longitudinal direction of wave propagation, could have been determined by taking the divergence of Equation 8.37 and by using the identity $\nabla \cdot (\nabla \times \vec{\vartheta}) = 0$. Solving for the speed of these compressional (pressure) waves leads to

$$c_L = \sqrt{\frac{(\lambda + 2\mu)}{\rho}}. \tag{8.43}$$

Note that in nearly incompressible soft tissues, because λ is typically several orders of magnitude larger than μ, shear waves (transverse direction) travel slower, typically in the range of 1 to 10 m/s in soft tissue [30], as compared with compressional waves (longitudinal direction), which typically travel at 1540 m/s. As a result, shear wave propagation can be monitored with ultrasonic imaging methods in soft tissue and used to derive elasticity information.

8.2.4 Force Field Geometry

The ARF field is spatially distributed throughout the geometric shadow of the aperture and is dependent on tissue properties along with characteristics of the transmitted excitation beam [19]. Recall Equation 8.32, which states that the applied ARF is a function of the attenuation and speed of sound of the tissue, along with the intensity of the applied excitation. The frequency of the transmitted pulse influences both the effects of tissue attenuation and also the focal geometry.

The dependence of attenuation on frequency has a significant impact on the ARF field geometry. Increased attenuation, which increases with frequency, leads to an increase in near field losses (generating radiation force) and a decrease in intensity at the focus. Therefore, higher frequencies are generally associated with a more evenly distributed ARF field. By contrast, lower frequencies are associated with fewer near field losses and more significant force generation at the focus. Therefore, as with diagnostic ultrasound, the optimal frequency used in ARF elasticity imaging methods involves a trade-off between attenuation losses in the near field and focal point gain [19].

The focal configuration of an acoustic beam dictates its beam width and depth of field, which determine the spatial distribution of the applied ARF. Pulse characteristics of the transmitted wave can be tailored for specific applications of ARF excitation. For example, a narrower force field geometry is associated with an increased transmit frequency and/or a larger aperture size, which is typically dictated by the transmit $F/\#$ defined as follows:

$$F/\# = \frac{z}{d},\tag{8.44}$$

where z is the focal length and d is the aperture size. Sources of error involved with actually monitoring the deformation response, discussed in Section 8.3, can also play an important role in the optimal excitation geometry and can thereby influence designed transmit pulse characteristics. For clinical applications, these design parameters are also dictated by safety considerations and exposure limits described in Section 8.5.

To gain a better understanding of the temporal behavior of deformed tissue in response to an impulse-like application of ARF, Figure 8.2 depicts the simulated axial displacements in a 3-D homogeneous isotropic elastic material at three separate time steps after excitation. Initially, the ARF is localized within the focal zone, resulting in a peak displacement response within the ROE as depicted in Figure 8.2a. After the initial excitation, shear waves created at the boundary of the ROE propagate orthogonally to the direction of acoustic wave propagation (Figure 8.2b and c). Figure 8.2d is a plot of the axial displacement versus time at the three different lateral locations located both within (pink X) and outside (red circle and green square) the ROE. Note that the peak axial displacements diminish with increasing distance from the ROE due to geometric spreading.

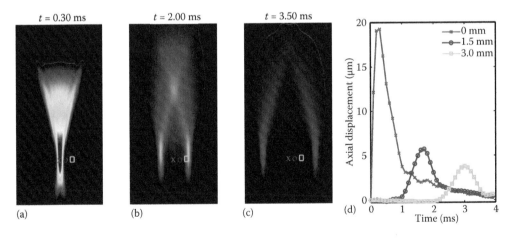

FIGURE 8.2 A finite element simulated deformation response in a 3-D homogeneous isotropic linear elastic solid with a Young's modulus of 4 kPa from the application of an impulsive ARF excitation. The axial displacements depicted at three different time steps after excitation in (a) through (c) show the propagation of shear waves away from the ROE. The displacement through time profiles depicted in (d) shows the axial displacements occurring at the focal depth for each of three separate lateral locations, located both on-axis (pink X) and off-axis (red circle and green square). These profiles reflect the decreased displacement amplitude with increased distance from the ROE due to geometric spreading.

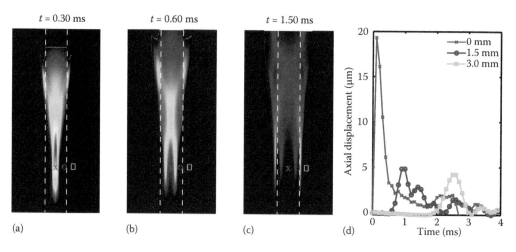

FIGURE 8.3 A finite element simulated deformation response from the application of an impulsive ARF excitation, in a 3-D linear elastic solid consisting of a soft material with a Young's modulus of 5 kPa centered between two stiffer materials with a Young's modulus of 20 kPa. The material boundaries are indicated by the white dashed lines in (a) through (c), which depict the axial displacements at 3 distinct times after excitation. In (b), the shear waves have not reached the layer boundaries and only the initial shear waves propagating from the ROE are observed. In (c), just after the shear waves reached the center material boundary, the reflected and transmitted waves are depicted. The displacement through time profiles in (d) shows the axial displacements occurring at the focal depth for each of three separate lateral locations, located both on-axis (pink X) and off-axis (red circle and green square). The multiple distinct peaks in these profiles are indicative of both the incident shear waves along with the reflected shear waves introduced by the material boundaries.

The presence of material boundaries, where acoustic impedance mismatches lead to wave reflection and transmission, can significantly complicate the dynamic response of soft tissue [31]. To demonstrate these boundary effects, Figure 8.3 shows the dynamic response in a 3-D linear elastic material consisting of a soft material centered between two stiffer materials. As shown in Figure 8.3b and c, the material boundaries (indicated by white dashed lines) lead to a much more complicated response compared with the homogeneous case depicted in Figure 8.2 due to wave reflections occurring at the boundaries. Figure 8.3d demonstrates this effect where the displacement through time profiles at different lateral locations consists of multiple distinct peaks indicative of not only the initial shear wave propagating away from the ROE but also the reflected shear waves that travel back toward the ROE.

8.3 Monitoring Tissue Deformation Response

To form meaningful measurements of stiffness, it is essential to accurately monitor the ARF-induced tissue deformation. Typically, this is performed using conventional pulse-echo ultrasound that monitors the tissue both temporally and spatially. Considering only the axial dimension, the deformation manifests itself in a relative shift and/or compression of a raw radio frequency (RF) signal collected after the excitation (tracking pulse) compared with an RF signal collected before the excitation (reference pulse). To form displacement estimates from collected data, the most common ultrasound signal-processing algorithms used in ARF-based elasticity imaging methods include cross-correlation and phase shift methods. The performance of these methods is typically quantified by bias (δ), which is the mean of the displacement error, and jitter (σ), which is the standard deviation of this error, defined as follows:

$$\delta = E[\acute{u} - u_0], \tag{8.45}$$

$$\sigma = E[(\acute{u} - u_0)^2] - \delta^2, \tag{8.46}$$

where u_0 is the true displacement, $\acute{u}$ is the estimated displacement, and $E[]$ denotes the expectation operator [32].

8.3.1 Cross-Correlation Methods

In ultrasound, time-delay estimation techniques are most commonly performed using methods based on cross-correlation. Using the cross-correlation method first proposed in blood velocity estimation [33], a window length of data is selected from the reference and delayed signals, and the cross-correlation value is computed for each time lag applied to the windowed region of the delayed signal. The maximum cross-correlation value indicates the time lag at which the two signals are most correlated (similar). This process is repeated through the entire axial extent of the data to find an estimated displacement through depth. The normalized cross-correlation method [34] functions on the same basic principal, but each signal is normalized by its standard deviation to reduce the impact of bright scatterers. In doing so, normalized cross-correlation has better performance and is typically considered the gold standard method for displacement estimation. The normalized cross-correlation function (c_n) for a reference signal (s_r) and a delay signal (s_d) as a function of the applied time lag (τ) over a given window length (T) can be written as follows [35]:

$$c_n(\tau) = \frac{\displaystyle\int_{-T/2}^{T/2} \left(s_r(t) s_d(t+\tau)\right) dt}{\sqrt{\displaystyle\int_{-T/2}^{T/2} \left(s_r(t)\right)^2 dt * \int_{-T/2}^{T/2} \left(s_d(t)\right)^2 dt}}. \tag{8.47}$$

Derived by Walker et al. [36], the theoretical limit of the jitter error of tracking algorithms based on cross-correlation for partially correlated speckle signals can be derived from the Cramer–Rao lower bound, which describes the minimum jitter of an unbiased ($\delta = 0$) estimator as follows:

$$\sigma = \sqrt{\frac{3}{2 f_c^3 \pi^2 T (B^3 + 12B)} \left(\frac{1}{\rho^2}\left(1 + \frac{1}{SNR^2}\right)^2 - 1\right)}, \tag{8.48}$$

where SNR is the signal-to-noise ratio, f_c is the transducer center frequency, T is the correlation window length, B is the fractional bandwidth, and ρ is the correlation value between the signals.

Chapter 8

8.3.2 Phase Shift Methods

As ultrasonic systems adopted the use of parallel-receive beamforming to achieve increased frame rates, memory requirements made it difficult to store the increased amount of RF channel data. For this reason, quadrature demodulation routines were implemented to save data in the less memory-intensive demodulated in-phase and quadrature (IQ) format. In 1985, Kasai et al. [37] made real-time 2-D color flow imaging possible with the development of the 1-D autocorrelation phase shift method for use with IQ data. As opposed to the methods based on cross-correlation that operate in the time domain, phase shift methods operate in the frequency domain using the autocorrelation function to search for the phase shift between the reference and the delayed signals. For detailed derivations of this method and similar Doppler techniques, the reader is referred to the vast library that exists on this topic (e.g., Jensen [38]). To calculate an estimate of the average axial displacement ($\bar{u}$) over a given axial range (M) and ensemble length (N) for a specific demodulation frequency (f_{dem}), Kasai's 1-D autocorrelator can be written as follows [39]:

$$\bar{u} = \frac{c}{4\pi} \frac{\tan^{-1}\left\{\dfrac{\sum_{m=0}^{M-1}\sum_{n=0}^{N-2}[Q(m,n)I(m,n+1)-I(m,n)Q(m,n+1)]}{\sum_{m=0}^{M-1}\sum_{n=0}^{N-2}[I(m,n)I(m,n+1)+Q(m,n)Q(m,n+1)]}\right\}}{f_{dem}}. \tag{8.49}$$

In general, the average axial displacement estimate can be calculated from a computed average phase shift ($\Delta\theta$) and known fundamental carrier frequency (f_0) of the RF signal using the following relation:

$$\bar{u} = \frac{c}{4\pi}\frac{\Delta\theta}{f_0}, \tag{8.50}$$

from which the use of the autocorrelation function to estimate the phase shift becomes apparent. Note that in Kasai's 1-D autocorrelation algorithm, the demodulation frequency is assumed to be the true carrier frequency of the RF signal. Often this demodulation frequency is depth dependent and can therefore account for attenuation effects; however, local frequency variations within the baseband of the transducer are neglected. For this reason, Kasai's algorithm is suited best for calculating an average displacement over a large axial range, in which local frequency variations average to zero. Over a small axial range, Kasai's algorithm suffers from increased jitter because the true carrier frequency of the RF signal may be different from the frequency used to demodulate the signal. For improved performance, Loupas et al. [39] developed the 2-D autocorrelation method, written as follows:

$$\bar{u} = \frac{c}{4\pi}\frac{\tan^{-1}\left\{\dfrac{\sum_{m=0}^{M-1}\sum_{n=0}^{N-2}[Q(m,n)I(m,n+1)-I(m,n)Q(m,n+1)]}{\sum_{m=0}^{M-1}\sum_{n=0}^{N-2}[I(m,n)I(m,n+1)+Q(m,n)Q(m,n+1)]}\right\}}{\left(f_{dem}+\dfrac{1}{2\pi t_s}\tan^{-1}\left\{\dfrac{\sum_{m=0}^{M-2}\sum_{n=0}^{N-1}[Q(m,n)I(m+1,n)-I(m,n)Q(m+1,n+1)]}{\sum_{m=0}^{M-2}\sum_{n=0}^{N-1}[I(m,n)I(m+1,n)+Q(m,n)Q(m+1,n)]}\right\}\right)} \tag{8.51}$$

Compared with Kasai's 1-D autocorrelator, the 2-D autocorrelator developed by Loupas includes extra terms in the denominator to calculate a shift in the carrier frequency for each displacement estimate to account for these local variations. In comparing the performance of these methods for the small displacements associated with ARF-induced deformation, Pinton et al. [40] found that although the computational time is faster with Kasai's algorithm, Loupas's algorithm is significantly faster than the methods based on cross-correlation and in most cases can perform reasonably well compared with normalized cross-correlation.

8.3.3 Sources of Bias and Jitter

On the basis of the Cramer–Rao lower bound (Equation 8.48), finite window length, poor SNR, and signal decorrelation can lead to shifts in the correlation function of two signals and result in displacement estimation errors [36]. Jitter errors, typically in the range of 0.2 to 0.4 μm, represent a slight shift in the displacement of the true peak in the correlation function that, unlike large false peak errors that result when a secondary peak has a higher correlation value than the true peak, cannot be easily filtered, placing a fundamental limit on the precision of displacement estimates [36]. These jitter errors result in noisy images and inaccurate moduli estimates. For elasticity imaging methods, decorrelation between reference and tracking signals resulting from the excitation source can degrade displacement estimates. Compared with external excitation methods, where large strains can lead to significant signal decorrelation [41], ARF methods are less corrupted by this noise mechanism because the induced displacements are so small [40]. However, ARF-based elasticity imaging methods can be compromised by scatterer shearing beneath the point spread function. Dependent on the tissue properties and also the focal point geometry of the excitation beam, this effect can introduce a displacement bias, in which there is an underestimation of the true displacement magnitude [42,43].

8.4 ARF-Based Elasticity Imaging Methods

Since first proposed in 1990 by Sugimoto et al. [44], a variety of ARF-based elasticity imaging methods have been developed. As depicted in Figure 8.4, these methods can be classified according to the temporal duration of the excitation (pushing) pulses, which can be applied (1) (quasi)statically to investigate the steady-state response of fluid and tissue, (2) transiently to measure the response of tissue to an impulse-like excitation, or (3) harmonically to excite tissues at specific frequencies.

In some cases, it helps to further divide these methods based on the spatial location of tracking beams used to monitor the tissue deformation. Tracking beams positioned within the ROE along the axis of applied excitation (on-axis) can be used to form qualitative images of stiffness, whereas quantitative data based on shear wave propagation speeds can be derived from tracking beams positioned outside the ROE (off-axis). To better understand the difference between these approaches, recall Equation 8.3, which states that for a given stress, the induced strain (deformation) is inversely related to the tissue stiffness. In the *in vivo* setting, where the magnitude of the applied ARF cannot be quantified because of unknown acoustic attenuation, measurements of on-axis

Chapter 8

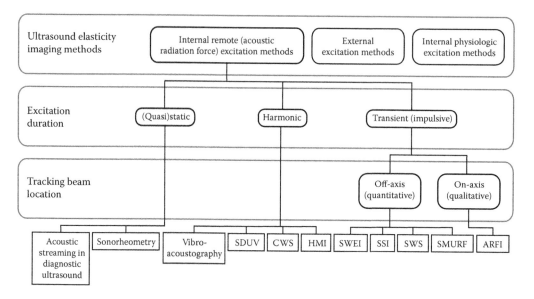

FIGURE 8.4 Summary of ARF-based elasticity imaging methods: acoustic streaming in diagnostic ultrasound, sonorheometry, vibroacoustography, shear wave dispersion ultrasound vibrometry (SDUV), crawling wave sonoelastography (CWS), harmonic motion imaging (HMI), shear wave elasticity imaging (SWEI), supersonic shear imaging (SSI), shear wave spectroscopy (SWS), spatially modulated radiation force (SMURF), and acoustic radiation force impulse (ARFI). As shown here, the methods have been classified according to the duration of the applied excitation and, in some cases, the location of tracking beams used to measure the deformation response.

displacements within the ROE cannot be used to provide a quantitative estimate of tissue stiffness but can only portray relative differences in tissue stiffness. On the other hand, by monitoring off-axis displacements at multiple lateral locations created by transversely propagating shear waves, as discussed in Section 8.2.3, an estimate of the shear wave velocity can be made. This shear wave velocity is independent of the applied force and can be related to the shear modulus via different material models, such as the linear elastic material model [45] or the more complex viscoelastic Voigt model [46], to form a quantitative estimate of tissue stiffness.

A schematic representation of many of the various ARF-based elasticity imaging methods discussed herein is provided in Figure 8.5. This figure not only depicts the different excitation sources used but also describes the relationship between the pushing beam and the tracking beam locations for each method.

8.4.1 (Quasi)Static Methods

The use of ARF to mechanically excite tissues in a (quasi)static fashion is generally accomplished by applying high-intensity acoustic pulses at a high pulse repetition frequency (PRF) for a duration that is long enough to approach a steady-state response. As shown in Figure 8.5a, these methods use tracking beams aligned with the pushing beam axis. Separate transducers can be used, or the pushing and tracking beams can be implemented in an interspersed manner using a single transducer. Because of safety considerations, described in Section 8.5, (quasi)static methods have generally

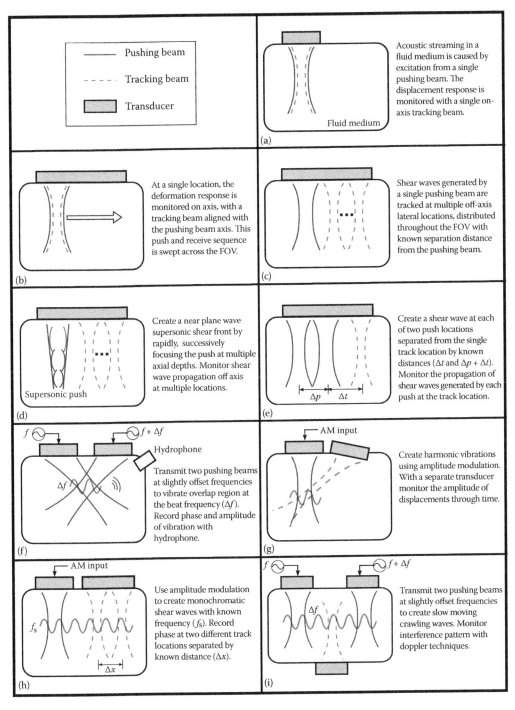

FIGURE 8.5 Overview of specific ARF methods. (a) Acoustic streaming and sonorheometry. (b) Acoustic radiation force impulse (ARFI). (c) Shear wave elasticity imaging (SWEI). (d) Supersonic shear imaging (SSI). (e) Spatially modulated radiation force (SMURF). (f) Vibroacoustography. (g) (AM) Harmonic motion imaging (HMI). (h) Shearwave dispersion ultrasound vibrometry (SDUV). (i) Crawling wave sonography (CWS).

been applied to fluids because fluids require smaller forces than solid tissues to achieve a steady-state ultrasonically detectable response.

8.4.1.1   Acoustic Streaming in Diagnostic Ultrasound

Acoustic streaming is the fluid flow that is induced in a viscous fluid arising from ARF, as discussed in Section 8.2.2. In the context of diagnostic ultrasonic imaging, acoustic streaming can be generated with diagnostic ultrasonic pulses, and the velocity of the induced fluid flow can be quantified using conventional Doppler techniques [47]. Clinically, this method was first applied by Nightingale et al. [48], who interspersed pulsed Doppler beams with the pushing beams to differentiate fluid-filled lesions (cysts) from solid lesions in the breast.

8.4.1.2   Sonorheometry

Using a model-based approach, the acoustic streaming response of a fluid can be used to estimate its underlying material properties. For use in identifying the onset of events associated with inappropriate blood coagulation during surgery, Viola et al. [49] developed the technique of sonorheometry. This technique uses a (quasi)static ARF excitation to generate a step in stress in the fluid and fits the resulting fluid response to a Voigt model to estimate the viscoelastic properties of blood during coagulation.

8.4.2   Transient (Impulsive) Methods

To induce a transient response, ARF is applied for a relatively short duration (tens to hundreds of microseconds) in an impulse-like fashion by exciting the soft tissue with a pulse that is several hundred cycles in length. The dynamic deformation response is monitored spatially and temporally using conventional diagnostic pulses. These pulses can be located within the ROE for on-axis tracking methods to depict the relative stiffness of tissues or outside the ROE for off-axis tracking methods to monitor shear wave propagation and derive a quantitative estimate of the tissue stiffness.

8.4.2.1   On-Axis Tracking (Qualitative) Methods

8.4.2.1.1 Acoustic Radiation Force Impulse Imaging   Acoustic radiation force impulse (ARFI) imaging [13,50] uses a conventional ultrasound scanner to create images of on-axis tissue displacements that portray relative differences in tissue mechanical properties. The method uses a single transducer to both generate and detect displacements within a localized ROE using an ensemble of acoustic pulses. In a single location, the transient deformation response is monitored through time by estimating the on-axis tissue displacements from returned echoes using either cross-correlation or phase shift motion tracking algorithms. As depicted in Figure 8.5b, this interrogation region can be swept across a lateral field of view, as performed in conventional B-mode imaging, to create 2-D (lateral vs. axial) data sets of the tissue displacement response as a function of time after excitation. Several parameters can be derived from these data to portray relative stiffness differences, including the tissue displacement at a particular time after excitation, the maximum tissue displacement, the time it takes the tissue to reach the maximum displacement, or the time it takes the tissue to recover [51]. In general, for a given impulsive radiation force excitation in a homogeneous material, softer tissues

displace farther, take longer to reach a peak displacement, and recover more slowly than stiffer tissues [19,31]. In heterogeneous media, the contrast and apparent lesion size varies dynamically with time after excitation due to shear wave reflections at structural boundaries [31]. As a result, the accuracy of structural details is the most accurate in ARFI images portraying relative displacement at an early time after force excitation, before appreciable shear wave propagation. The contrast transfer efficiency of ARFI images (i.e., the ratio of lesion contrast in the image as compared with the underlying mechanical contrast) varies as a function of ROE geometry and lesion size, where contrast is maximized when the ROE cross-sectional width is smaller than the structure being imaged [52].

A typical ARFI imaging sequence consists of three pulse types: reference pulses that establish a baseline reference tissue location before the applied ARF, excitation (pushing) pulses that deform the tissue, and an ensemble of tracking pulses that monitor the deformation response immediately after the excitation and for a period exceeding full recovery. Motion not induced by ARF (i.e., physiological and transducer based) is removed using motion filters [53]. These filters estimate the underlying motion by deriving a fit between preexcitation and postrecovery displacement estimates, which is then subtracted from the raw displacement estimates. For improved image quality, increased frame rates, and decreased transducer face heating, novel pulse sequencing techniques have been developed [54,55]. More recently, these techniques have been combined with Doppler methods in multimodal systems [56] and also implemented to provide for near real-time characterization of the mechanical properties of tissue during freehand imaging [57].

Clinically, ARFI imaging provides 2-D images that allow for visualizing the relative mechanical properties of tissue, often with higher contrast than corresponding B-mode ultrasound images. This imaging approach has been optimized for several different clinical applications. As demonstrated by Fahey et al. [58], and shown in Figure 8.6, ARFI methods have been used to visualize liver masses with better contrast than B-mode and CT imaging. ARFI methods have also been applied to visualize the relative stiffness of arterial plaques, toward the goal of identifying rupture prone regions [59] (Figure 8.7). Thermal ablation procedures have been monitored with ARFI imaging [60], providing significantly improved thermal lesion contrast as compared with B-mode imaging. Cardiovascular ARFI imaging has been used to assess the relative stiffness differences occurring in vascular [61–63] and cardiac tissue [64,65] throughout the cardiac cycle. The method has also been used for tumor characterization in breast imaging [66], gastrointestinal imaging [67], and prostate imaging [68,69].

8.4.2.2 Off-Axis Tracking (Quantitative) Methods

8.4.2.2.1 Shear Wave Elasticity Imaging   Using a high-intensity focused ultrasound (HIFU) piston to generate impulsive ARF and MRI methods to monitor shear wave propagation, Sarvazyan et al. [14] first proposed shear wave elasticity imaging (SWEI) for quantifying the stiffness of tissues. Using the same excitation as described for ARFI imaging, Nightingale et al. [70] generated and monitored the propagation of shear waves using a single ultrasound transducer in vivo. With multiple tracking beams located outside the ROE, as depicted in Figure 8.5c, displacements are estimated using

Chapter 8

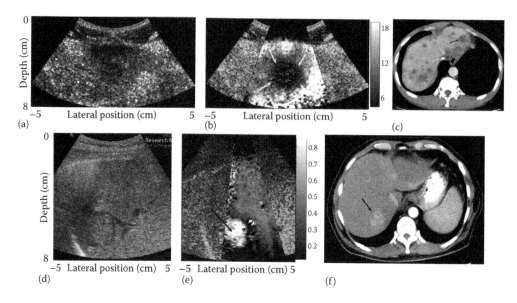

FIGURE 8.6 B-mode, ARFI, and CT images of in vivo liver lesions. Arrows indicate the lesions in the ARFI and CT images. In the top row ARFI image (b), where the grayscale color bar represents displacements in microns, the lesion displaces less, suggesting it is stiffer compared with the healthy background tissue and is less easily observed in either the B-mode (a) or the CT image (c). In the bottom row is a hepatocellular carcinoma in a fibrotic liver, which appears to displace more in the ARFI image (e) compared with the stiffer diseased liver tissue. Again, the lesion is more easily observed in the ARFI image (e) compared with either the B-mode (d) or the CT image (f). (Reprinted from Fahey, B. J. et al., *Phys. Med. Biol.*, 53:279–293, 2008. With permission.)

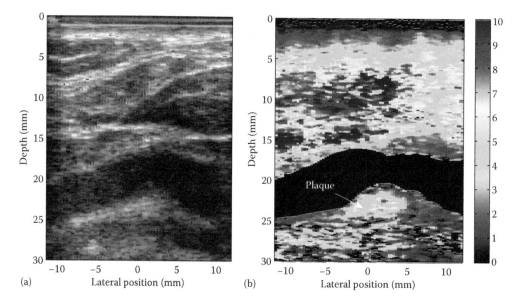

FIGURE 8.7 Matched B-mode (a) and ARFI (b) images of an in vivo carotid artery plaque, where the ARFI image color bar represents displacements in microns. The heterogeneous carotid artery plaque (indicated by arrow) displaces more compared with the surrounding tissue, suggesting that it is a soft lipid pool with a thin fibrous cap. (Reprinted from Dahl, J. J. et al., *Ultras. Med. Biol.*, 5(5), 707–716, 2009. With permission.)

cross-correlation or phase shift techniques at each location to monitor shear wave propagation as a function of time and distance traveled from the ROE.

To quantify the tissue stiffness from collected off-axis displacement information, early methods were based on algebraic inversion of the Helmholtz equation (Equation 8.41) as described by Nightingale et al. [70] and Bercoff et al. [71]. However, because of the jitter inherent in ultrasonically monitored displacement estimates, the second-order differentiation of the data required considerable smoothing, leading to poor resolution. For this reason, time-of-flight (TOF) methods are now commonly used, which perform linear regression between the wave arrival time and the lateral position data [72–74]. The concept of this approach is depicted in Figure 8.8, which portrays a finite element method simulation of the axial displacements occurring in a 3-D homogeneous isotropic elastic material as a function of both lateral position from the excitation source and time after excitation.

There is an inherent trade-off between the regression kernel size and the precision of the shear wave speed estimate obtained with TOF methods. Larger kernels include more data and can result in improved estimate precision, assuming that local homogeneity is satisfied. However, larger kernels are also associated with decreased spatial resolution [75]. Thus, optimal processing is determined by the clinical application. For liver fibrosis staging, which is a diffuse stiffening throughout the gland, large regression kernels have generally been used to achieve higher precision [73,74]. This approach has demonstrated clinical utility in the context of noninvasively staging liver fibrosis [76,77].

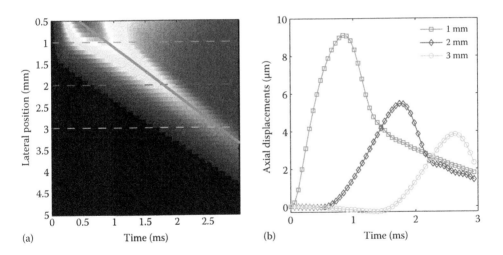

(a) Time (ms)

(b) Time (ms)

FIGURE 8.8 A finite element simulation showing the propagation of shear waves in a 3-D homogeneous isotropic linear elastic material with a Young's modulus of 4 kPa, generated by an impulsive excitation of ARF. (a) The image depicts the axial displacements as a function of lateral position from the excitation source (applied in the upper left corner) and the time after excitation. (b) The image depicts the displacement through time response at three specific lateral positions located 1, 2, and 3 mm from the excitation source, at positions indicated by the corresponding horizontal dashed lines in the left image. At each lateral location, the time to maximum displacement can be used as the time of arrival of the shear wave at that particular location, which can be used to obtain the linear regression line shown on the left image (gray line). The slope of this line provides an estimate of the shear wave speed in time-of-flight (TOF) methods.

Chapter 8

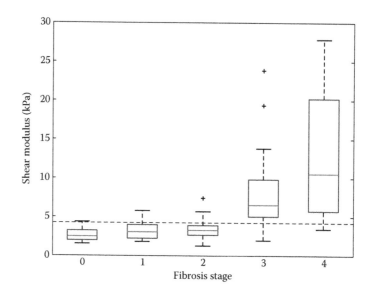

FIGURE 8.9 Quantification of liver stiffness as a function of the fibrosis stage in patients evaluated for nonalcoholic fatty liver disease using SWEI methods. The RANSAC method was implemented to solve for the shear wave speed, which was then used to calculate the shear modulus. The dotted line was chosen as a threshold to distinguish between no fibrosis (stages 0–2) and mild fibrosis (stages 3 and 4) with 90% sensitivity and specificity. (Reprinted from Palmeri, M. L. et al., *J Hepatol*, 55(3), 666–672, 2011. With permission.)

8.4.2.2.2 Supersonic Shear Imaging   By successively focusing ARF pushing beams at increasing focal depths, Bercoff et al. [71] introduced a shear wave technique, supersonic shear imaging (SSI), capable of providing 2-D image maps of quantitative shear modulus estimates. The method relies on transmitting multiple ARF excitations at increasing depths in rapid succession to create an axially extended shear wave source. By applying the ARF at increasing depths faster than the shear waves can propagate away from each excitation, the shear waves sum constructively along a Mach cone to create a cylindrical shear wave front propagating from the ROE, as depicted in Figure 8.5d. Compared with a single foci excitation, cylindrical shear excitation is advantageous in that it satisfies TOF assumptions over a larger axial depth. SSI methods use shear wave monitoring by extensive parallel beamforming, enabling kilohertz frame rates over relatively large FOVs. Recent implementations of this approach use TOF-based shear wave speed estimation algorithms using small regression kernels to achieve spatial resolution in the order of millimeters (Figure 8.9) [78].

SSI has been demonstrated for multiple clinical applications, including the musculoskeletal system [79], mapping hepatic viscoelasticity [80], monitoring thermal ablation procedures [81], and characterizing breast lesions [78] (Figure 8.10).

8.4.2.2.3 Shear Wave Spectroscopy   Shear wave spectroscopy is a transient ARF elasticity method that provides an evaluation of dispersion, the frequency-dependent nature of shear wave speed propagation in viscoelastic materials such as soft tissue, by monitoring off-axis displacements. The frequency content of the shear waves generated by an impulsive ARF excitation is dictated by the spatial extent of the ROE. By Fourier transforming the propagating shear wave and by evaluating phase differences

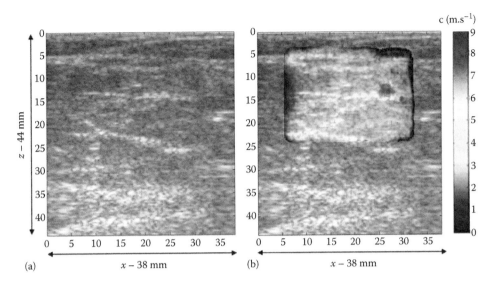

FIGURE 8.10   Matched B-mode (a) and shear wave velocimetry map (b) images of a 5 mm grade III infiltrating ductal carcinoma using SSI methods. Unlike in the B-mode image, the velocimetry map clearly distinguishes the carcinoma (bright red) compared with the surrounding tissue. (Reprinted from Tanter M. et al., *Ultrasound Med Biol*, 34(5), 2008. With permission.)

as a function of frequency, dispersion can be assessed in the frequency range from 75 to 600 Hz [82].

8.4.2.2.4 Spatially Modulated Radiation Force   The use of a spatially modulated ARF excitation has been investigated by McAleavey et al. [83,84] to generate a disturbance of known spatial frequency for shear modulus reconstruction in a technique termed spatially modulated radiation force (SMURF). In this method, depicted in Figure 8.5e, sequential pushing pulses are offset spatially to design shear waves with a wavelength equal to the spatial frequency of the applied excitations. Shear wave propagation is tracked off-axis at a single location, and methods based on cross-correlation are used to estimate the difference in arrival times between successive pulses to solve for the temporal frequency of the traveling waves. Modulated by the underlying shear stiffness, this frequency (f) is related to the shear modulus (μ), the tissue density (ρ), and the designed shear wavelength (λ) in a linear elastic model according to the following equation [83]:

$$\mu = \rho(\lambda f)^2. \tag{8.52}$$

8.4.3   Harmonic Methods

High-intensity ARF excitation pulses can also be applied in a pulsed manner to create sinusoidal tissue excitations. Methods using this approach are often termed harmonic methods because they drive tissue at one or more specific frequencies. Unlike SMURF, where the spatial frequency of the applied excitation was varied, harmonic methods modulate the temporal frequency of the applied excitation by varying the PRF

Chapter 8

of pushing pulses applied at a single location. The vibratory response of the interrogated tissue is dependent on its mechanical properties and can be used either qualitatively to differentiate structures of varying stiffness or quantitatively to form estimates of material properties.

8.4.3.1 Vibroacoustography

By using two ultrasonic beams focused at the same spatial location, but separated spatially and with slightly offset frequencies, oscillating radiation force excitations at the difference (beat) frequency of the two beams are generated in vibroacoustography, a method developed by Fatemi and Greenleaf [15,85]. As depicted in Figure 8.5f, tissue response is monitored using a hydrophone to detect the amplitude and phase of the resulting sinusoidal acoustic (compressive wave) vibrations that are generated within the overlapping focal regions of the two beams. To depict relative differences in stiffness, the excitation beams are mechanically translated in a raster scan across a 2-D scanning plane to create a C-scan type image of the tissue vibration response. The method has been applied both in vitro and in vivo, including applications in breast [86], vascular [87], and prostate imaging [88]. Recent efforts involve the implementation of the method on a commercially available scanner with a single linear array transducer [89].

8.4.3.2 Harmonic Motion Imaging

In harmonic motion imaging (HMI), a method developed by Konofagou and Hynynen [90], sinusoidal excitations can be generated in tissue using either the same two beam excitation source developed in vibroacoustography or by using the amplitude modulation (AM) [91] of a single excitation beam at a desired beat frequency [92], as depicted in Figure 8.5g. Unlike vibroacoustography, however, the harmonic motion is monitored using techniques based on cross-correlation from data collected with a separate ultrasonic imaging transducer using standard diagnostic pulse echo techniques. With the ability to monitor displacements during the application of the ARF, the AM HMI technique has shown promise for imaging thermal ablation lesions created during HIFU or focused ultrasound therapy [93]. An example AM HMI image of the relative stiffness changes that occur during HIFU is shown in Figure 8.11 [93].

8.4.3.3 Shear Wave Dispersion Ultrasound Vibrometry

In the method called shear wave dispersion ultrasound vibrometry (SDUV), developed by Chen et al. [94], ARF is applied in a single location to generate harmonic shear waves with a temporal frequency equal to the PRF of the applied excitation pulses. Tissue response is monitored off-axis using diagnostic imaging pulses. The phase shift of the monochromatic propagating shear waves between two off-axis tracking locations, depicted in Figure 8.5h, is determined as a function of the distance between the locations to form an estimate of shear wave speed. The narrowband shear waves generated with SDUV can be used to assess dispersion. This is accomplished by estimating the shear wave speeds over a range of different excitation frequencies. By using a theoretical Voigt dispersion model, the method can provide a quantitative estimate of tissue stiffness and viscosity [94]. The method has been implemented to study dispersion in the liver [95] and blood vessels [96].

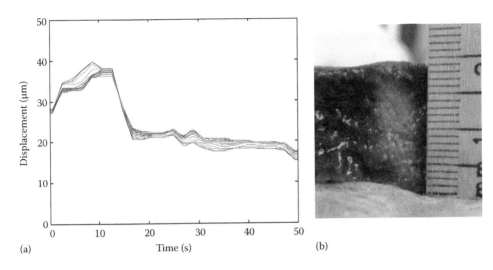

FIGURE 8.11 Displacement profiles (a) as a function of sonification time at 10 different locations (multicolored lines) separated by 0.5 mm obtained using HMI methods during a liver ablation procedure, representing the decrease in tissue stiffness resulting from the ablation. The ablated region is visualized in the accompanying pathology image (b). (Reprinted from Maleke C. and E. E. Konofagou, *Phys Med Biol*, 53(6), 1773–1793, 2008. With permission.)

8.4.3.4 Crawling Wave Sonoelastography

By applying amplitude modulated ARF excitations at two spatially offset positions, each driven at slightly offset frequencies as depicted in Figure 8.5i, crawling waves are created between the two shear-vibration sources in a method developed by Parker et al. termed crawling wave sonoelastography (CWS) [97]. Dependent on the mechanical properties of the tissue, the speed of the slow moving crawling waves can be estimated using conventional Doppler techniques. As with other quantitative methods, a mathematical model can be used to form an estimate of the tissue modulus. CWS methods have been implemented to provide a quantitative estimate of the viscoelastic properties of cancerous and normal tissues in excised human prostate glands [98].

8.5 Safety Considerations

As with all diagnostic ultrasound imaging techniques, the safety of ARF-based elasticity imaging methods must be monitored to minimize patient risk. In particular, ARF techniques used *in vivo* are regulated by the thermal index (TI)—an estimate of expected heating, and the mechanical index (MI)—a measure of the potential for cavitation [99,100]. With configurable parameters, including spatial location, excitation duration, pulse length, PRF, and intensity of the excitation beams used, there exist many trade-offs for all ARF-based elasticity imaging methods that balance frame rates and image quality versus patient safety.

The generation of ARF results from the absorption of energy, which also leads to the generation of heat within the tissue. Fortunately, for ARF-based elasticity imaging methods, differences between the mechanical and the thermal properties of soft tissue are such that it takes less energy to displace tissue several microns than to increase its

Chapter 8

temperature a fraction of a degree Celsius. Nevertheless, energy deposition that occurs during application of the excitation beams must be quantified to ensure that peak tissue heating is kept within diagnostic limits (i.e., peak tissue heating <6°C) [101]. On the basis of the linear bioheat transfer equation and neglecting heat loss due to perfusion and thermal conduction, the temperature rise in tissue for a single impulsive radiation force excitation can be estimated using the following equation [102,103]:

$$\Delta T = \frac{q_v}{c_v}t = \frac{2\alpha\langle I\rangle}{c_v}t,$$

(8.53)

where q_v is the rate of heat production per unit volume, c_v is the heat capacity per unit volume of the tissue, $\langle I\rangle$ is the *in situ* pulse average intensity, and t is the duration of the excitation. On the basis of this equation, tissue heating is a function of the focal configuration of the transducer as well as material properties. For a single impulsive interrogation of the focused beams used in ARF-based elasticity imaging methods, the maximum temperature increase is largest at the focal point and typically ranges from 0.02°C to 0.2°C [103]. In highly attenuating media that primarily absorb energy in the near field, the increase in temperature at the focus is smaller. For methods using repeated excitations, issues including conduction and near field beam overlap must be accounted for. In these cases, the maximum temperature increase shifts from the focal point to the location of maximum spatial and thermal overlap between sequential excitation locations [103]. ARF-based elasticity imaging methods are also vulnerable to transducer surface heating [104], which can be especially problematic for high excitation beam density and multiframe acquisitions.

Typically, the generation of sufficient ARF capable of deforming soft tissue in the order of 10 μm can be achieved for short durations while maintaining an MI within diagnostic limits (i.e., MI < 1.9). Although no bioeffect has been reported to date in vivo using impulsive ARF excitation pulses, the effects of this type of pulse (i.e., much longer duration than other diagnostic pulses) when used in conjunction with contrast agents or in obstetric imaging have not been evaluated.

8.6 Summary

ARF-based elasticity imaging methods have been under development for the past 15 years. Advantages of ARF excitation in the context of elasticity imaging include the coupling of applied force directly within the region of interest, the relatively small tissue deformations reducing the potential for complications from signal decorrelation and out of plane motion, and the relatively high-frequency content of the shear excitation (70–600 Hz). Several novel approaches have been developed that use (quasi)static, transient (impulsive), and harmonic modes of ARF excitation, providing both qualitative and quantitative stiffness information about soft tissues. Recently, commercial versions of ARF-based elasticity imaging have been released by both Supersonic Imagine [105] and Siemens [106], with promising initial clinical findings for both hepatic fibrosis staging [107–109] and breast tumor characterization [110].

As ultrasonic imaging systems become more flexible in their beamforming and acoustic output capabilities, the development of ARF-based elasticity imaging methods is likely to expand. With increased parallel beamforming and 3-D volumetric imaging, we are likely to see the development of multimodal systems capable of acquiring both on-axis and off-axis displacement information simultaneously. With increased transmit pulse programmability, the development of novel beam geometries may prove useful. In addition, the investigation of more complex material models and tissue structures is already under way, which may lead to additional diagnostic information based on parameters characterizing, for example, tissue nonlinearity, viscoelasticity, and anisotropy. Because of the availability of commercial realizations of ARF elasticity methods, it is likely that large-scale clinical investigations will be performed, which will facilitate a better understanding of the physiologic changes that generate differences in tissue stiffness during disease progression and treatment. Perhaps, such advancements and clinical studies will lead to the identification of specific clinical applications of ARF elasticity imaging that will become part of standard clinical examinations in the 21st century.

References

1. P. N. Wells and H. Liang, Medical ultrasound: Imaging of soft tissue strain and elasticity. *Journal of the Royal Society Interface*, 2011;8(64):1521–1549.
2. E. Tohno, D. O. Cosgrove, and J. P. Sloane, *Ultrasound Diagnosis of Breast Diseases* 1994, New York: Churchill Livingstone Inc.
3. J. N. Wolfe, *Xeroradiography of the Breast* 1972, Springfield, IL: Thomas.
4. L. Bassett, R. Gold, and C. Kimme-Smith, *Hand-Held and Automated Breast Ultrasound* 1986, Thorafore, NJ: Slack Inc.
5. W. L. Donegan and J. S. Spratt, *Cancer of the Breast* 2002, Philadelphia: Saunders.
6. J. Ophir, I. Cespedes, H. Ponnekanti, Y. Yazdi, and X. Li, Elastography: A quantitative method for imaging the elasticity of biological tissues. *Ultrasonic Imaging*, 1991;13(2):111–134.
7. R. M. Lerner, S. R. Huang, and K. J. Parker, Sono-elasticity: Medical elasticity images derived from ultrasound signals in mechanically vibrated targets. *Acoustic Imaging*, 1988;16(317–327).
8. L. S. Wilson and D. E. Robinson, Ultrasonic measurement of small displacements and deformations of tissue. *Ultrasonic Imaging*, 1982;4(1):71–82.
9. R. J. Dickinson and C. R. Hill, Measurement of soft tissue motion using correlation between A-scans. *Ultrasound in Medicine and Biology*, 1982;8(3):263–271.
10. J. J. Mai and M. F. Insana, Strain imaging of internal deformation. *Ultrasound in Medicine and Biology*, 2002;28(11–12):1475–1484.
11. C. L. de Korte and A. F. v. der Steen, Intravascular ultrasound elastography: An overview. *Ultrasonics*, 2002;40(1–8):859–865.
12. J. D'hooge, B. Bijnens, J. Thoen, F. Van de Werf, G. R. Sutherland, and P. Suetens, Echocardiographic strain and strain-rate imaging: A new tool to study regional myocardial function. *IEEE Transactions on Medical Imaging*, 2002;21(9):1022–1030.
13. K. R. Nightingale, M. L. Palmeri, R. W. Nightingale, and G. E. Trahey, On the feasibility of remote palpation using acoustic radiation force. *Journal of the Acoustical Society of America*, 2001;110(1):625–634.
14. A. P. Sarvazyan, O. V. Rudenko, S. D. Swanson, J. B. Fowlkes, and S. Y. Emelianov, Shear wave elasticity imaging: A new ultrasonic technology of medical diagnostics. *Ultrasound in Medicine and Biology*, 1998;24(9):1419–1435.
15. M. Fatemi and J. F. Greenleaf, Ultrasound-stimulated vibro-acoustic spectrography. *Science*, 1998;280:82–85.
16. J. F. Greenleaf, M. Fatemi, and M. Insana, Selected methods for imaging elastic properties of biological tissues. *Annual Review of Biomedical Engineering*, 2003;5(1):57–78.
17. K. J. Parker, L. S. Taylor, and S. Gracewski, A unified view of imaging the elastic properties of tissue. *Journal of the Acoustical Society of America*, 2005;117(5):2705–2712.
18. G. S. Kino, *Acoustic Waves: Devices, Imaging, and Analog Signal Processing* 1987, Englewood Cliffs, NJ: Prentice-Hall Inc.

Chapter 8

19. M. L. Palmeri, A. C. Sharma, R. R. Bouchard, R. W. Nightingale, and K. R. Nightingale, A finite-element method model of soft tissue response to impulsive acoustic radiation force. *IEEE Transactions on Ultrasonics, Ferroelectrics and Frequency Control*, 2005;52(10):1699–1712.

20. Y. C. Fung, *Biomechanics: Mechanical Properties of Living Tissue, 2nd Edition* 1993, New York: Springer-Verlag.

21. W. M. Lai, D. Rubin, and E. Krempel, *Introduction to Continuum Mechanics* 1999, Burlington, MA: Butterworth-Heinmann.

22. T. A. Krouskop, T. M. Wheeler, F. Kallel, B. S. Garra, and T. Hall, Elastic moduli of breast and prostate tissue under compression. *Ultrasonic Imaging*, 1998;20:260–274.

23. W. Nyborg, *Acoustic Streaming*, edited by P. Mason, 1965, New York: Academic Press.

24. C. Eckart, Vortices and streams caused by sound waves. *American Physical Society*, 1948;73(1):68–76.

25. R. Fox and A. McDonald, *Introduction to Fluid Mechanics* 1992, New York: Wiley.

26. J. H. Ginsberg, *Perturbation Methods, in Nonlinear Acoustics*, edited by M. F. Hamilton and D. T. Blackstock, 1998, New York: Academic Press.

27. D. A. Christensen, *Ultrasonic Bioinstrumentation* 1988, New York: Wiley.

28. J. P. Westervelt, The theory of steady forces caused by sound waves. *Journal of the Acoustical Society of America*, 1951;23(4):312–315.

29. J. D. Achenbach, *Wave Propagation in Elastic Solids* 1999, Elsevier Science.

30. J. Bishop, G. Poole, and D. Plewes, Magnetic resonance imaging of shear wave propagation in excised tissue. *Journal of Magnetic Resonance Imaging*, 1998;8(6):1257–1265.

31. M. L. Palmeri, S. A. Mcaleavey, G. E. Trahey, and K. R. Nightingale, Dynamic mechanical response of elastic spherical inclusions to impulsive acoustic radiation force excitation. *IEEE Transactions on Ultrasonics, Ferroelectrics and Frequency Control*, 2006;53(11):2065–2079.

32. G. Casella and R. Berger, *Statistical Inference* 2001, Duxbury Press.

33. O. Bonnefous and P. Pesque, Time domain formulation of pulse-Doppler ultrasound and blood velocity estimation by cross correlation. *Ultrasonic Imaging*, 1986;8:73–85.

34. P. Embree, *The Accurate Ultrasonic Measurement of the Volume Flow of Blood by Time-Domain Correlation* 1985, Urbana, IL: University of Illinois.

35. F. Viola and W. F. Walker, A comparison of the performance of time-delay estimators in medical ultrasound. *IEEE Transactions on Ultrasonics, Ferroelectrics and Frequency Control*, 2003;50(4):392–401.

36. W. F. Walker and G. E. Trahey, A fundamental limit on delay estimation using partially correlated speckle signals. *IEEE Transactions on Ultrasonics, Ferroelectrics and Frequency Control*, 1995;42(2):301–308.

37. C. Kasai, K. Namekawa, A. Koyano, and R. Omoto, Real-time two-dimensional blood flow imaging using an autocorrelation technique. *IEEE Transactions on Ultrasonics, Ferroelectrics and Frequency Control*, 1985;SU-32(3):458–464.

38. J. A. Jensen, *Estimation of Blood Velocities Using Ultrasound* 1996, Cambridge University Press.

39. T. Loupas, J. Powers, and R. Gill, An axial velocity estimator for ultrasound blood flow imaging, based on a full evaluation of the Doppler equation by means of a two-dimensional autocorrelation approach. *IEEE Transactions on Ultrasonics, Ferroelectrics and Frequency Control*, 1995;42(4):672–688.

40. G. F. Pinton, J. J. Dahl, and G. E. Trahey, Rapid tracking of small displacements with ultrasound. *IEEE Transactions on Ultrasonics, Ferroelectrics and Frequency Control*, 2006;53(6):1103–1117.

41. M. Bilgen and M. Insana, Error analysis in acoustic elastography. I. Displacement estimation. *Journal of the Acoustical Society of America*, 1997;101(2):1139–1146.

42. M. L. Palmeri, S. A. McAleavey, G. E. Trahey, and K. R. Nightingale, Ultrasonic tracking of acoustic radiation force-induced displacements in homogeneous media. *IEEE Transactions on Ultrasonics, Ferroelectrics and Frequency Control*, 2006;53(7):1300–1313.

43. S. A. McAleavey, K. R. Nightingale, and G. E. Trahey, Estimates of echo correlation and measurement bias in acoustic radiation force impulse imaging. *IEEE Transactions on Ultrasonics, Ferroelectrics and Frequency Control*, 2003;50(6):631–641.

44. T. Sugimoto, S. Ueha, and K. Itoh, Tissue hardness measurement using the radiation force of focused ultrasound radiation force. *Proceedings of the IEEE Ultrasonics Symposium*, 1990:1377–1380.

45. W. F. Walker, Internal deformation of a uniform elastic solid by acoustic radiation force. *Journal of the Acoustical Society of America*, 1999;105(4):2508–2518.

46. W. F. Walker, F. J. Fernandez, and L. A. Negron, A method of imaging viscoelastic parameters with acoustic radiation force. *Physics in Medicine and Biology*, 2000;45(6):1437–1447.

47. S. O. Dymling, H. W. Persson, T. G. Hertz, and K. Lindstrom, A new ultrasonic method for fluid property measurements. *Ultrasound in Medicine and Biology*, 1991;17(5):497–500.

48. K. R. Nightingale, P. J. Kornguth, W. F. Walker, and B. A. McDermott, A novel ultrasonic technique for differentiating cysts from solid lesions: Preliminary results in the breast. *Ultrasound in Medicine and Biology*, 1995;21(6):745–751.

49. F. Viola, M. D. Kramer, M. B. Lawrence, J. P. Oberhauser, and W. F. Walker, Sonorheometry: A non-contact method for the dynamic assessment of thrombosis. *Annual Review of Biomedical Engineering*, 2004;32(5):696–705.

50. K. Nightingale, M. S. Soo, R. Nightingale, and G. Trahey, Acoustic radiation force impulse imaging: In vivo demonstration of clinical feasibility. *Ultrasound in Medicine and Biology*, 2002;28(2):227–235.

51. K. R. Nightingale, R. C. Bentley, and G. E. Trahey, Observations of tissue response to acoustic radiation force: Opportunities for imaging. *Ultrasonic Imaging*, 2002;24(3):100–108.

52. K. R. Nightingale, M. L. Palmeri, and G. E. Trahey, Analysis of contrast in images generated with transient acoustic radiation force. *Ultrasound in Medicine and Biology*, 2006;32(1):61–72.

53. S. J. Hsu, R. R. Bouchard, D. M. Dumont, C. W. Ong, P. D. Wolf, and G. E. Trahey, Novel acoustic radiation force impulse imaging methods for visualization of rapidly moving tissue. *Ultrasonic Imaging*, 2009;31(3):183–200.

54. J. J. Dahl, G. F. Pinton, M. L. Palmeri, V. Agrawal, K. R. Nightingale, and G. E. Trahey, A parallel tracking method for acoustic radiation force impulse imaging. *IEEE Transactions on Ultrasonics, Ferroelectrics and Frequency Control*, 2007;54(2):301–312.

55. R. R. Bouchard, J. J. Dahl, S. J. Hsu, M. L. Palmeri, and G. E. Trahey, Image quality, tissue heating, and frame rate trade-offs in acoustic radiation force impulse imaging. *IEEE Transactions on Ultrasonics, Ferroelectrics and Frequency Control*, 2009;56(1):63–76.

56. D. M. Dumont, *Assessment of Mechanical and Hemodynamic Vascular Properties Using Radiation-Force Driven Methods* 2011, Duke University.

57. J. R. Doherty, D. M. Dumont, D. Hyun, J. J. Dahl, and G. E. Trahey, Development and evaluation of pulse sequences for freehand ARFI imaging. *Proceedings of the 2011 IEEE International Ultrasonics Symposium*, 2011.

58. B. J. Fahey, R. C. Nelson, D. P. Bradway, S. J. Hsu, D. M. Dumont, and G. E. Trahey, In vivo visualization of abdominal malignancies with acoustic radiation force elastography. *Physics in Medicine and Biology*, 2008;53:279–293.

59. J. J. Dahl, D. M. Dumont, J. D. Allen, E. M. Miller, and G. E. Trahey, Acoustic radiation force impulse imaging for noninvasive characterization of carotid artery atherosclerotic plaques: A feasibility study. *Ultrasound in Medicine and Biology*, 2009;35(5):707–716.

60. S. A. Eyerly, S. J. Hsu, S. H. Agashe, G. E. Trahey, L. Yang, and P. D. Wolf, An in vitro assessment of acoustic radiation force impulse imaging for visualizing cardiac radiofrequency ablation lesions. *Journal of Cardiovascular Electrophysiology*, 2010;21(5):557–563.

61. D. M. Dumont, J. J. Dahl, E. M. Miller, J. D. Allen, B. Fahey, and G. E. Trahey, Lower-limb vascular imaging with acoustic radiation force elastography: Demonstration of in vivo feasibility. *IEEE Transactions on Ultrasonics, Ferroelectrics and Frequency Control*, 2009. 56(5):931–944.

62. R. H. Behler, T. C. Nichols, H. Zhu, E. P. Merricks, and C. M. Gallippi, ARFI imaging for noninvasive material characterization of atherosclerosis. Part II: Toward in vivo characterization. *Ultrasound in Medicine and Biology*, 2009;35(2):278–295.

63. G. E. Trahey, M. L. Palmeri, R. C. Bentley, and K. R. Nightingale, Acoustic radiation force impulse imaging of the mechanical properties of arteries: In vivo and ex vivo results. *Ultrasound in Medicine and Biology*, 2004;30(9):1163–1171.

64. S. J. Hsu, J. L. Hubert, S. W. Smith, and G. E. Trahey, Intracardiac echocardiography and acoustic radiation force impulse imaging of a dynamic ex vivo ovine heart model. *Ultrasonic Imaging*, 2008;30(2):63–77.

65. S. J. Hsu, R. R. Bouchard, D. M. Dumont, P. D. Wolf, and G. E. Trahey, In vivo assessment of myocardial stiffness with acoustic radiation force impulse imaging. *Ultrasound in Medicine and Biology*, 2007;33(11):1706–1719.

66. A. C. Sharma, M. S. Soo, G. E. Trahey, and K. R. Nightingale, Acoustic radiation force impulse imaging of in vivo breast masses. *Proceedings of the IEEE Ultrasonics Symposium*, 2004.

67. M. L. Palmeri, K. D. Frinkley, R. C. Bentley, K. Ludwig, M. Gottfried, and K. R. Nightingale, Acoustic radiation force impulse (ARFI) imaging of the gastrointestinal tract. *Ultrasonic Imaging*, 2005;27:75–88.

68. L. Zhai, T. J. Polascik, W. C. Foo, S. J. Rosenzweig, M. L. Palmeri, J. Madden, K. R. Nightingale, and V. Mouraviev, Acoustic radiation force impulse imaging of human prostates ex vivo. *Ultrasound in Medicine and Biology*, 2010;36(4):576–588.

Chapter 8

69. L. Zhai, T. J. Polascik, W. C. Foo, S. J. Rosenzweig, M. L. Palmeri, J. Madden, K. R. Nightingale, and V. Mouraviev, Acoustic radiation force impulse imaging of human prostates: Initial in vivo demonstration. *Ultrasound in Medicine and Biology*, 2012;38(4):50–61.

70. K. R. Nightingale, S. A. McAleavey, and G. E. Trahey, Shear-wave generation using acoustic radiation force: In vivo and ex vivo results. *Ultrasound in Medicine and Biology*, 2003;29(12):1714–1723.

71. J. Bercoff, M. Tanter, and M. Fink, Supersonic shear imaging: A new technique for soft tissue elasticity mapping. *IEEE Transactions on Ultrasonics, Ferroelectrics and Frequency Control*, 2004;51(4):396–409.

72. J. McLaughlin and D. Renzi, Shear wave speed recovery in transient elastography and supersonic imaging using propagating fronts. *Inverse Problems*, 2006;22:681–706.

73. M. H. Wang, M. L. Palmeri, V. M. Rotemberg, N. C. Rouze, and K. R. Nightingale, Improving the robustness of time-of-flight based shear wave speed reconstruction methods using RANSAC in human liver in vivo. *Ultrasound in Medicine and Biology*, 2010;36(5):802–813.

74. N. C. Rouze, M. H. Wang, M. L. Palmeri, and K. R. Nightingale, Robust estimation of time-of-flight shear wave speed using a radon sum transformation. *IEEE Transactions on Ultrasonics, Ferroelectrics and Frequency Control*, 2010;57(12):2662–2670.

75. N. C. Rouze, M. H. Wang, M. L. Palmeri, and K. R. Nightingale, Parameters affecting the resolution and accuracy of 2D quantitative shear wave images. *IEEE Transactions on Ultrasonics, Ferroelectrics and Frequency Control*, 2012;59(8):1729–1740.

76. M. L. Palmeri, M. H. Wang, J. J. Dahl, K. D. Frinkley, and K. R. Nightingale, Quantifying hepatic shear modulus in vivo using acoustic radiation force. *Ultrasound in Medicine and Biology*, 2008;34(4): 546–558.

77. M. L. Palmeri, M. H. Wang, N. C. Rouze, M. F. Abdelmalek, C. D. Guy, B. Moser, A. M. Diehl, and K. R. Nightingale, Noninvasive evaluation of hepatic fibrosis using acoustic radiation force-based shear stiffness in patients with nonalcoholic fatty liver disease. *Journal of Hepatology*, 2011;55(3):666–672.

78. M. Tanter, J. Bercoff, A. Athanasiou, T. Deffieux, J.-L. Gennisson, G. Montaldo, M. Muller, A. Tardivon, and M. Fink, Quantitative assessment of breast lesion viscoelasticity initial clinical results using supersonic shear imaging. *Ultrasound in Medicine and Biology*, 2008;34(5):1373–1386.

79. T. Deffieux, J. L. Gennisson, M. Tanter, and M. Fink, Assessment of the mechanical properties of the musculoskeletal system using 2-D and 3-D very high frame rate ultrasound. *IEEE Transactions on Ultrasonics, Ferroelectrics and Frequency Control*, 2008;55(10):2177–2190.

80. M. Muller, J.-L. Gennisson, T. Deffieux, M. Tanter, and M. Fink, Quantitative viscoelasticity mapping of human liver using supersonic shear imaging: Preliminary in vivo feasibility study. *Ultrasound in Medicine and Biology*, 2009;35(2):219–229.

81. J. Bercoff, M. Pernot, M. Tanter, and M. Fink, Monitoring thermally-induced lesions with supersonic shear imaging. *Ultrasonic Imaging*, 2004;26(2):71–84.

82. T. Deffieux, G. Montaldo, M. Tanter, and M. Fink, Shear wave spectroscopy for in vivo quantification of human soft tissues viscoelasticity. *IEEE Transactions on Medical Imaging*, 2009;28(3):313–322.

83. S. A. McAleavey, M. Menon, and J. Orszulak, Shear-modulus estimation by application of spatially-modulated impulsive acoustic radiation force. *Ultrasonic Imaging*, 2007;29(2):87–104.

84. S. A. McAleavey, E. Collins, K. Johanna, E. Elegbe, and M. Menon, Validation of SMURF estimation of shear modulus in hydrogels. *Ultrasonic Imaging*, 2009;31(2):131–150.

85. M. Fatemi and J. F. Greenleaf, Vibro-acoustography: An imaging modality based on ultrasound-stimulated acoustic emission. *Proceedings of the National Academy of Sciences of the United States of America*, 1999;96(6603–6608).

86. M. Fatemi, L. E. Wold, and J. F. Greenleaf, Vibro-acoustic tissue mammography. *IEEE Transactions on Medical Imaging*, 2002;21(1):1–8.

87. C. Pislaru, B. Kantor, R. R. Kinnick, J. L. Anderson, C. Aubry, M. W. Urban, M. Fatemi, and F. James, In vivo vibroacoustography of large peripheral arteries. *Investigative Radiology*, 2008;43(4):243–252.

88. F. G. Mitri, B. J. Davis, A. Alizad, J. F. Greenleaf, T. M. Wilson, L. A. Mynderse, and M. Fatemi, Prostate cryotherapy monitoring using vibroacoustography: Preliminary results of an ex vivo study and technical. *IEEE Transactions on Biomedical Engineering*, 2008;55(11):2584–2592.

89. M. W. Urban, C. Chalek, R. R. Kinnick, T. M. Kinter, B. Haider, J. F. Greenleaf, K. E. Thomenius, and M. Fatemi, Implementation of vibro-acoustography on a clinical ultrasound system. *IEEE Transactions on Ultrasonics, Ferroelectrics and Frequency Control*, 2011;58(6):1169–1181.

90. E. E. Konofagou and K. Hynynen, Localized harmonic motion imaging: Theory, simulations, and experiments. *Ultrasound in Medicine and Biology*, 2003;29(10):1405–1413.

91. M. W. Urban, M. Fatemi, and J. F. Greenleaf, Modulation of ultrasound to produce multifrequency radiation force. *Journal of the Acoustical Society of America*, 2010;127(3):1228–1238.

92. C. Maleke, M. Pernot, and E. E. Konofagou, Single-element focused ultrasound transducer method for harmonic motion imaging. *Ultrasonic Imaging*, 2006;28:144–158.

93. C. Maleke and E. E. Konofagou, Harmonic motion imaging for focused ultrasound (HMIFU): A fully integrated technique for sonification and monitoring of thermal ablation in tissues. *Physics in Medicine and Biology*, 2008;53(6):1773–1793.

94. S. G. Chen, M. Fatemi, and J. F. Greenleaf, Quantifying elasticity and viscosity from measurement of shear wave speed dispersion. *Journal of the Acoustical Society of America*, 2004;115(6):2781–2785.

95. S. Chen, M. W. Urban, C. Pislaru, R. Kinnick, Y. Zheng, A. Yao, and J. F. Greenleaf, Shear wave dispersion ultrasound vibrometry (SDUV) for measuring tissue elasticity and viscosity. *IEEE Transactions on Ultrasonics, Ferroelectrics and Frequency Control*, 2009;56(1):55–62.

96. X. Zhang, R. R. Kinnick, M. Fatemi, and J. F. Greenleaf, Noninvasive method for estimation of complex elastic modulus of arterial vessels. *IEEE Transactions on Ultrasonics, Ferroelectrics and Frequency Control*, 2005;52(4):642–652.

97. Z. Hah, C. Hazard, Y. T. Cho, D. Rubens, and K. Parker, Crawling waves from radiation force excitation. *Ultrasonic Imaging*, 2010;32(3):177–189.

98. L. An, B. Mills, Z. Hah, S. Mao, J. Yao, J. Joseph, D. J. Rubens, J. Strang, and K. J. Parker, Crawling wave detection of prostate cancer: Preliminary in vitro results. *Medical Physics*, 2011;38(5):2563–2571.

99. W. D. O'Brien, Ultrasound biophysics mechanisms. *Progress in Biophysics and Molecular Biology*, 2007;93(1–3):212–255.

100. F. A. Duck, Medical and non-medical protection standards for ultrasound and infrasound. *Progress in Biophysics and Molecular Biology*, 2006;93(2007):176–191.

101. Standard, I.E.C.S., *Particular Requirements for the Safety of Ultrasonic Medical Diagnostic and Monitoring Equipment.* Vol. Tech. Rep. 60601-2-37. 2011, Geneva, Switzerland: International Electrotechnical Committee.

102. W. Nyborg, Solutions of the bio-heat transfer equation. *Physics in Medicine and Biology*, 1988;33:785–792.

103. M. L. Palmeri and K. R. Nightingale, On the thermal effects associated with radiation force imaging of soft tissue. *IEEE Transactions on Ultrasonics, Ferroelectrics and Frequency Control*, 2004;51(5):551–565.

104. J. Wu, J. D. Chase, Z. Zhu, and T. P. Holzagfel, Temperature rise in tissue-mimicking material generated by unfocused and focused ultrasound transducers. *Ultrasound in Medicine and Biology*, 1992;18(5):495–512.

105. J. Bercoff, *Shearwave Elastography.* SuperSonic Imagine, 2008. White Paper.

106. R. S. Lazebnik, *Tissue Strain Analytics Virtual Touch Tissue Imaging and Quantification.* Siemens Healthcare, 2008. White Paper.

107. M. Haque, C. Robinson, D. Owen, E. M. Yoshida, and A. Harris, Comparison of acoustic radiation force impulse imaging (ARFI) to liver biopsy histologic scores in the evaluation of chronic liver diseases: A pilot study. *Annals of Hepatology*, 2010;9(3):289–293.

108. C. Fierbinteanu-Braticevici, D. Andronescu, R. Usvat, D. Cretoiu, C. Baicus, and G. Marinoschi, Acoustic radiation force imaging sonoelastography for noninvasive staging of liver fibrosis. *World Journal of Gastroenterology*, 2009;15(4):5525–5532.

109. J. Boursier, G. Isselin, I. Fouchard-Hubert, F. Oberti, N. Dib, J. Lebigot, S. Bertrais, Y. Gallois, P. Cales, and C. Aube, Acoustic radiation force impulse: A new ultrasonographic technology for the widespread noninvasive diagnosis of liver fibrosis. *European Journal of Gastroenterology and Hepatology*, 2010;22(9):1074–1084.

110. D. O. Cosgrove, C. C. Bacrie, C. Dore, B. Cavanaugh, C. Balu-Maestro, H. Madjar, A. Cossi, and A. Athanasion, *Shearwave elastography improves the specificity of the BI-RADS classification of breast masses by ultrasound: Results on 1000 cases in the breast elastography (BE1) investigators multicenter study.* Presented at the RSNA North America 2010 Scientific Assembly and Annual Meeting.

Chapter 8

Therapeutic and Interventional Ultrasound Imaging

9. Three–Dimensional Ultrasound–Guided Prostate Biopsy

Aaron Fenster, Jeff Bax, Vaishali Karnik, Derek Cool, Cesare Romagnoli, and Aaron Ward

Chapter 9

Ultrasound Imaging and Therapy. Edited by Aaron Fenster and James C. Lacefield © 2015 CRC Press/Taylor & Francis Group, LLC. ISBN: 978-1-4398-6628-3

9.1   Introduction: The Clinical Problem

Prostate disease is predominantly a disease of older men, and its relative importance in the overall health of the population has increased in the last three decades as the population has aged. The two most important diseases are prostate cancer (PCa) and benign prostatic hypertrophy. PCa is the most commonly diagnosed malignancy in men, and it is found at autopsy in 30% of men at age 50 years, 40% at age 60 years, and almost 90% at age 90 years (McNeal et al. 1986; Garfinkel and Mushinski 1994). Worldwide, it is the second leading cause of death due to cancer in men, accounting for between 2.1% and 15.2% of all cancer deaths (Silverberg et al. 1990; Abbas and Scardino 1997). In Canada, approximately 20,700 new PCa cases were diagnosed (26% of all new cancers in men) and approximately 4200 died from this disease in 2006 (Jemal et al. 2007). Symptoms due to carcinoma of the prostate are generally absent until extensive local growth or metastases develop, accounting for the fact that only approximately 65% of patients are diagnosed with locally confined disease. When diagnosed at an early stage, the disease is curable (Shinohara et al. 1989; Terris et al. 1992), and treatment can be effective even at later stages (Rifkin 1997). However, once the tumor has extended beyond the prostate, the risk of metastases and locally aggressive cancer increases. Clearly, early diagnosis, accurate staging of PCa, and appropriate therapies are critical to the patient's well-being.

In managing patients with possible PCa, the challenges facing physicians are as follows: (a) to diagnose clinically relevant cancers at a curable stage, (b) to stage the disease accurately, (c) to apply appropriate therapy accurately to optimize destruction of cancer cells while preserving normal tissues and function, and (d) to follow patients to assess side effects and therapy effectiveness. This chapter focuses on improving early PCa diagnosis and staging using three-dimensional (3-D) ultrasound-guided prostate biopsy and on helping to prescribe and deliver appropriate therapy through the use of 3-D ultrasound-based therapy planning.

9.2   Treatment Options for PCa

The controversies related to the decision of how best to manage early stage PCa are intensely debated by medical professionals as well as concerned patients (Jani and Hellman 2003; Matlaga et al. 2003). Management options for early stage PCa include active surveillance, hormone therapy, radiotherapy, cryosurgery, high-frequency ultrasound, and surgery. Some have argued that active surveillance is an appropriate management for patients older than 65 years diagnosed with PCa because the expected natural history of PCa progression can occur for decades and many in this age-group are at greater risk of dying from other causes first (e.g., stroke and myocardial infarction). However, patients with intermediate-grade and high-grade cancers are considered to have a more aggressive cancer and are at a substantial risk of early local failure if an intervention is not undertaken (Goto et al. 1996). Clearly, an accurate prostate biopsy is crucial for obtaining sufficient tissue to detect the presence of cancer, to identify the appropriate grade or aggressiveness of the malignancy, and to provide the patient with the most appropriate treatment option. Furthermore, as PCa is detected at earlier stages, when the tumor volumes are small, there is impetus to develop safe and efficacious focal therapy techniques that will decrease the morbidity (e.g., incontinence and impotence) that is associated with invasive therapies such as radiotherapy or surgery. Critical to these treatments is an accurate biopsy, including the capability to reliably reassess the same intraprostatic location during subsequent biopsies (for monitoring and treatment result verification).

9.3 Prostate Biopsy

Not all cancers are palpable, and even if hypoechoic tumor is seen on transrectal ultrasound (TRUS), they are not sufficiently diagnostic. Thus, PCa diagnosis is established by a histological examination of prostate tissue obtained most commonly by TRUS-guided biopsy and sometimes by transurethral resection procedures. Prostate needle biopsy is the only definitive diagnostic modality capable of confirming malignancy and is now always performed with TRUS guidance (Figure 9.1).

Indications for initial prostate needle biopsy are well established, consisting of any abnormality on digital rectal examination (DRE) or an abnormal prostate-specific antigen (PSA) result (Eastham and Scardino 2001; Punglia et al. 2003; Schroder and Kranse 2003). Typically, the physician will attempt to obtain two or three core biopsy samples from each prostate lesion detected with DRE and/or TRUS. Because many small tumors are not detected by TRUS or DRE, biopsy samples are obtained from predetermined regions of the prostate known to have a high probability of harboring cancer. These are typically in the peripheral zone (which harbors 80% of all PCas and a higher proportion of clinically significant ones) and close to the capsule, as most cancers are thought to start within 5 mm of the prostate capsule. Most centers are now taking 8 to 12 cores or more as part of their routine assessment (Djavan et al. 2000, 2002; Matlaga et al. 2003; Presti et al. 2003).

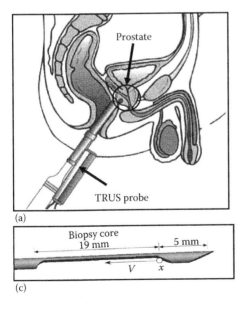

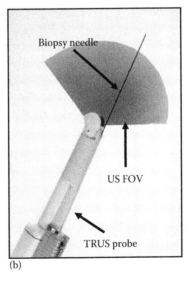

FIGURE 9.1   (a) Schematic diagram of the TRUS transducer, biopsy needle, and guide in the rectum during a prostate biopsy. (b) Schematic diagram of a TRUS transducer with an attached biopsy guide. Note that the needle is inserted into the plane of the TRUS image so that it is continuously visible in the US image. (c) Diagram of biopsy needle with 19 mm notch for biopsy core.

Chapter 9

The introduction of TRUS has revolutionized prostate biopsy techniques and has greatly increased the accuracy of biopsy, as biopsies were previously guided using prostate palpation without direct visualization. TRUS biopsies are performed with a thin, 18-gauge needle mounted on a spring-loaded gun connected to the TRUS probe, forcing the needle to stay in the imaging plane so it is always visible in the US image (Figure 9.1). Each core is separately identified as to the prostate region from which it was drawn, so that the pathologist can report the extent and grade of the cancer within each region. This is especially important if the histological result is equivocal and the pathologist requests a repeat biopsy. It is therefore vital to know exactly where the initial sample was obtained to target more relevant tissue if a repeat biopsy is performed.

Widespread screening for PCa using the serum PSA test has greatly increased the use of TRUS-guided biopsy (Djavan et al. 2000), although the threshold PSA indicator for biopsy is now being questioned. Approximately 250,000 men are diagnosed with PCa each year in North America using TRUS-guided biopsy, and another factor of 4 undergo prostate biopsy to rule out carcinoma (Matlaga et al. 2003). Although TRUS-guided prostate biopsy has become a commonly performed urological procedure, it is not without limitations. Chief among the problems facing urologists relates to the management of patients in whom a first set of prostate biopsies was negative for cancer. Because prostate volume sampled by the biopsy is small, and PCa is often multifocal, involving only a small volume of the prostate in the early stages of the disease (Jemal et al. 2002; Nelson et al. 2003), the probability for obtaining a sample of the tumor on biopsy is small. Thus, a negative biopsy may be in fact false, and the patient may be harboring cancer at an early and curable stage. The management of these patients, in addition to those diagnosed with early stage, low-grade disease that is typically considered to be indolent, is currently a major challenge, which requires improved prostate biopsy techniques to resolve.

9.3.1 Repeat Prostate Biopsy

Various reports have shown that the detection rate on repeat biopsy (after the first biopsy was negative) ranges from 10% to 25% (Djavan et al. 2000, 2002). Because cancer is still present in one-tenth to one-fourth of patients with a negative first biopsy, the current biopsy procedure is still suboptimal (Djavan et al. 2002; Park et al. 2003). Although advances in technology and understanding of the disease have produced improvements in the biopsy procedure, significant clinical and technical challenges remain. For example, if an initial biopsy fails to detect cancer, who should undergo a repeat biopsy? How should a repeat biopsy be directed? Should the repeat (and initial) biopsy be lesion directed, random, or based on the details of the patient's anatomy (e.g., prostate regions, volume, shape) (Djavan et al. 2002; Presti et al. 2003)? Clearly, an improved procedure with improved planning and recording of biopsy locations is necessary to resolve these issues.

9.3.2 PSA Indication for Prostate Biopsy

Improvements in PCa mortality have resulted from improved early detection and treatment. Detection has traditionally relied on DRE screening, but the increased detection rate has been primarily due to the widespread use of PSA screening. The PSA blood test is well established for the early detection of PCa and particularly for monitoring of PCa after treatment. The wide availability of the PSA test and the public's increased awareness of PCa together have increased the proportion of PCa diagnosed at an early stage. Currently, 77% of men are diagnosed with early stage PCa compared with only 57% between 1975 and 1979 (Mettlin et al. 1993; Middleton et al. 1995).

9.3.3 Morbidity and Complications Due to Prostate Biopsy

Although prostate biopsy is a safe and effective method to diagnose PCa in general, no procedure is risk free. Significant discomfort is reported during the procedure by 65% to 90% of men. After biopsy, common side effects include hematuria, hematospermia, and hematochezia in approximately a third to a half of patients (Clements et al. 1993; Collins et al. 1993). Although these are relatively minor, other less frequent postbiopsy morbidity includes sepsis (0.2%–0.6%), urinary tract infection (0.1%–4.5%), and urinary retention (0.2%–1.2%) (Rodriguez and Terris 1998). Thus, reducing the number of repeat biopsies would minimize the total number of patient complications.

Given the large and increasing number of men undergoing this procedure, the low (15%–30%) positive yield on first biopsy, and the reported frequency of morbidity, great effort is being made to increase the specificity of PSA screening and to increase the positive yield of the first and even the second biopsy.

9.4 Limitations of Current TRUS-Guided Prostate Biopsy

Although TRUS-guided core needle prostate biopsy is routinely used for diagnosis of PCa, evaluations in the past decade have identified several important challenges, described as follows.

9.4.1 Repeat TRUS-Guided Prostate Biopsy

With more men undergoing PSA testing and the potentially lowered PSA threshold for prostate biopsy, physicians commonly face the dilemma of the patient with a negative prostate biopsy who still has a suspicious clinical examination or serum PSA results. Because a negative result does not preclude the possibility of a missed cancer (Coplen et al. 1991), patients undergo repeat biopsies when indicated by clinical suspicion and in cases when a positive biopsy for cancer would have therapeutic consequences. Because there is an appreciable number of men with false-negative biopsy who in fact harbor curable PCa, the physician is faced with a difficult challenge. Sometimes these patients are imaged with other modalities and undergo a second and sometimes a third biopsy.

Many investigators have examined the positive yields on repeated biopsies of men with elevated PSA, suspicious DRE, or TRUS findings. Results have shown that on the first biopsy, approximately 15% to 40% of men had PCa (Presti et al. 2000; Altman and Resnick 2001), approximately 15% to 23% on the second biopsy, and 8% to 10% on the third (Coplen et al. 1991; Keetch et al. 1994). In some of the patients with false-negative biopsy, the cancer might be clinically insignificant, warranting no therapy, but some of these patients might benefit from detection and subsequent treatment.

Another important challenge facing physicians involves men diagnosed on biopsy to have premalignant lesions, for example, high-grade prostatic intraepithelial neoplasia

(PIN), and particularly atypical small acinar proliferation (ASAP) (Iczkowski et al. 2002). These are challenging to manage as there is a 40% to 50% chance of finding cancer on repeat biopsy with ASAP (Iczkowski et al. 1998). Because coexisting cancer might be present, especially with ASAP, where the pathologist finds only a small amount of histological "atypia" but not enough material to confidently diagnose cancer, these patients require a repeat biopsy soon after the first. In these situations, it is vital to rebiopsy the same area (Iczkowski et al. 2002). However, the current biopsy procedure only records the general region from which a biopsy core is taken. Therefore, it is not possible to be certain that the same area has been sampled on the repeat biopsy (see Figures 9.2 and 9.3).

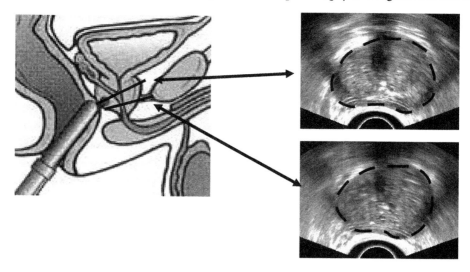

FIGURE 9.2   Diagram showing the problem with targeting using 2-D TRUS imaging. Because many views of the prostate appear similar, it is very difficult to identify a previously biopsied target in the prostate. Here we show two TRUS views from different locations in the prostate that appear similar.

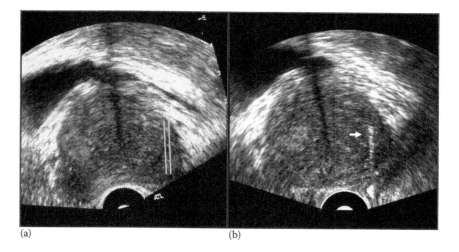

(a)   (b)

FIGURE 9.3   (a) A 2-D TRUS image obtained just before the biopsy needle was inserted into the prostate. The concave boundary of the rectal wall and prostate is caused by the pressure of the tip of the end-firing TRUS transducer. The graphical overlay corresponds to the planned trajectory of the biopsy needle. (b) A 2-D TRUS prostate image with the biopsy needle (white arrow) in the prostate.

Because of the increasing number of younger men with early and potentially curable PCa undergoing repeated prostate biopsy, it is therefore vital not to rebiopsy the same area if the original biopsy was negative, and it is particularly vital to rebiopsy the same area if a possible abnormal area was detected on first biopsy as ASAP (Iczkowski et al. 2002). Thus, the locations of the cores obtained from the prostate must be known accurately to help guide the physician during the repeat biopsy (Thorson and Humphrey 2000; San Francisco et al. 2003), to help in correlating any imaging evidence of the disease, and to provide improved planning for subsequent therapy.

9.4.2   Multimodality Directed Prostate Biopsy

A variety of imaging techniques and molecular probes are being investigated to improve the early detection of PCa. Different magnetic resonance imaging (MRI) techniques have been evaluated using body and endorectal coils, contrast enhancement, and different pulse sequences (Futterer et al. 2006; Hricak et al. 2007; Manenti et al. 2007), resulting in disease detection sensitivity and specificity of 80% to 88% and 75% to 95%, respectively (Heijmink et al. 2007; Hricak et al. 2007; Morgan et al. 2007). Positron emission tomography (PET) combined with computed tomography (CT) is used to detect early disease, with the newer PET imaging probes proving to be the more promising (Farsad et al. 2005; Martorana et al. 2006; Schoder and Gonen 2007). Although progress has been made with improved PET and MRI techniques, they do not yet have ideal specificity or sufficient accuracy to assess the grade of the cancer. Thus, a biopsy of suspicious lesions on MRI or PET is required to provide a definitive diagnosis and grade of the disease. Systems have been developed to perform biopsies in the MRI suite; however, the cost of the equipment and prolonged use of the MRI is extremely expensive and likely prohibitive given the large number of patients requiring biopsy. However, conventional two-dimensional (2-D) TRUS guidance of the biopsy procedure limits the physician's ability to target locations identified as suspicious on other modalities (Figure 9.2).

As we currently do not have a highly sensitive and specific imaging test for local staging of PCa, there is a growing belief that the optimal method to guide prostate biopsy will involve not just one but a combination of imaging modalities. Three-dimensional TRUS imaging combined with functional or molecular imaging from another imaging modality such as radiopharmaceutical imaging (PET, SPECT) or magnetic resonance imaging (MRS, MRI) may provide the best approach for guiding prostate biopsy.

9.4.3   Optimal Prostate Sampling

The optimal distribution of cores within the prostate has been studied extensively, and it has been shown that uniform biopsy approaches such as sextant methods are subject to sampling limitations in view of the wide variations in gland sizes (Karkiewicz et al. 1998; Chon et al. 2002). This issue has been explored using computer simulations of the biopsy procedure and prostate anatomy, with probability distribution of location, frequency, and volume of PCa obtained from radical prostatectomy specimens (Shen et al. 2001; Zeng et al. 2001). Results from studies that explored different systematically distributed cores demonstrated that the positive biopsy yield depends on

Chapter 9

the magnitude of gland sampling (Zeng et al. 2001). Increasing the number of biopsy cores increases the biopsy yield, and this effect is most pronounced in larger prostates. Using the same number of cores regardless of individual prostate characteristics may lead to the oversampling of small glands and less extensive and potentially inadequate sampling of large glands. These sampling differences may account for the reported variation in biopsy yields in many studies (Bazinet et al. 1996; Karakiewicz et al. 1997). Thus, prostate volume information is essential at the time of biopsy for determining the number and distribution of biopsy cores required for optimal clinical results (Middleton et al. 1995).

9.5 Three-Dimensional TRUS-Guided Prostate Biopsy

Given the clinical need for 3-D image-guided navigation, several systems have been developed using CT or MRI to identify potential malignant tumors and to guide a needle for a biopsy (Fichtinger et al. 2002; Barnes et al. 2005; Krieger et al. 2005). Because ultrasound imaging is the clinical standard for image-guided biopsy of the prostate, not only the use of CT or MRI would be costly but also the resulting workflow would differ significantly from the current protocol. Thus, we have developed a 3-D TRUS-based navigation system that provides a reproducible record of the 3-D locations of the biopsy targets throughout the procedure.

For the past few years, we have developed tools for 3-D prostate TRUS imaging and have explored their application to prostate cryosurgery, brachytherapy, and biopsy guidance (Downey and Fenster 1995, 1998; Chin et al. 1996, 1998; Elliot et al. 1996; Onik et al. 1996; Tong et al. 1996, 1998a,b; Bax et al. 2008; Cool et al. 2008). During that time, we focused on improving the management of PCa by developing a new prostate biopsy approach, which promises to overcome some of its current limitations. This is being accomplished by combining 3-D TRUS and novel image processing tools into a 3-D TRUS-guided prostate biopsy system. This system uses 3-D TRUS imaging, with the capability to be registered with images from other modalities, to help the physician to optimally plan and sample the prostate and to automatically record the biopsy locations for use in guiding subsequent biopsies or therapy.

The system we have developed is a mechanical 3-D biopsy system that maintains the procedural workflow, minimizing costs and physician retraining. This mechanical system has 4 degrees of freedom (DF) and has an adaptable cradle that supports commercially available end-firing TRUS transducers used for prostate biopsy (Bax et al. 2008). It also enables real-time tracking and recording of the 3-D position and orientation of the biopsy needle as the physician manipulates the TRUS transducer. The following describes the components of the system, including hardware, modeling and segmentation algorithms, and system validation using a multimodal US/CT prostate phantom.

9.5.1 Mechanical System Design

Our approach involves the use of a device composed of two mechanisms (Figures 9.4 and 9.5): an articulated multijointed stabilizer and a transducer tracking mechanism.

The end-firing TRUS transducer with the biopsy needle guide in place is mounted to the mechanical tracking mechanism in a manner where the US probe is free to rotate

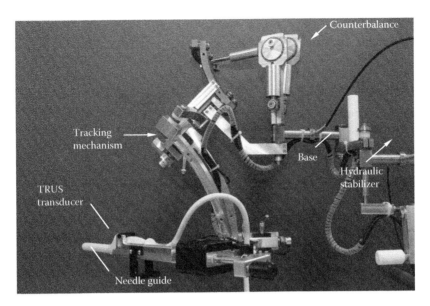

FIGURE 9.4 Photograph of the mechanical tracking system to be used for 3-D TRUS-guided prostate biopsy. The system is mounted at the base of a stabilizer while the linkage enables the TRUS transducer to be manually manipulated about an RCM, to which the center of the probe tip is aligned. The spring-loaded counterbalance is used to fully support the weight of the system throughout its full range of motion about the RCM.

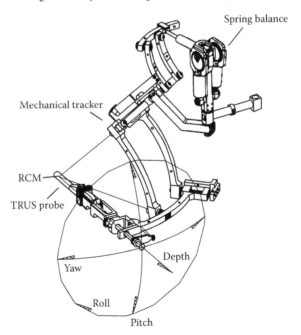

FIGURE 9.5 A rear perspective view of the mechanical tracker which supports the TRUS probe and attached cradle about a fixed point in space. This configuration constrains the TRUS probe motion to 3 DF (roll, pitch, and yaw) and one degree of translation along the axis of the probe. The mechanical guidance system uses a spring-loaded counterbalance mechanism to support the payload, which prevents the probe from drifting when the physician is not holding the probe and enables the physician to control its movements with a light touch of the hand. The circle in the diagram illustrates the working space of the mechanism where the roll and yaw movements enable a maximum of ±45° of movement from the orientation illustrated.

around its longitudinal axis (Figures 9.4 and 9.5). The tracking assembly is attached to a stabilizer, which is mounted on the patient's examination bed or a free-standing cart (preferred). In this way, the physician can manipulate the tracking mechanism freely, insert the transducer through the anus, and rotate the transducer to acquire a 3-D image of the prostate. The tracking linkage contains angle-sensing encoders mounted to each joint to transmit to the computer the angles between the arms (Figure 9.5). This arrangement enables the computer to determine the relative position of the transducer as it is being manipulated. The needle location can be calculated because the biopsy gun is mounted onto the transducer and its position relative to the transducer is calibrated.

The mechanical tracking device is a spherical linkage assembly consisting of three elements and three (pinned) connections. The axis of each revolute joint converges to a common point on the remote center of motion (RCM) (see Figure 9.5). This linkage is a nonparametric planar joint with 2 DF. The base link (Figure 9.5, L1) defines the reference axis of the proposed coordinate system and is fixed to a stabilizer. This RCM is critical in the linkage design, as the RCM minimizes targeting errors within the prostate. As the TRUS transducer is constrained through a stationary point, the physician's movements are replicated at a scaled down rate (minified through the RCM), minimizing changes in morphology and dislocation of the prostate. In addition, the RCM enables a precision equivalent to that of robotic assisted machines. Thus, the system improves the physician's ability to accurately biopsy a point of interest within the patient's prostate. In addition, stabilization is also accomplished using a mechanical spring-loaded counterbalancing system that maintains the orientation of the probe and attached biopsy gun even when the physician removes his hand.

9.5.2 Prostate Biopsy Workflow and System User Interface

The end-firing transducer is mounted onto the tracking assembly such that the tip of the probe is initially set to the RCM point of the tracker linkage. The multijointed stabilizer is unlocked and the physician is able to manipulate the transducer. The physician begins the procedure by inserting the TRUS transducer into the patient's rectum and aligns the prostate to the center of the 2-D TRUS image. The physician then acquires a 3-D image of the prostate by rotating the transducer 180 degrees about its longitudinal axis (Figure 9.5) (Fenster et al. 2001). A graphical model of the prostate is then generated by a semi-automatic 3-D segmentation algorithm (Ladak et al. 2000; Hu et al. 2003; Wang et al. 2003; Cool et al. 2006). This segmentation algorithm requires that the physician selects four points around the boundary of a 2-D prostate cross section (Figures 9.6 and 9.7). The algorithm then performs a stepwise rotational 2-D segmentation by fitting a dynamic deformable contour to each image slice and the result is then propagated to each succeeding image slice. In addition, stabilization is also accomplished using a mechanical spring-loaded counterbalancing system that maintains the orientation of the probe and attached biopsy gun even when the physician removes his hand.

After the prostate model has been constructed, the physician can then manipulate the 3-D image on the computer screen and select locations to biopsy. Other images or information (e.g., MRI or PET/CT images), if available, are registered to the 3-D TRUS image

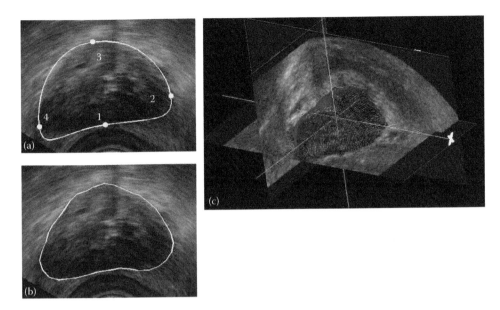

FIGURE 9.6   Steps of the 2-D and 3-D prostate segmentation algorithm. (a) The user initializes the algorithm by placing four points on the boundary to generate an initial contour. (b) The deformable dynamic contour approach is used to refine the initial contour until it matches the prostate boundary. (c) The contour is propagated to adjacent 2-D slices of the 3-D TRUS image and refined. The process is repeated until the complete prostate is segmented as shown.

and displayed as an overlay on the computer screen. After all biopsy targets have been selected, the system then displays the 3-D needle guidance interface (see Figure 9.8), which facilitates the systematic targeting of each biopsy location previously selected.

As the physician manually manipulates the TRUS transducer, the 3-D location and orientation of the transducer and needle trajectory are tracked in real time throughout the procedure on the computer screen. Figure 9.8 illustrates the biopsy interface, which is composed of four windows: the live 2-D TRUS video stream, the 3-D TRUS image, and two 3-D model views. The 2-D TRUS window displays the real-time 2-D TRUS image from the US machine. The 3-D TRUS window contains a 2-D slice of the 3-D static model in real time to reflect the expected orientation and position of the TRUS probe. This correspondence enables the physician to compare the 3-D image with the real-time 2-D image to determine if the prostate has moved or deformed to a prohibitive extent. Finally, the two 3-D model windows show: (i) orthogonal views (sagittal and coronal) of the 3-D prostate model, (ii) the real-time position of the 2-D TRUS image plane, (iii) and the expected trajectory of the biopsy needle, which is illustrated by the line intersecting the circle. The targeting circle on the screen illustrates all accessible needle trajectories by rotating the US probe about its longitudinal axis. When the needle path intersects the chosen target, the tracker is locked in place and a biopsy is performed. The biopsy location is then recorded in 3-D from the tracker orientation, and the system is ready for the next biopsy. After the needle is withdrawn, a 3-D image may be obtained to determine if there is any movement or swelling of the prostate.

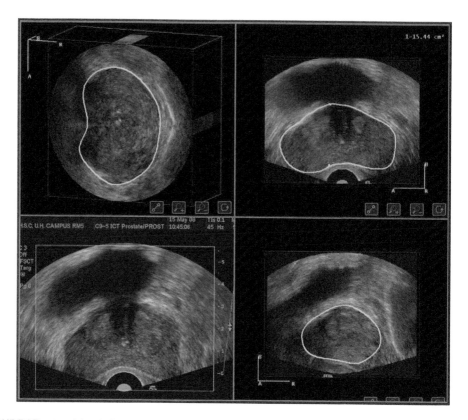

FIGURE 9.7   Axial (top left), sagittal (top right), and coronal views (bottom right) of the 3-D TRUS prostate image. The coronal view of the prostate is not possible using the current 2-D TRUS biopsy procedure. The image within the green box (bottom left) shows the live video stream from the US machine. The image within the green bounding box was digitized by a frame grabber as the physician rotated the TRUS transducer and then reconstructed into the 3-D image shown in the other three windows.

9.5.3   Mock Biopsy System Testing

The accuracy of the system was tested in a mock biopsy procedure. Using our 3-D TRUS-guided biopsy system, a traditional sextant biopsy was performed within a prostate phantom with contralateral biopsy targets at the base, mid-gland, and apex of the prostate. The accuracy of the biopsy was then validated using CT. The prostate test phantom consisted of two components: a simulated prostate and surrounding background tissue. Both the background tissue and the simulated prostate were constructed from a colloidal agar solution composed of cellulose scattering agent (15% by weight) to mimic the appearance of tissue in US (Rickey et al. 1995; Cool et al. 2008). In addition, the simulated prostate consisted of tungsten powder (1% by weight), which was used to provide image contrast in CT images. The simulated prostate was embedded in the background tissue, which assisted in the stabilization of the simulated prostate within the phantom. Furthermore, stainless steel ball bearings (1 mm diameter) were placed within the background tissue material and positioned in a configuration surrounding the simulated prostate. These fiducials were used for coordinate registration between the TRUS and the CT reference images. To determine the location of the fiducials within the US and CT images, an intensity center-of-mass algorithm was used to localize the fiducials.

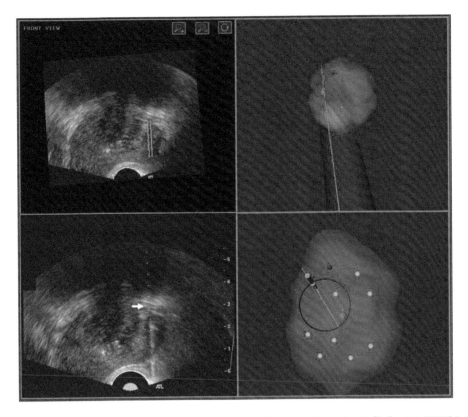

FIGURE 9.8 The 3-D biopsy system interface is composed of 4 windows: (top left) the 3-D TRUS image dynamically sliced to match the real-time TRUS probe 3-D orientation, (bottom left) the live 2-D TRUS video stream, (right side) and the 3-D location of the biopsy core is displayed within the 3-D prostate models. The targeting ring in the bottom right window shows all the possible needle paths that intersect the preplanned target by rotating the TRUS about its long axis. This enables the physician to maneuver the TRUS probe to the target (highlighted by the red dot) in the shortest possible distance. The biopsy needle (arrow) is visible within the real-time 2-D TRUS image.

A standard six-core biopsy procedure using a 9 MHz TRUS probe (Philips Medical Systems, Seattle, WA) was performed on the prostate phantom using our 3-D tracking system. Before needle insertion, the prostate phantom was scanned to reconstruct a 3-D prostate model. The 6-core biopsy targets were then selected and displayed in the user interface. After the biopsy procedure, the phantom was imaged using a preclinical high-resolution micro-CT scanner (eXplore Locus, GE Healthcare, London, ON) to determine the morphology of the prostate and the location of the 3-D biopsy cores within. With isometric voxel dimensions of 0.15 mm, the CT image was used as a reference to determine the targeting error of the biopsy procedure.

The accuracy of the biopsy system was evaluated by guiding a biopsy needle to a target and recording the 3-D biopsy core location using the mechanical tracking apparatus (Bax et al. 2008; Cool et al. 2008). For a given biopsy site, the needle guidance error (NGE) is defined as the minimum Euclidean distance between the biopsy target and the CT biopsy core, which has a 19 mm core length resulting from the 18-gauge biopsy needle notch used. The NGE can be further decomposed into the human guidance error and the needle trajectory error (NTE). The needle guidance human error is defined as the distance between the biopsy target and the expected needle path, and the NTE is the minimum distance between the actual needle trajectory and the recorded location of the biopsy core. The error in recording the 3-D biopsy core location is quantified by three metrics: BLE^{min}, BLE^{center}, and BLE^{θ}. BLE^{min} is defined as the minimum distance between the CT and the TRUS biopsy cores. BLE^{θ} is the angle between the cores when projected on a plane perpendicular to BLE^{min}. BLE^{center} is the center-to-center distance between the coregistered biopsy cores.

9.5.4   Results of Mock Biopsy System Testing

The reconstructed axial and coronal image slices of the 3-D TRUS image using the biopsy system (see Figure 9.9) of a certified industrial US phantom (CIRS, phantom volume = 21.50 cm³) was scanned and segmented with a mean volume error of 4.7%.

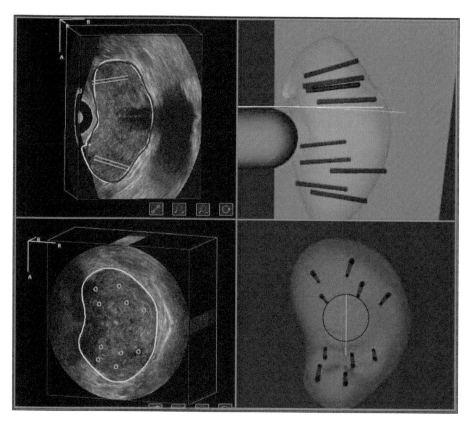

FIGURE 9.9   Previous biopsy locations can be viewed to plan the rebiopsy. (Top left) This image also enables the physician to compare the previously segmented prostate boundary (red) with the currently segmented boundary (green). (Bottom left) The coronal view of the patient's prostate with the segmented boundary and the location of the previous biopsy cores as circles. (Top and bottom right) The 3-D locations of the biopsy core are displayed within the 3-D prostate models. The targeting ring in the bottom right window shows all the possible needle paths that intersect the selected target by rotating the TRUS about its long axis. The vertical white line through the targeting ring shows the orientation of the US imaging plane.

Table 9.1   A 3-D Biopsy System Accuracy Based on the Biopsy Core Analysis Described in Section 9.5.4 and on Bax et al. (2011)

Experiment	System Accuracy			Localization Error Metrics		
	NGE (mm)	NGHE (mm)	NTE (mm)	BLE^{min} (mm)	BLE^{center} (mm)	BLE$^{\theta}$ (°)
1	3.79 ± 0.86	0.37 ± 0.51	4.41 ± 0.27	1.93 ± 0.75	3.98 ± 1.45	5.65 ± 2.04
2	0.70 ± 0.32	0.27 ± 0.12	1.44 ± 0.46	1.08 ± 0.61	3.82 ± 2.52	5.61 ± 3.30
3	1.95 ± 0.39	0.19 ± 0.09	1.97 ± 0.59	2.39 ± 0.98	4.07 ± 1.22	7.76 ± 2.01
4	2.21 ± 1.39	0.99 ± 0.08	2.06 ± 1.94	1.20 ± 0.67	4.34 ± 3.17	6.46 ± 2.55
5	1.99 ± 0.77	0.90 ± 0.11	0.50 ± 0.51	0.97 ± 0.92	3.13 ± 2.00	7.93 ± 1.20
Mean	2.13 ± 1.28[a]	0.54 ± 0.41	2.08 ± 1.59	1.51 ± 0.92	3.87 ± 1.81	6.68 ± 2.23
95% CI	1.67–2.59	0.39–0.69	1.51–2.65	1.18–1.84	3.22–4.52	5.88–7.48

Note: NGE, needle guidance error; NGHE, needle guidance human error; NTE, and needle trajectory error; all evaluate biopsy targeting accuracy. The biopsy localization metrics (BLE^{min}, BLE^{center}, and BLE$^{\theta}$) indicate the errors in recording the 3-D location of the biopsy cores; 95% CI represents the confidence interval of the mean. All values are reported as mean ± SD.

[a] Statistically less than 5 mm using a one-sided *t*-test ($p < 0.001$, $B = 0.2$, $n = 29$).

The quantitative results for 3-D biopsy NGE and recording errors of biopsy core locations are summarized in Table 9.1. Results from the US targeting experiments revealed a mean NGE = 2.13 ± 1.28 mm. NHGE resulted in a mean error of 0.54 ± 0.41 mm, and the trajectory error of the biopsy needle deviating from the expected needle path, NTE, was 2.08 ± 1.59 mm. The biopsy core localization error was found to be BLE^{min} = 1.51 ± 0.92 mm and with a mean angulation difference of BLE$^{\theta}$ = 6.68° ± 2.23°. The center-to-center distance (BLE^{center}) between corresponding biopsy cores was larger than BLE^{min}, at 3.87 ± 1.81 mm.

9.6   Prostate Motion and Deformation

9.6.1   Intrasession Motion and Deformation Correction

During a biopsy procedure, the prostate may shift and/or deform due to patient motion and/or transducer probe pressure. This may lead to a misalignment between the targets in the prebiopsy 3-D image and their correct locations in the real-time image, thus making it challenging to establish a consistent reference frame for needle guidance and procedure planning. Therefore, biopsy targets determined based on a previous biopsy procedure must be transformed during the procedure to compensate for intraprocedure prostate motion and deformation. For a registration method to be suitable for clinical use, the postregistration misalignment between corresponding points in the transformed source image and target image must be no greater than 2.5 mm. The performance of iterative closest point and thin-plate spline surface-based registration methods and block matching and B-spline image-based methods was evaluated by measuring the TRE as the postregistration misalignment of 60 manually marked, corresponding, intrinsic fiducials (Karnik et al. 2010). These fiducials were obtained

Chapter 9

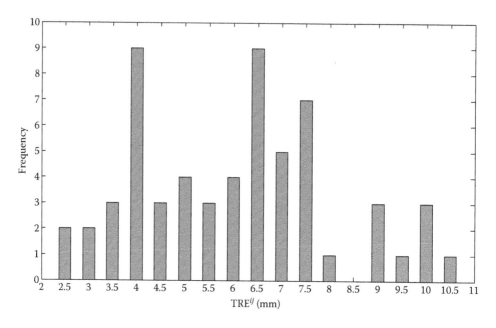

FIGURE 9.10   Frequency distribution of preregistration distances between all 60 corresponding fiducials from intrabiopsy 3-D TRUS scans.

from 3-D TRUS images of 16 subjects who either had previous negative biopsies or were suspect for premalignant cells. Two 3-D images of each prostate were collected. The first 3-D TRUS image was obtained at the start of the procedure, before any biopsy cores were collected, and was used for planning the biopsy procedure. The second 3-D TRUS image was obtained near the end of the procedure to maximize potential prostate shifting and deformation. In addition to calculating the TRE, we measured the fiducial localization error and the effects on TRE of segmentation variability and fiducial distance from the transducer probe tip.

The frequency distributions of the distances between corresponding fiducials in the images obtained at the start and near the end of the biopsy procedure are shown in Figures 9.10 and 9.11. The values of the TRE before registration ranged from 2.35 to 10.1 mm, with a mean of 6.11 ± 2.02 mm. The mean TRE values for all 60 fiducials were 2.13 ± 0.80 mm and 2.09 ± 0.77 mm for surface-based rigid and nonrigid registration, respectively, and the mean image-based rigid and nonrigid TRE values for all 60 fiducials were 1.74 ± 0.84 mm and 1.50 ± 0.83 mm, respectively. This shows that the best correction was obtained with nonrigid image-based registration giving a mean TRE = 1.50 ± 0.83 mm.

9.6.2   Intersession Motion and Deformation Correction

In the case of a repeat biopsy, image registration is required to map biopsy targets planned on a previous session's 3-D TRUS image into the context of the current session. The performance of surface- and image-based rigid and nonrigid registration algorithms was measured for this task, as shown in Section 9.6.1 (Karnik et al. 2011). Two 3-D TRUS images were collected for each of 13 subjects, where each image was collected in separate

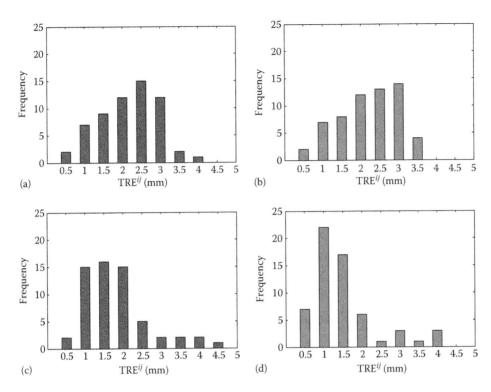

FIGURE 9.11   Frequency distribution of TRE^{ij}: (a) rigid surface-based registration, TRE = 2.13 ± 0.80 mm; (b) nonrigid surface-based registration, TRE = 2.09 ± 0.77 mm; (c) rigid image-based registration, TRE = 1.74 ± 0.84 mm; (d) nonrigid image-based registration, TRE = 1.50 ± 0.83 mm.

biopsy sessions 1 week apart. We evaluated the accuracy and variability of iterative closest point and thin-plate spline surface-based registration methods, and block matching and B-spline image-based methods. One hundred and nine manually marked intrinsic fiducials (calcifications) were used to calculate a TRE for each of the tested methods for the whole gland (WG). A similar analysis was performed separately for the peripheral zone (PZ) because this area harbors up to 80% of all PCa.

Figure 9.12 shows the frequency distribution of the preregistration distances between the corresponding fiducial pairs for the WG and PZ, with RMS distances of 7.27 and 6.00 mm, respectively, and the values ranging from 0.28 to 18.00 mm and from 0.41 to 10.87 mm, respectively.

Table 9.2 summarizes the computed TRE values for all tested methods. The surface-based rigid and nonrigid TREs differed by 0.07 mm, whereas the image-based TREs differed by 0.31 mm. A two-tailed paired t-test showed statistically significant differences ($\alpha = 0.05$ for all statistical significance tests in this paper, and the normality of all such tested data was verified using the D'Agostino and Pearson omnibus (K2) normality test between the rigid and the nonrigid approaches for both the surface-based ($p = 0.046$) and the image-based ($p \ll 0.001$) methods. The rigid registration TREs (for surface- and image-based methods) differed significantly ($p \ll 0.001$) by 0.70 mm, with the rigid image-based method yielding the better performance. The nonrigid image-based method outperformed the surface-based method by 1.08 mm ($p \ll 0.001$).

Chapter 9

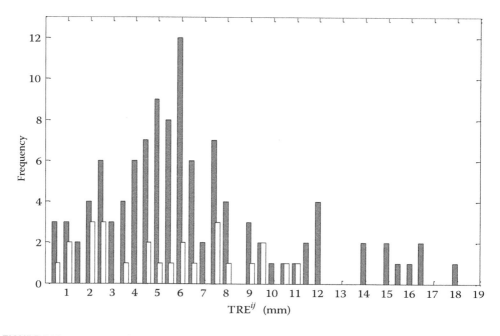

FIGURE 9.12 Frequency distribution of whole gland (gray) and peripheral zone (white) preregistration fiducial misalignments for the case of inter-biopsy session.

Table 9.2 Summary of TRE before Registration and after Surface- and Image-Based Registration for Whole Gland (WG) and Peripheral Zone (PZ) for the Case of Interbiopsy Session

Registration	Rigid (mm)	Nonrigid (mm)	*N* (fiducials)
Preregistered, WG		7.27	*N* = 109
Surface based, WG	3.16	3.23	
Image based, WG	2.46	2.15	
Preregistered, PZ		6.00	*N* = 26
Surface based, PZ	3.42	3.48	
Image based, PZ	2.61	2.36	

9.7 Clinical Evaluation of 3-D TRUS/MRI-Guided Biopsy

In collaboration with the University Hospital of the London Health Sciences Centre (London, Ontario, Canada), we are conducting clinical studies to evaluate the clinical effect of fusion of MRI to intrabiopsy 3-D TRUS for 3-D US-guided targeted biopsy of suspicious MRI lesions on PCa detection and grading. Prostate MR imaging was performed on 31 patients with clinical suspicion for PCa before their 3-D TRUS-guided biopsy. T2, diffusion-weighted, and dynamic contrast–enhanced MR sequences were collected in a 3T magnet with an endorectal RF coil. All suspicious lesions in the MR images were then identified and delineated on the images, which were then registered to the 3-D TRUS image obtained during the biopsy procedure (see Figure 9.13). Using the 3-D TRUS-guided biopsy system, prostate biopsy cores were targeted toward each

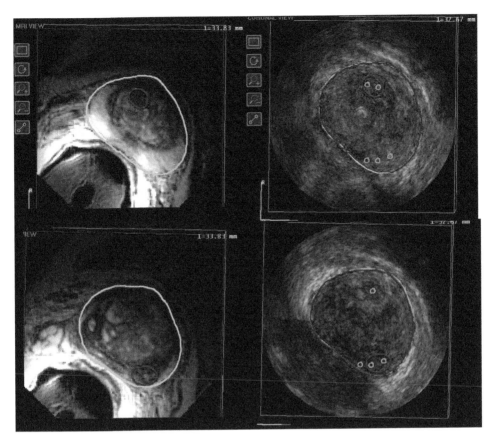

FIGURE 9.13 Three-dimensional TRUS and MRI images of the same patient showing two delineated suspicious lesions identified in the MR images (left two panels). The MR images were then registered with the 3-D TRUS images (right two panels) and the delineated regions superimposed on the 3-D TRUS images. These regions were then targeted with the 3-D TRUS-guided biopsy system, showing that one lesion was targeted with two cores and the second lesion was targeted with 3 cores.

suspicious delineated MRI lesion, which were displayed on the 3-D TRUS image (see Figure 9.13). A standard 12-core set of random biopsies was also performed on each patient and used as an internal control.

The results of this study showed that MRI–3-D TRUS fusion was successfully performed and the targeted biopsy needle cores had a significantly higher rates of prostate malignancy (30.0%) compared with random, sextant cores (10.0%). In total, PCa was biopsy confirmed in 11 patients; however, only 7 of these patients had abnormal MRI findings (even in retrospective analysis) and were sampled with targeted MRI–3-D TRUS fusion. Random sampling detected the remaining four patients. A significantly higher percentage of the targeted biopsy cores (47% ± 26%) contained cancer compared with the randomly sampled cores (28% ± 26%), and the MRI-targeted cores detected a higher Gleason cancer grade than the random cores for three patients, modifying potential treatment modalities. This study showed that MRI–3-D TRUS fusion enables superior sampling of PCa visible on MRI. This technology may benefit both cancer detection and accurate malignancy grading for appropriate therapeutic management; however, further testing is needed to establish the full utility of this technology.

Chapter 9

9.8   Conclusions

The mechanical tracking system to guide and record prostate biopsy locations is easy to maneuver with a 4 DF range of motion of the tracker. The tracker permits manual motions identical to the current conventional procedure, where restricted movements are produced by the US probe in the patient's rectum. The procedural workflow for the 3-D TRUS-guided biopsy system is also similar to current procedure. The only additions are a 10-second scan needed to produce a 3-D TRUS image, the segmentation of the prostate in 3-D, and the selection of several 3-D biopsy targets (optional). Using the 3-D TRUS imaging tools and protocols of current biopsy procedures, the clinical integration of the mechanical 3-D TRUS-guided biopsy system should be easily performed into most centers.

The reconstruction of a 3-D TRUS image of a prostate phantom demonstrated that the rotational approach to 3-D TRUS imaging can produce 3-D images without significant visible discontinuity or probe alignment artifacts. Volume calculations from the 3-D TRUS image were within 4.7% of the CT standard, demonstrating that the 3-D TRUS system can generate accurate volume measurements. Experiments with a mock biopsy procedure using a prostate phantom demonstrated that the system is able to guide the biopsy needle to predefined targets to within 2.13 mm. This error is primarily due to deflection of the biopsy needle from the expected path of the biopsy guide (Wan et al. 2005). Despite this error, the targeting accuracy is still within the 5 mm radius of the smallest tumors considered clinically significant. Recording the location of the biopsy in 3-D within the prostate will increase the physician's ability to accurately guide the biopsy needle to selected targets in cases of rebiopsy and active surveillance.

The patient studies have demonstrated that it is possible to minimize the effects of prostate motion through a variety of mechanical and software mechanisms. By providing mechanical means to stabilize and support the US transducer, the pressure between the transducer tip and the rectal wall is maintained and the probe insertion depth kept constant. The remote fulcrum of the mechanical device (i.e., RCM) further maintains the constant pressure on the rectal wall by minimizing motion between the RCM and the probe tip.

Although hardware and software solutions can be provided to overcome prostate motion and deformation, it is not possible to control all patient motion during the biopsy, particularly if the patient moves a significant distance after the firing of the biopsy needle. In these cases, the real-time 2-D TRUS image would not correspond to the 2-D coronal slice displayed in the 3-D static prostate model in our software interface, visually informing the physician that the preselected targets are no longer valid. To overcome this problem, a software module would have to be developed to inform the physician that the prostate has moved and then correct for the motion and deformation. This task must be performed quickly, possibly in real time, using an implementation of the software in a graphical processing unit.

Acknowledgments

The authors gratefully acknowledge the financial support of the Canadian Institutes of Health Research, the Ontario Institute for Cancer Research, the Ontario Research Fund, the National Science and Engineering Research Council, and the Canada Research Chair program.

References

Abbas, F. and P. T. Scardino (1997). The natural history of clinical prostate carcinoma [editorial; comment]. *Cancer* **80**(5): 827–33.

Altman, A. L. and M. I. Resnick (2001). Ultrasonographically guided biopsy of the prostate gland. *Journal of Ultrasound Medicine* **20**(2): 159–67.

Barnes, A. S., S. J. Haker, R. V. Mulkern, M. So, A. V. D'Amico, and C. M. Tempany (2005). Magnetic resonance spectroscopy-guided transperineal prostate biopsy and brachytherapy for recurrent prostate cancer. *Urology* **66**(6): 1319.

Bax, J., D. Cool, L. Gardi, K. Knight, D. Smith, J. Montreuil, S. Sherebrin, C. Romagnoli, and A. Fenster (2008). Mechanically assisted 3D ultrasound guided prostate biopsy system. *Medical Physics* **35**(12): 5397–410.

Bazinet, M., P. I. Karakiewicz, and J. A. Hanly (1996). Prediction of cancer detection rates with sector biopsy of the prostate: A three dimensional dynamic model. *Journal of Urology* **155**: 259A.

Chin, J. L., D. B. Downey, M. Mulligan, and A. Fenster (1998). Three-dimensional transrectal ultrasound guided cryoablation for localized prostate cancer in nonsurgical candidates: A feasibility study and report of early results. *Journal of Urology* **159**(3): 910–4.

Chin, J. L., D. B. Downey, G. Onik, and A. Fenster (1996). Three-dimensional prostate ultrasound and its application to cryosurgery. *Techniques in Urology* **2**(4): 187–93.

Chon, C. H., F. C. Lai, J. E. McNeal, and J. C. Presti, Jr. (2002). Use of extended systematic sampling in patients with a prior negative prostate needle biopsy. *Journal of Urology* **167**(6): 2457–60.

Clements, R., O. U. Aideyan, G. J. Griffiths, and W. B. Peeling (1993). Side effects and patient acceptability of transrectal biopsy of the prostate. *Clinical Radiology* **47**(2): 125–6.

Collins, G. N., S. N. Lloyd, M. Hehir, and G. B. McKelvie (1993). Multiple transrectal ultrasound-guided prostatic biopsies—True morbidity and patient acceptance. *British Journal of Urology* **71**(4): 460–3.

Cool, D., D. Downey, J. Izawa, J. Chin, and A. Fenster (2006). 3D prostate model formation from non-parallel 2D ultrasound biopsy images. *Medical Image Analysis* **10**(6): 875–87.

Cool, D., S. Sherebrin, J. Izawa, J. Chin, and A. Fenster (2008). Design and evaluation of a 3D transrectal ultrasound prostate biopsy system. *Medical Physics* **35**(10): 4695–707.

Coplen, D. E., G. L. Andriole, J. J. Yuan, and W. J. Catalona (1991). The ability of systematic transrectal ultrasound guided biopsy to detect prostate cancer in men with the clinical diagnosis of benign prostatic hyperplasia. *Journal of Urology* **146**(1. 1): 75–7.

Djavan, B., M. Remzi, C. C. Schulman, M. Marberger, and A. R. Zlotta (2002). Repeat prostate biopsy: Who, how and when? A review. *European Urology* **42**(2): 93–103.

Djavan, B., A. R. Zlotta, S. Ekane, M. Remzi, G. Kramer, T. Roumeguere, M. Etemad, R. Wolfram, C. C. Schulman, and M. Marberger (2000). Is one set of sextant biopsies enough to rule out prostate Cancer? Influence of transition and total prostate volumes on prostate cancer yield. *European Urology* **38**(2): 218–24.

Downey, D. B. and A. Fenster (1995). Three-dimensional power Doppler detection of prostate cancer [letter]. *American Journal of Roentgenology* **165**(3): 741.

Downey, D. B. and A. Fenster (1998). Three-dimensional ultrasound: A maturing technology. *Ultrasound Quarterly* **14**(1): 25–40.

Eastham, J. A. and P. T. Scardino (2001). Early diagnosis and treatment of prostate cancer. *Disease-a-Month* **47**(9. 9): 421–59.

Elliot, T. L., D. B. Downey, S. Tong, C. A. McLean, and A. Fenster (1996). Accuracy of prostate volume measurements in vitro using three-dimensional ultrasound. *Academic Radiology* **3**(5): 401–6.

Farsad, M., R. Schiavina, P. Castellucci, C. Nanni, B. Corti, G. Martorana, R. Canini, W. Grigioni, S. Boschi, M. Marengo, C. Pettinato, E. Salizzoni, N. Monetti, R. Franchi, and S. Fanti (2005). Detection and localization of prostate cancer: Correlation of (11)C-choline PET/CT with histopathologic step-section analysis. *Journal of Nuclear Medicine* **46**(10): 1642–9.

Fenster, A., D. B. Downey, and H. N. Cardinal (2001). Topical review: Three-dimensional ultrasound imaging. *Physics in Medicine and Biology* **46**(5): R67–99.

Chapter 9

Fichtinger, G., T. L. DeWeese, A. Patriciu, A. Tanacs, D. Mazilu, J. H. Anderson, K. Masamune, R. H. Taylor, and D. Stoianovici (2002). System for robotically assisted prostate biopsy and therapy with intraoperative CT guidance. *Academic Radiology* **9**(1. 1): 60–74.

Futterer, J. J., S. W. Heijmink, T. W. Scheenen, J. Veltman, H. J. Huisman, P. Vos, C. A. Hulsbergen-Van de Kaa, J. A. Witjes, P. F. Krabbe, A. Heerschap, and J. O. Barentsz (2006). Prostate cancer localization with dynamic contrast-enhanced MR imaging and proton MR spectroscopic imaging. *Radiology* **241**(2): 449–58.

Garfinkel, L. and M. Mushinski (1994). Cancer incidence, mortality and survival: Trends in four leading sites. *Statistical Bulletin (Metropolitan Life Insurance Company)* **75**(3): 19–27.

Goto, Y., M. Ohori, A. Arakawa, M. W. Kattan, T. M. Wheeler, and P. T. Scardino (1996). Distinguishing clinically important from unimportant prostate cancers before treatment: Value of systematic biopsies. *Journal of Urology* **156**(3. 3): 1059–63.

Heijmink, S. W., J. J. Futterer, T. Hambrock, S. Takahashi, T. W. Scheenen, H. J. Huisman, C. A. Hulsbergen-Van de Kaa, B. C. Knipscheer, L. A. Kiemeney, J. A. Witjes, and J. O. Barentsz (2007). Prostate cancer: Body-array versus endorectal coil MR imaging at 3 T—Comparison of image quality, localization, and staging performance. *Radiology* **244**(1): 184–95.

Hricak, H., P. L. Choyke, S. C. Eberhardt, S. A. Leibel, and P. T. Scardino (2007). Imaging prostate cancer: A multidisciplinary perspective. *Radiology* **243**(1): 28–53.

Hu, N., D. Downey, A. Fenster, and H. Ladak (2003). Prostate boundary segmentation from 3D ultrasound images. *Medical Physics* **30**: 1648–59.

Iczkowski, K. A., T. J. Bassler, V. S. Schwob, I. C. Bassler, B. S. Kunnel, R. E. Orozco, and D. G. Bostwick (1998). Diagnosis of "suspicious for malignancy" in prostate biopsies: Predictive value for cancer. *Urology* **51**(5): 749–57; discussion 757–8.

Iczkowski, K. A., H. M. Chen, X. J. Yang, and R. A. Beach (2002). Prostate cancer diagnosed after initial biopsy with atypical small acinar proliferation suspicious for malignancy is similar to cancer found on initial biopsy. *Urology* **60**(5): 851–4.

Jani, A. B. and S. Hellman (2003). Early prostate cancer: Clinical decision-making. *Lancet* **361**(9362): 1045–53.

Jemal, A., R. Siegel, E. Ward, T. Murray, J. Xu, and M. J. Thun (2007). Cancer statistics, 2007. *CA: A Cancer Journal for Clinicians* **57**(1): 43–66.

Jemal, A., A. Thomas, T. Murray, and M. Thun (2002). Cancer statistics, 2002. *CA: A Cancer Journal for Clinicians* **52**(1): 23–47.

Karakiewicz, P. I., M. Bazinet, A. G. Aprikian, C. Trudel, S. Aronson, M. Nachabe, F. Peloquint, J. Dessureault, M. S. Goyal, L. R. Begin, and M. M. Elhilali (1997). Outcome of sextant biopsy according to gland volume. *Urology* **49**(1. 1): 55–9.

Karkiewicz, P. I., J. A. Hanley, and M. Bazinet (1998). Three-dimensional computer-assisted analysis of sector biopsy of the prostate. *Urology* **52**: 208–212.

Karnik, V. V., A. Fenster, J. Bax, D. W. Cool, L. Gardi, I. Gyacskov, C. Romagnoli, and A. D. Ward (2010). Assessment of image registration accuracy in three-dimensional transrectal ultrasound guided prostate biopsy. *Medical Physics* **37**(2): 802–13.

Karnik, V. V., A. Fenster, J. Bax, C. Romagnoli, and A. D. Ward (2011). Evaluation of intersession 3D-TRUS to 3D-TRUS image registration for repeat prostate biopsies. *Medical Physics* **38**(4): 1832–43.

Keetch, D. W., W. J. Catalona, and D. S. Smith (1994). Serial prostatic biopsies in men with persistently elevated serum prostate specific antigen values. *Journal of Urology* **151**(6. 6): 1571–4.

Krieger, A., R. C. Susil, C. Menard, J. A. Coleman, G. Fichtinger, E. Atalar, and L. L. Whitcomb (2005). Design of a novel MRI compatible manipulator for image guided prostate interventions. *IEEE Transactions on Biomedical Engineering* **52**(2): 306–13.

Ladak, H. M., F. Mao, Y. Wang, D. B. Downey, D. A. Steinman, and A. Fenster (2000). Prostate boundary segmentation from 2D ultrasound images. *Medical Physics* **27**(8): 1777–88.

Manenti, G., M. Carlani, S. Mancino, V. Colangelo, M. Di Roma, E. Squillaci, and G. Simonetti (2007). Diffusion tensor magnetic resonance imaging of prostate cancer. *Investigative Radiology* **42**(6): 412–9.

Martorana, G., R. Schiavina, B. Corti, M. Farsad, E. Salizzoni, E. Brunocilla, A. Bertaccini, F. Manferrari, P. Castellucci, S. Fanti, R. Canini, W. F. Grigioni, and A. D'Errico Grigioni (2006). 11C-choline positron emission tomography/computerized tomography for tumor localization of primary prostate cancer in comparison with 12-core biopsy. *Journal of Urology* **176**(3): 954–60; discussion 960.

Matlaga, B. R., L. A. Eskew, and D. L. McCullough (2003). Prostate biopsy: Indications and technique. *Journal of Urology* **169**(1): 12–9.

McNeal, J. E., D. G. Bostwick, R. A. Kindrachuk, E. A. Redwine, F. S. Freiha, and T. A. Stamey (1986). Patterns of progression in prostate cancer. *Lancet* **1**(8472): 60–3.

Mettlin, C., G. W. Jones, and G. P. Murphy (1993). Trends in prostate cancer care in the United States, 1974–1990: Observations from the patient care evaluation studies of the American College of Surgeons Commission on Cancer. *CA: A Cancer Journal for Clinicians* **43**(2): 83–91.

Middleton, R. G., I. M. Thompson, M. S. Austenfeld, W. H. Cooner, R. J. Correa, R. P. Gibbons, H. C. Miller, J. E. Oesterling, M. I. Resnick, S. R. Smalley et al. (1995). Prostate cancer clinical guidelines panel summary report on the management of clinically localized prostate cancer. The American Urological Association. *Journal of Urology* **154**(6): 2144–8.

Morgan, V. A., S. Kyriazi, S. E. Ashley, and N. M. DeSouza (2007). Evaluation of the potential of diffusion-weighted imaging in prostate cancer detection. *Acta Radiologica* **48**(6): 695–703.

Nelson, W. G., A. M. De Marzo, and W. B. Isaacs (2003). Prostate cancer. *New England Journal of Medicine* **349**(4): 366–81.

Onik, G. M., D. B. Downey, and A. Fenster (1996). Three-dimensional sonographically monitored cryosurgery in a prostate phantom. *Journal of Ultrasound Medicine* **15**(3): 267–70.

Park, S. J., H. Miyake, I. Hara, and H. Eto (2003). Predictors of prostate cancer on repeat transrectal ultrasound-guided systematic prostate biopsy. *International Journal of Urology* **10**(2): 68–71.

Presti, J. C., Jr., G. J. O'Dowd, M. C. Miller, R. Mattu, and R. W. Veltri (2003). Extended peripheral zone biopsy schemes increase cancer detection rates and minimize variance in prostate specific antigen and age related cancer rates: Results of a community multi-practice study. *Journal of Urology* **169**(1): 125–9.

Presti, J. C. J., J. J. Chang, V. Bhargava, and K. Shinohara (2000). The optimal systematic prostate biopsy scheme should include 8 rather than 6 biopsies: Results of a prospective clinical trial. *Journal of Urology* **163**(1. 1): 163–6; discussion 166–7.

Punglia, R. S., A. V. D'Amico, W. J. Catalona, K. A. Roehl, and K. M. Kuntz (2003). Effect of verification bias on screening for prostate cancer by measurement of prostate-specific antigen. *New England Journal of Medicine* **349**(4): 335–42.

Rickey, D. W., P. A. Picot, D. A. Christopher, and A. Fenster (1995). A wall-less vessel phantom for Doppler ultrasound studies. *Ultrasound in Medicine and Biology* **21**(9): 1163–76.

Rifkin, M. D. (1997). *Ultrasound of the Prostate-Imaging in the Diagnosis and Therapy of Prostatic Disease.* Philadelphia, New York, Lippincott-Raven Publishers.

Rodriguez, L. V. and M. K. Terris (1998). Risks and complications of transrectal ultrasound guided prostate needle biopsy: A prospective study and review of the literature. *Journal of Urology* **160**(6 Pt 1): 2115–20.

San Francisco, I., W. DeWolf, S. Rosen, M. Upton, and A. Olumi (2003). Extended prostate needle biopsy improves concordance of Gleason grading between prostate needle biopsy and radical prostatectomy. *Urology* **169**: 136–140.

Schoder, H. and M. Gonen (2007). Screening for cancer with PET and PET/CT: Potential and limitations. *Journal of Nuclear Medicine* **48**(Suppl 1): 4S–18S.

Schroder, F. H. and R. Kranse (2003). Verification bias and the prostate-specific antigen test—Is there a case for a lower threshold for biopsy? *New England Journal of Medicine* **349**(4): 393–5.

Shen, D., Z. Lao, J. Zeng, E. H. Herskovits, G. Fichtinger, and C. Davatzikos (2001). A statistical atlas of prostate cancer for optimal biopsy. *Proceedings of the Medical Image Computing and Computer-Assisted Intervention, MICCAI '01*: 416–424.

Shinohara, K., P. T. Scardino, S. S. Carter, and T. M. Wheeler (1989). Pathologic basis of the sonographic appearance of the normal and malignant prostate. *Urologic Clinics of North America* **16**(4): 675–91.

Silverberg, E., C. C. Boring, and T. S. Squires (1990). Cancer statistics, 1990 [see comments]. *CA: A Cancer Journal for Clinicians* **40**: 9–26.

Terris, M. K., J. E. McNeal, and T. A. Stamey (1992). Estimation of prostate cancer volume by transrectal ultrasound imaging. *Journal of Urology* **147**(3 Pt 2): 855–7.

Thorson, P. and P. A. Humphrey (2000). Minimal adenocarcinoma in prostate needle biopsy tissue. *American Journal of Clinical Pathology* **114**(6): 896–909.

Tong, S., H. N. Cardinal, D. B. Downey, and A. Fenster (1998a). Analysis of linear, area and volume distortion in 3D ultrasound imaging. *Ultrasound in Medicine and Biology* **24**(3): 355–73.

Tong, S., H. N. Cardinal, R. F. McLoughlin, D. B. Downey, and A. Fenster (1998b). Intra- and inter-observer variability and reliability of prostate volume measurement via two-dimensional and three-dimensional ultrasound imaging. *Ultrasound in Medicine and Biology* **24**(5): 673–81.

Tong, S., D. B. Downey, H. N. Cardinal, and A. Fenster (1996). A three-dimensional ultrasound prostate imaging system. *Ultrasound in Medicine and Biology* **22**(6): 735–46.

Chapter 9

Wan, G., Z. Wei, L. Gardi, D. B. Downey, and A. Fenster (2005). Brachytherapy needle deflection evaluation and correction. *Medical Physics* **32**(4): 902–9.

Wang, Y., H. N. Cardinal, D. B. Downey, and A. Fenster (2003). Semiautomatic three-dimensional segmentation of the prostate using two-dimensional ultrasound images. *Medical Physics* **30**(5): 887–97.

Zeng, J., J. Bauer, W. Zhang, I. Sesterhenn, R. Connelly, J. Lynch, J. Moul, and S. K. Mun (2001). Prostate biopsy protocols: 3D visualization-based evaluation and clinical correlation. *Computer Aided Surgery* **6**(1): 14–21.

10. Ultrasound Applications in the Brain

Meaghan A. O'Reilly and Kullervo Hynynen

10.1 Introduction

Ultrasound is an appealing modality for both diagnostic and therapeutic applications in the brain. As an imaging modality, it is nonionizing and thus less harmful than computed tomography (CT), and it is inexpensive and portable by comparison with both CT and magnetic resonance imaging (MRI). For therapeutic purposes, ultrasound has the ability to focus deep within tissue and can be applied noninvasively. This is particularly important in the brain, where conventional surgical approaches involve penetrating through healthy brain tissue and where the response to drug therapy is typically poor. Despite the attractive qualities of ultrasound for brain applications, its use in the brain is greatly limited by the unique challenges of the skull bone. This chapter examines both diagnostic and

Chapter 10

Ultrasound Imaging and Therapy. Edited by Aaron Fenster and James C. Lacefield © 2015 CRC Press/Taylor & Francis Group, LLC. ISBN: 978-1-4398-6628-3

therapeutic applications in the brain and the successful realization of transcranially applied ultrasound.

10.2 Skull Bone

The use of ultrasound in the brain is limited by the skull bone because of several difficulties associated with transmission of ultrasound through bone. First, ultrasound passing through the skull bone suffers from high losses, which increase with ultrasound frequency. The insertion loss of the skull is in part due to large reflective losses, which dominate at lower frequencies (<approximately 500 kHz) (Fry and Barger 1978). These reflective losses can be attributed to the acoustic impedance mismatch that occurs at the bone/tissue interface and can vary from 30% to 80% at normal incidence (Fry and Barger 1978). When the bone is of a resonance thickness for the ultrasound frequencies being transmitted (e.g., thickness = 1/2 λ), the reflective losses are minimized and the ultrasound transmission is a maximum (Fry and Barger 1978; Aarnio et al. 2005). However, other mechanisms, including scattering and absorption, contribute to the total insertion loss and are of greater significance at higher frequencies. In the cancellous bone layer, the porous internal bone structure scatters the ultrasound and has been reported as the dominant mechanism for losses at frequencies higher than 0.7 MHz (Barger and Linzer 1979).

The absorption of ultrasound energy by bone occurs at a much higher rate than that in water or soft tissues. For example, the attenuation coefficient in the human skull bone is an order of magnitude higher than that reported for brain tissue (Goldman and Hueter 1956; Fry and Barger 1978; Duck 1990; Pichardo et al. 2011). As a result, there is the potential for significant bone heating to occur from transcranial ultrasound. When high intensities are used to induce therapeutic effects, there is the potential for the temperature rise at the scalp/bone and skull/brain interfaces to exceed that at the ultrasound focus (Connor and Hynynen 2004), thus requiring careful array design and treatment planning consideration to make transcranial ultrasound therapy feasible. These considerations will be discussed in detail later in this chapter. For now, it is sufficient to say that the high losses experienced by ultrasound passing through the skull can result in low SNR in diagnostic ultrasound and low achievable intensities in therapeutic ultrasound, and that the potential for skull heating prevents overcoming these losses by simply increasing the transducer driving power.

The skull is irregularly shaped and very heterogeneous. The longitudinal speed of sound in skull bone is almost double that in water, with an average speed of approximately 2900 m/s (Fry and Barger 1978; Pichardo et al. 2011); however, the speed of sound in bone varies with density (Clement and Hynynen 2002a; Padilla et al. 2006; Pichardo et al. 2011). Thus, sound passing through the skull undergoes location-specific changes in speed. The distortion of the ultrasound beam by the skull is minimal at low frequencies; however, the phase shifts become significant relative to the wavelength of the ultrasound at higher frequencies, producing high distortion (Fry and Barger 1978; Hynynen and Jolesz 1998). Beam distortion is further complicated by refraction effects because of the nonnormal incidence of the beam on the skull and by multiple reflections between the skull bone and the transducer (Clement et al. 2001).

Finally, standing waves can be generated in the skull cavity (Azuma et al. 2004; Baron et al. 2009) or bone (Connor and Hynynen 2004). Standing waves form when

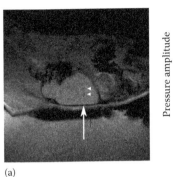 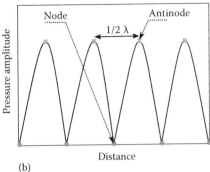

(a)   (b)

FIGURE 10.1 (a) T2*-weighted sagittal MR image of a rat brain (supine) after FUS exposure. The white arrow indicates the direction of the ultrasound propagation. The arrowheads mark two regions of damage separated by 1/2 λ. (b) Illustration of the characteristic pressure profile of a purely standing wave. Spacing between antinodes or between nodes is 1/2 λ. In practice, some travelling wave components will also contribute to the pressure profile.

long insonations reflect from the bone interfaces (Figure 10.1), which can cause secondary foci (Deffieux and Konofagou 2011) and bone heating (Connor and Hynynen 2004). Standing waves most easily form for loosely focused transducers (Deffieux and Konofagou 2011; Song et al. 2012), and in one clinical trial using unfocused, low-frequency transcranial ultrasound to enhance treatment of stroke (Daffertshofer et al. 2005), standing waves are thought to have contributed to adverse effects in the contralateral hemisphere (Baron et al. 2009).

Despite the many difficulties associated with transcranial ultrasound, both diagnostic and therapeutic ultrasound exposures of the brain have been achieved. The various techniques used to overcome the challenges posed by the skull bone are discussed in the following sections.

10.3 Brain Imaging

Despite the attractiveness of ultrasound as an inexpensive, portable, and nonionizing imaging modality, the use of diagnostic ultrasound in the brain is limited. As described, this is primarily due to the beam distortion and attenuation that occurs as the ultrasound passes through the skull (White et al. 1968). However, there remains a place for ultrasound in brain diagnostics.

The brain is most easily imaged in the absence of the skull bone. This can be achieved intraoperatively through burr holes or a craniotomy window (Andrews et al. 1990; Auer and van Velthoven 1990). Intraoperative brain ultrasonography can be used to visualize brain lesions (Auer and van Velthoven 1990) and guide interventions (van Velthoven and Auer 1990; Guthkelch et al. 1991). However, the creation of a bone window is highly invasive, and thus this approach is ill-suited for many patients.

An alternative approach is to image through soft tissue windows in the skull or thinner areas of the bone where attenuation and distortion effects will be reduced. Acoustic windows in the skull that can be used for ultrasound imaging include the anterior fontanel in newborns (Edwards et al. 1981; Sauerbrei et al. 1981) and the temporal and suboccipital bone windows in adults (Aaslid et al. 1982; Kirkham et al. 1986; Berland

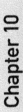

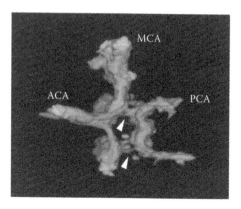

FIGURE 10.2 Contrast-enhanced 3-D transcranial color-coded sonography rendering of the circle of Willis. The anterior cerebral artery (ACA), posterior cerebral artery (PCA), and middle cerebral artery (MCA) are indicated. (Reprinted from Wessels, T., Bozzato, A., Mull, M., and Klötzsch, C., *Ultrasound Med. Biol*, 31(11), 1435–1440, Copyright 2004, with permission from Elsevier.)

et al. 1988; Ringelstein et al. 1990; Smith et al. 2009). Transtemporal Doppler imaging of the middle cerebral artery was first demonstrated in the early 1980s (Aaslid et al. 1982). Transcranial Doppler has applications for the diagnosis of stroke and assessment of recanalization (Spencer et al. 1990; Burgin et al. 2000; Smith et al. 2009) and is a particularly attractive imaging modality for stroke diagnosis because of the speed at which patients can access ultrasound compared with other imaging modalities. However, each imaging window has a limited field of view (Smith et al. 2009), which prevents imaging of the whole brain, and the imaging windows are insufficient in some patients. Contrast-enhanced transcranial sonography (Figure 10.2) has been used to improve image quality (Gahn et al. 2000; Wessels et al. 2004). Using a contrast agent increases the signal-to-noise ratio and improves imaging through poor acoustic windows. However, some recent studies have shown that short insonations are capable of disrupting the blood–brain barrier (BBB) (Bing et al. 2009; O'Reilly et al. 2011; Choi et al. 2011), although not in all studies (Hynynen et al. 2003; Jungehulsing et al. 2008), and thus techniques to improve image quality that do not rely on the use of a contrast agent may prove safer. To improve the field of view, the registration of simultaneous Doppler scans through both temporal windows has also been proposed (Smith et al. 2009). Using this approach, a three-dimensional visualization in both hemispheres of blood flow in vessels of the Circle of Willis has been demonstrated, which could have great benefit in the rapid diagnosis of stroke (Lindsey et al. 2011). This technique has been demonstrated with and without the use of a contrast agent (Lindsey et al. 2011).

Imaging through the temporal and occipital windows still suffers from beam dephasing, and refraction artifacts and aberration correction is required to produce high quality images.

10.3.1 Adaptive Focusing

Diagnostic ultrasound images of the brain can be improved using adaptive focusing techniques to account for aberrations caused by the skull. Smith et al. (1986) used a planar layer model to calculate refraction effects and applied the results to improve transcranial image quality. However, the model is only valid for regions of the skull that have fairly constant skull thickness. Other adaptive focusing techniques include cross-correlation and speckle brightness algorithms. In cross-correlation techniques, the signals received by neighboring receiver elements are assumed to be very similar but with a phase shift. Using cross correlation, the phase shift between elements can be estimated to adjust the focus (O'Donnell and Flax 1988). This technique works very well for a single emitting point source. However, when multiple or diffuse scatterers are present, cross correlation can still be used to adaptively focus by iterating on the results (O'Donnell and Flax 1988). Cross-correlation algorithms have been implemented in real-time three-dimensional Doppler imaging through the temporal window, resulting in an improvement in image quality (Ivancevich et al. 2008). Aberration correction by maximizing speckle brightness is an alternative technique to estimate phase delays (Nock et al. 1989). However, in one study, the direct comparison of cross-correlation and speckle brightness techniques for correcting phase aberrations through the temporal bone found that cross-correlation produced better results (Ivancevich et al. 2009).

If an acoustic source is available within the skull, time reversal acoustics can be used to correct for phase aberrations (Fink 1992). Signals emitted from a source, or scattered from a reflector, within the skull can be recorded by the transmit/receive array. The signals can then be time reversed and transmitted from the array to create a focus at the location of the source or scatterer and to correct the received echoes (Dorme and Fink 1995). This method works well for determining the phase shifts and timing of the bursts, but not for determining the magnitude of the signal because the time reversal does not account for tissue attenuation unless inverse modeling is used. Alternatively, to gain a greater control over the acoustic field, an inverse filter approach can be used to control the beam profile at several control points in the focal region (Tanter et al. 2000, 2001). Neither of these approaches can be easily realized in brain imaging when no acoustic source or scatterer exists in the brain and when the field in the brain cannot be directly measured. However, a variation on the spatiotemporal inverse filter, which uses a second transducer placed outside the skull to capture the transmitted signals and calculate the corrected wavefronts, has been proposed (Vignon et al. 2006). As an alternative, a cavitating source within the brain could be used to calculate phase corrections (Pernot et al. 2006; Gâteau et al. 2010). However, inducing inertial cavitation within the brain poses safety concerns. Potentially less harmful, in situ vaporization of liquid nanodroplets has been proposed, which may make it possible to create an acoustic source for beam aberration correction (Kripfgans et al. 2000).

10.3.2   Shear Wave Imaging

For nonnormal angles of incidence, longitudinal sound waves undergo conversion to longitudinal and shear components as they enter the skull bone. Upon exiting the skull, the shear waves are converted back to longitudinal form. The sound field in the skull cavity is thus a combination of the longitudinal mode from longitudinal transmission and the longitudinal mode from shear transmission (Figure 10.3). The fraction of transmitted sound that undergoes shear mode conversion increases with incident angle up to approximately 30°, beyond which purely shear wave transmission occurs (Hayner and Hynynen 2001; Clement et al. 2004).

Chapter 10

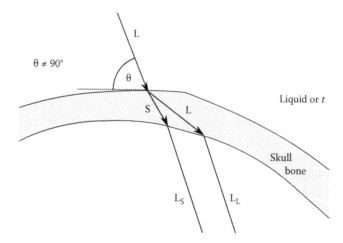

FIGURE 10.3   Illustration of shear mode conversion through the skull bone. For nonnormal angles of incidence, sound passing through the skull results in both shear (S) and longitudinal (L) components, which propagate at different speeds. Exiting the bone, two longitudinal components are present: that generated from longitudinal propagation in the bone (L_L) and that generated from shear propagation in the bone (L_S).

There are several advantages to using shear mode conversion for transcranial imaging. First, although the longitudinal sound speed in bone is approximately twice that in water, the shear speed of sound is close to the speed of sound in water at room temperature at approximately 1500 m/s (White et al. 2006). Thus, beam distortion due to refraction and skull bone inhomogeneities is significantly less for shear waves compared with longitudinal waves (Clement et al. 2004). Second, by using higher incident angles on the skull, large reflections from the skull are not captured by the transducer. This eliminates reflections between the skull and the transducer, which can complicate echoes from the intracranial target.

One disadvantage to shear mode imaging is the substantially higher attenuation in bone of shear waves compared with longitudinal waves (Wu and Cubberley 1997). However, the acoustic impedance match with water of shear mode propagation is much better than for longitudinal mode propagation. This reduces the reflective losses and allows reasonable transmission to be achieved (White et al. 2006).

Shear mode imaging is not routinely used in clinics. However, it is a topic of increasing research interest, and preliminary investigations have shown potential for brain imaging. Transcranial shear mode imaging has been demonstrated in ex vivo skulls, using scanned A-mode images to produce a B-mode image (Yousefi et al. 2009). In humans, A-mode imaging has been used to monitor brain shift with good results (White et al. 2009).

10.4   Brain Therapy

10.4.1   Early Studies

Investigation of focused ultrasound (FUS) as a neurosurgical tool dates to the 1940s when Lynn et al. (1942) and Lynn and Putnam (1944) applied FUS exposures to the brains of laboratory animals. In the 1950s, Fry and Fry treated more than 100 patients using FUS for neurosurgical intervention. However, because of the challenges associated with insonating through the skull bone (heating, attenuation, and beam distortion), these treatments were performed through a craniotomy window. The creation of a craniotomy window is highly invasive, and the treatable brain region can still be limited (Guthkelch et al. 1991; Ram et al. 2006). A small number of studies in humans were performed between the 1950s and early 1990s (Fry and Fry 1960; Heimburger 1985; Shimm et al. 1988; Guthkelch et al. 1991), and some positive effects of the ultrasound exposures were observed. However, until the late 1990s, transcranial FUS was believed impossible, and the necessity of an invasive bone window greatly limited the interest in FUS for brain applications.

10.4.2   Transcranial Therapeutic Ultrasound

The renewal of interest in transcranial FUS applications came in 1998 when Hynynen and Jolesz demonstrated that an intact focus could be achieved through an ex vivo skull using clinically relevant large multielement two-dimensional arrays (Hynynen and Jolesz 1998). This was achieved without phase correction for frequencies lower than 0.5 MHz, where distortion effects were much less. At higher frequencies, phase correction was used, and a sharp focus was demonstrated at 1.58 MHz. In a separate numerical study, the same group proposed that the skull heating problem could also be solved by using very large aperture arrays and low frequencies (Sun and Hynynen 1998). At lower frequencies, the absorption by the skull is significantly less, and the ultrasound energy is distributed over a larger skull area by using large aperture arrays, reducing the energy passing through any particular point on the skull while achieving high focal gains. This concept was demonstrated experimentally by Clement et al. (2000) with the first hemispherical phased array (64 elements). This array was used to insonate rabbit thigh through a human skull and create lesions. Temperature rises of 12°C to 18°C were observed, suggesting that transcranial FUS ablation could be feasible with active cooling of the scalp to minimize temperature elevations in the bone (Clement et al. 2000).

Phase correction through the skull can be achieved by several means. Because of the large aperture of therapy transducers and larger element spacing, cross-correlation and speckle brightness techniques, which work reasonably well for diagnostics with small arrays, are generally not used. The gold standard for phase correction through ex vivo skulls is placing a hydrophone at the desired focus and transmitting from each element, one at a time (Smith et al. 1977; Thomas and Fink 1996; Hynynen and Jolesz 1998; Clement et al. 2000). On the basis of the differences in the time of arrival of the ultrasound signals, appropriate phase delays can be applied to each element. As discussed previously, time reversal can also be used, placing an emitter at the focus and recording with the array elements (Fink 1992; Dorme and Fink 1995), or a spatiotemporal filter can be used for greater control, if an array of receive elements is available at the focus (Tanter et al. 2001).

The most commonly used technique for phase correction is now image-based correction (Hynynen and Sun 1999). Using image-derived information, the skull bone can be modeled, and the sound propagation from a virtual source at the focus can be simulated. From the simulation results, the required phase delays can then be calculated as with experimental phase correction techniques. This approach was first proposed using MRI images and bulk properties for water, bone, and tissue to create a model (Hynynen and Sun 1999). This technique has been expanded upon to use CT-based bone density information to assign bone properties (Clement and Hynynen 2002b; Connor et al. 2002; Aubry et al. 2003), and recent models account for both shear and longitudinal wave propagation (Pulkkinen et al. 2011). Simulation-based methods are entirely noninvasive and are the technique used for current transcranial clinical trials (Martin et al. 2009; McDannold et al. 2010a; Elias et al. 2011).

Amplitude corrections can also be applied to the driving signals to compensate for the changes in insertion loss through the different areas of the skull. The most straightforward technique for amplitude correction is to apply higher powers through those areas that experience higher losses to produce equal amplitude at the focus from each individual driving element (White et al. 2005). However, this approach can lead to hot spots in the skull, as areas that absorb more are subject to higher powers. In addition, White et al. (2005) found this approach to be electrically inefficient as more power is being devoted to elements in areas with high losses. These shortcomings can be avoided if, instead, the objective is to obtain equal absorption through the skull from each driving element. Referred to as an "inverse amplitude correction" by White et al. (2005), this approach assigns lower powers to areas of the skull that absorb more and higher powers to areas that more easily transmit the ultrasound. This approach avoids hot spots and is the most efficient amplitude correction approach, powerwise. Current clinical thermal ablation treatments (McDannold et al. 2010a) use a simpler implementation of this technique, where uniform intensity on the skull surface is sought (Hynynen et al. 2006).

10.4.3 Large Aperture Array Design

For the past decade, several large aperture arrays have been developed for transcranial FUS research purposes (Clement et al. 2000; Pernot et al. 2003; Hynynen et al. 2004; Pernot et al. 2007; Song and Hynynen 2010; Aubry et al. 2010). In addition to these, two clinical prototype systems are commercially available (ExAblate 4000, InSightec, Haifa, Israel) (Figure 10.4). These prototypes both have 1024 elements at either a low frequency (220 kHz) or a high frequency (660 kHz). The high-frequency ExAblate 4000 system and its previous 512 kHz version have been used in three centers to date to perform clinical brain ablation (Martin et al. 2009; McDannold et al. 2010a; Elias et al. 2011). These trials will be discussed in greater detail later in this chapter.

Large aperture, multielement arrays provide several advantages over small aperture transducers, in addition to the already discussed reduction in skull heating. Large aperture arrays produce shaper focal spots (Figure 10.5), allowing small volumes to be targeted, and also reduce the potential for standing waves (Song et al. 2012). Small aperture arrays can result in significant standing waves within the skull cavity and between the skull and the transducer (Song et al. 2012). Standing waves have the most impact at lower frequencies (Deffieux and Konofagou 2010), which are often required because of their skull penetrating abilities. However, when large aperture arrays are used, standing wave effects become negligible, even at frequencies as low as 220 kHz (Song et al. 2012). Standing waves can be further reduced if linear chirps (Deffieux and Konofagou 2010), sweep frequencies (Mitri et al. 2005), or random phase shifts (Tang and Clement 2009) are used.

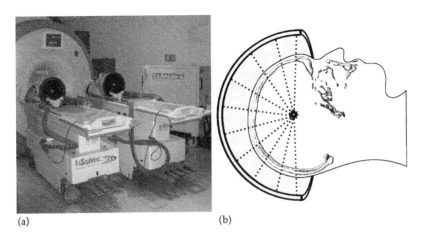

(a)   (b)

FIGURE 10.4   (a) ExAblate 4000 low and high-frequency clinical prototype arrays. (b) Illustration of treatment geometry using a large aperture array. The incident ultrasound beam is distributed across the maximum available surface area to reduce skull heating.

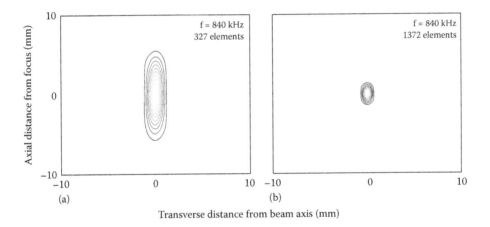

FIGURE 10.5   Simulated pressure profiles for an f-number 0.75 array (a) and a full hemispherical array (b). Results are based on the array presented by Song and Hynynen (2010) operating at 840 kHz. The full array consists of 1372 elements, whereas only 327 elements were used to simulate the smaller aperture array.

Optimal center-to-center element spacing for reducing grating lobes is $1/2 \, \lambda$ (von Ramm et al. 1983). However, for large aperture arrays, full array population at optimal spacing is often not practical because of the large number of elements and associated hardware required to drive them. Sparse random arrays provide one alternative to fully populated arrays. Random configurations will reduce grating lobes compared with sparse symmetric configurations. However, larger numbers of elements typically can deliver higher focal pressures through superior aberration correction. Pernot et al. (2003) used elements capable of generating high acoustic powers to compensate for the reduced number of elements in their 200-element random sparse array. If sparse arrays are to be used, it is important to consider that too sparse an array may revive skull heating issues, as the effective area over which the sound penetrates the skull is once again reduced.

Chapter 10

10.4.4 Thermal Effects

Thermal treatments using ultrasound can be broadly classified into two categories: hyperthermia and ablation. Hyperthermia generally refers to moderate temperature elevations (42°C–45°C) that are achieved more gradually using long insonations and are maintained for minutes to hours. Conversely, thermal ablation occurs at higher temperatures (55°C–60°C) using high intensity to quickly necrotize tissue. Both hyperthermia and ablation techniques have been applied in clinical investigations in the brain.

The popularity of ultrasound hyperthermia in early studies was likely due to the success of other heating techniques in inducing a therapeutic response in brain tissue. Investigations in the 1960s using other heating modalities identified a temperature threshold of approximately 44°C beyond which detrimental effects were observed in the brains of laboratory animals (Harris et al. 1962; Burger and Fuhrman 1964). Cancer cells have been shown to be particularly sensitive to heat, particularly when hyperthermia is used as an adjuvant to other interventions (e.g., chemotherapy, radiation therapy, and immunotherapy) (Marmor and Hahn 1978; Larkin 1979). Ultrasound hyperthermia has several advantages over other heating techniques. FUS is much more localized than whole-body hyperthermia and can therefore minimize deleterious effects outside the treatment volume. By comparison with microwave or RF heating, ultrasound is capable of focusing much deeper in the tissue. Thus, ultrasound was an intuitive choice for brain hyperthermia investigations.

Clinical studies using ultrasound hyperthermia for the treatment of brain tumors were performed through a craniotomy window (Shimm et al. 1988; Guthkelch et al. 1991). The transducer focus during these treatments was mechanically scanned (Hynynen et al. 1987) to allow control of the shape and volume of the treatment area. Control of the treatment volume could also be achieved using a phased array and electronic steering.

A single phase I clinical trial was performed in 1991 by Guthkelch et al. (1991). Fifteen patients with primary brain tumors participated in the trial and received ultrasound hyperthermia (target temperature = 42.5°C) in combination with radiation therapy. In this study, the temperature rise in the treatment volume was monitored by in situ temperature probes (Guthkelch et al. 1991). The invasiveness of the craniotomy window and the temperature probes limited the appeal of the FUS procedure, despite observation of some positive effects of the combined treatment. In addition to the invasiveness of the craniotomy window, the treatable regions of the brain were still limited. This could only be avoided if a very large craniotomy window were used. The study by Guthkelch et al. (1991) was continued using ultrasound hyperthermia in combination with the chemotherapy agent carboplatin (Hynynen 2011, personal communication). However, the results of that investigation were never published.

The investigations by Guthkelch et al. (1991) marked the end of hyperthermia investigations in the brain. Research focus was instead turned to high intensity ablation of brain tissue. As with the hyperthermia investigations, early ablation studies were performed through a skull window. Clinical ablation studies began in the 1950s. Fry and Fry (1960) treated patients for several neurological disorders, including Parkinson's disease, essential tremor, intractable pain, and phantom limb pain. Positive responses were observed after many of the treatments, although these responses were sometimes only temporary. In 1985, Heimburger published a paper describing the treatment of both metastatic and primary brain cancers using high intensity FUS. The metastatic tumors did not respond to the treatment. The majority of the treatments were performed through a bone window, although a few transcranial treatments were attempted at more moderate intensities and lower frequencies. The transcranial treatments were discontinued because of skin burns.

Two major developments led to the current transcranial FUS ablation clinical trials. The first of these was the introduction of MRI for guiding and monitoring FUS treatments (Cline et al. 1993; Hynynen et al. 1993a,b). MRI can be used for targeting the treatment area and to monitor temperature rise during treatments. MRI thermometry can be performed sufficiently quickly to allow treatment exposures to be controlled in real time (Smith et al. 2001; Quesson et al. 2002). MR thermometry marked a huge improvement in monitoring over temperature probes. In addition to its noninvasiveness, MRI thermometry allows large areas of the brain to be monitored, instead of just discreet points. This means that areas near the skull, which are the most likely location for unwanted heating, can be easily monitored (McDannold et al. 2004). MRI thermometry has been shown in the brain to be sufficiently sensitive to detect temperature elevations below the threshold for irreversible damage (Hynynen et al. 1997) and can be used to predict tissue damage (Vykhodtseva et al. 2000). The second development that revolutionized FUS brain ablation was the demonstration of the feasibility of transcranial FUS (Hynynen and Jolesz 1998), which has already been discussed in this chapter.

Several investigations of brain ablation have been performed in primate models (McDannold et al. 2003; Hynynen et al. 2006; Marquet et al. 2006) and in cadaver studies (Aubry et al. 2010). Since 2005, there have been a total of five reports of FUS brain ablation in humans, of which three studies were performed transcranially using the ExAblate systems (InSightec).

In 2005, the treatment of a single patient with anaplastic glioma using FUS through a craniotomy window was reported (Park et al. 2006). An additional three patients with glioma were treated by another group, using a craniotomy window and an in-bed FUS system designed for uterine fibroids (ExAblate 2000, InSightec) (Ram et al. 2006). In the first study, the full study details were never published, although some improvement in patient symptoms and tumor shrinkage was reported (Park et al. 2006). The second study was largely unsuccessful (Ram et al. 2006). Three patients were treated. In one case, a technical malfunction prevented therapeutic exposures from being reached. In another patient, therapeutic temperatures were achieved without complications. However, the treatment volume was limited by the craniotomy window. In the final patient, a secondary focus formed in the brain, resulting in hemiparesis. This was thought to have formed because of reflections of the ultrasound from the skull base, a complication that was not observed in preliminary feasibility studies in pigs (Cohen et al. 2007). A more likely explanation is the low ultrasound attenuation in the fluid-filled ventricles, which can lead to larger than expected intensities at depth.

Three studies have now been performed using clinical transcranial FUS prototype arrays, either the high-frequency (660 kHz) ExAblate 4000 or the previous generation 512 element device operating at a similar frequency. In Boston, three patients with glioblastomas were treated using the 512 element array (McDannold et al. 2010a). With the 512 element array, the deliverable power was limited. However, in two patients, temperature elevations to 48°C to 51°C were recorded. MRI thermometry measurements during treatment illustrated that ablation would be feasible if more power were available, as the temperature rise at the focus was sufficiently high compared with those at the skull surface (McDannold et al. 2010a).

Another group in Zurich has reported unilateral and bilateral ablations in nine patients for the treatment of chronic pain (Martin et al. 2009). These treatments used the upgraded 1024 element ExAblate 4000 and achieved temperature elevations from 50°C to 60°C. T2w MRI 24 to 48 hours posttreatment showed precisely located lesions with sharp margins (Figure 10.6). Early pain relief after the treatment was reported.

Finally, and most recently, ablation for the treatment of essential tremor has been performed in 15 patients in Virginia (Elias et al. 2011) using the ExAblate 4000. Only preliminary results have been reported, but a marked reduction in tremor was observed posttreatment, and early reports are positive.

Unfortunately, thermal FUS ablation is limited when performed transcranially. Treatments are only feasible in the midbrain because treatments close to the skull

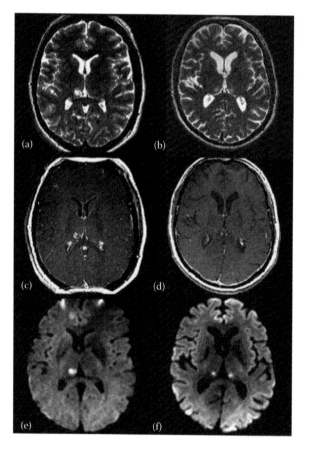

FIGURE 10.6 Posttreatment magnetic resonance (MR) images from two patients treated in Zurich with transcranial HIFU. (a, c, e) Axial T2-weighted image (T2WI), postcontrast T1-weighted image (T1WI), and isotropic diffusion tensor image, respectively, of a patient obtained immediately after treatment. (b, d, f) Axial T2WI, postcontrast T1WI, and isotropic diffusion tensor images of another patient obtained 48 hours posttreatment. (Reprinted from Martin, E. et al., *Ann. Neurol.*, 66(6), 858–861, 2009. With permission.)

result in significant temperature elevations in the skull bone, close to or beyond those achievable at the focus (Connor and Hynynen 2004; Pulkkinen et al. 2011).

10.4.5 Sonothrombolysis and Stroke Treatments

Ultrasound also has therapeutic potential for the treatment of stroke. Stroke has a large global impact, with more than 15 million individuals worldwide affected each year (Mackay and Mensah 2004). In addition, the treatment window associated with positive outcomes in ischemic stroke is narrow at around 3 hours from the onset of symptoms. Thus, there is great interest in the development of treatment strategies that can be rapidly administered. The gold standard for the treatment of ischemic stroke is the administration of the lytic agent tissue plasminogen activator (tPA) to dissolve the clot (N.I.N.D.S. 1995), the effects of which can be enhanced by the application of ultrasound (Francis et al. 1992).

Ultrasound has long been known to affect clot lysis (Trübestein et al. 1976). When combined with tPA, transcranial Doppler ultrasound has been shown in humans to enhance tPA effects and improve recanalization rates (Alexandrov et al. 2004; Molina et al. 2006; Tsivgoulis et al. 2010). In most studies, there was no indication of increased hemorrhagic risk from the combination of ultrasound and tPA (Alexandrov et al. 2004; Molina et al. 2006; Tsivgoulis et al. 2010). However, one clinical trial was prematurely terminated after abnormally high hemorrhage rates in the contralateral hemisphere (Daffertshofer et al. 2005). The study used 300 kHz planar Doppler ultrasound, and the contralateral hemorrhage could have been caused by skull reflections and standing waves arising from the use of such a low-frequency, unfocused beam (Baron et al. 2009). Because tPA alone is associated with an increased risk of intracranial hemorrhage (N.I.N.D.S. 1995), the development of alternative stroke treatment approaches that do not rely on the administration of tPA is of great importance.

Recent research has demonstrated the feasibility of using ultrasound alone (Holscher et al. 2010; Burgess et al. 2011a,b; Wright et al. 2012a,b), or in combination with circulating microbubbles (Culp et al. 2011), to treat ischemic stroke (Figure 10.7). In 2010,

Poststroke   Postsonication

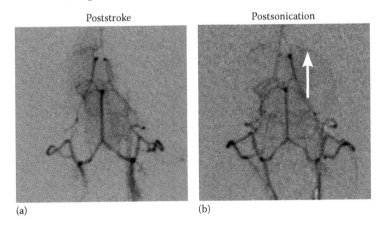

(a)   (b)

FIGURE 10.7 Fluoroscopy images of a rabbit brain (a) after the induction of a stroke in the left middle cerebral artery and (b) after sonication with FUS alone. The images are mirrored left/right. The arrow indicates the restoration of flow through the middle cerebral artery. (From Burgess et al., 2011a, with permission.)

Chapter 10

Holscher et al. presented preliminary results demonstrating in vitro clot lysis using transcranially applied ultrasound alone. The ultrasound was applied through ex vivo human skulls using the ExAblate 4000 (InSightec) transcranially FUS system. More recently, two studies have demonstrated in vivo clot lysis in the brain in animal models. The first, by Culp et al. (2011), used ultrasound and microbubbles to dissolve clots in a rabbit model. An improvement in outcome was seen when ultrasound and microbubbles were used by comparison with control groups of tPA alone and ultrasound combined with tPA. Importantly, the use of ultrasound and microbubbles was not associated with an increased risk of hemorrhage. The technique used in the second study is based on the in vitro work and in vivo studies in rabbit femoral arteries of Wright et al. (2012a,b) and uses ultrasound alone. The mechanism for clot dissolution using ultrasound without microbubbles or lytic agent is thought to be inertial cavitation within the clot (Wright et al. 2012a). In a rabbit model of stroke, Burgess et al. (2011a) showed that this approach, using FUS alone, can be used for clot dissolution in the brain in vivo.

10.4.6   BBB Disruption

The delivery of pharmacological agents to the brain is greatly inhibited by the presence of the BBB. The BBB is characterized by reduced vesicular transport and the presence of tight junctions between brain endothelial cells (Rubin and Staddon 1999). These characteristics greatly restrict both active transcellular transport and paracellular delivery routes. The delivery of agents to the brain parenchyma is thus limited to small molecules (<500 Da) with high lipid solubility, which describes less than 2% of known neurotherapeutics (Pardridge 2005).

Techniques for circumventing the BBB are either highly invasive or disrupt the BBB globally, exposing whole regions to potential pathogens. For example, drugs can be delivered by direct injection into the brain tissue or ventricles, which is invasive and damages the tissue that is penetrated (Pardridge 2005). Because of the limited diffusion of therapeutics within the parenchyma (Blasberg et al. 1975), direct injection of drugs into the brain also results in poor drug dispersion unless convection-enhanced methods, which use pressure gradients during infusion to induce bulk flow in the extracellular space, are used (Bobo et al. 1994; Lieberman et al. 1995; Lidar et al. 2004; Kunwar et al. 2010). The BBB can be disrupted by chemical means through an intracarotid injection of a hyperosmotic solution, such as Mannitol (Salahuddin et al. 1988). However, this disrupts the BBB in the whole region fed by the artery. In addition, Mannitol has been shown to cause permanent neurological changes in animal models (Salahuddin et al. 1988). Ultrasound provides a noninvasive, transient, and localized alternative for disrupting the BBB.

It was first observed in the 1950s that ultrasound-induced lesions in the brain were accompanied by a reversible zone of BBB disruption (BBBD) at their periphery (Bakay et al. 1956; Patrick et al. 1990). Early studies were able to disrupt the BBB without causing a lesion, either with (Ballantine et al. 1960) or without (Vykhodtseva et al. 1995; Mesiwala et al. 2002) tissue damage, but not consistently. Early studies did, however, demonstrate that different acoustic parameters resulted in differing degrees of histological damage (Vykhodtseva et al. 1995) and that the threshold exposures for inducing BBBD varied between different brain structures (Patrick et al. 1990). It was not until 2001 that Hynynen et al. showed that by using ultrasound in combination with circulating preformed microbubbles, the BBB could be consistently and safely disrupted at powers two orders of magnitude less than those used for ablation. Pulsed ultrasound combined with microbubbles is now the accepted technique for inducing BBBD (Figure 10.8) in animal models.

FUS-induced BBBD is associated with increased vesicular transport and disruption of tight junctions, which is observable via electron microscopy (Sheikov et al. 2004, 2008). The upregulation of transcellular transport proteins caveolin 1 (Xia et al. 2009; Deng et al. 2011) and caveolin 2 (Xia et al. 2009) has been documented after FUS in normal and tumor brain tissue, in addition to the downregulation of tight junctional proteins claudin 1 and claudin 5 (Sheikov et al. 2008; Zhang et al. 2009, 2010; Shang et al. 2011), occludin (Zhang et al. 2009), and zonula occludin-1 (Sheikov et al. 2008; Zhang et al. 2010; Shang et al. 2011). The exact physical mechanism for ultrasound-induced disruption is unknown, but stretching of the tight junctions due to radial expansion of the microbubbles (Hynynen et al. 2001; Caskey et al. 2007), vessel wall shear stress arising from radiation force (Borgnis 1953) and/or acoustic microstreaming (Krasovitski and Kimmel 2004), and a microthermal mechanism (Klotz et al. 2010) are all possibilities.

The threshold for BBBD is related to frequency by the mechanical index (McDannold et al. 2008). Thus, investigations at higher frequencies suitable for small animal models can be related to frequencies more appropriate for use in humans (generally <1 MHz). When appropriate exposures are used, damage is avoided and the BBB is restored in less than 24 hours, and usually less than 6 hours (Hynynen et al. 2001, 2006; Sheikov et al. 2008; Xie et al. 2008; Mei et al. 2009; Yang et al. 2009; Wang et al. 2009; Zhang et al. 2010; Shang et al. 2011; Deng et al. 2011; Wang et al. 2011). A wide range of acoustic parameters have been investigated, and the use of ultrasound combined with microbubbles appears to be a robust technique for disrupting the BBB, with varying pulse lengths, pulse repetition frequency, and sonication duration all succeeding in inducing BBBD. Generally, the microbubbles used are commercially available ultrasound

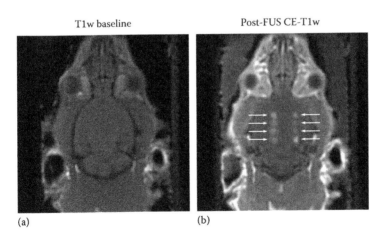

T1w baseline   Post-FUS CE-T1w

(a)   (b)

FIGURE 10.8 (a) Baseline T1w image of a rat brain before 1.18 MHz FUS with microbubbles. (b) Contrast enhanced T1w image of the same rat brain after FUS with microbubbles in eight locations. Enhancement indicating disruption is visible at the 8 sonication locations (indicated by the arrows). The direction of the ultrasound propagation is into the page.

Chapter 10

contrast agents (e.g., Definity, Optison, SonoVue, etc.). However, some groups have used custom bubbles, either monodisperse bubbles, to better understand the effects of bubble size (Choi et al. 2010; Vlachos et al. 2010, 2011; Samiotaki et al. 2011; Tung et al. 2011a) or payload bubbles carrying a drug or gene payload to be released in situ (Ting et al. 2012; Huang et al. 2011).

The disruption of the BBB can be assessed using contrast enhanced MRI. However, this cannot be performed in real time as leakage from the vessel can occur after the completion of the sonication (Cho et al. 2011). In 2006, McDannold et al. (2006b) demonstrated that an increase in harmonic emissions occurs when the BBB is successfully disrupted. Other groups have found similar changes in acoustic emissions with successful BBBD (Tung et al. 2010). Recently, it has been shown that treatment exposures can be modulated in real time using microbubble ultraharmonic emissions as a control parameter (O'Reilly and Hynynen 2012). In small animal models, it thus appears that BBBD can be consistently and safely induced using microbubble emissions to monitor treatments.

A wide range of agents that do not normally cross the BBB have been delivered to the brain by FUS-induced BBBD. The largest of these agents are antibodies (Herceptin [Kinoshita et al. 2006a], D4-receptor antibodies [Kinoshita et al. 2006b], and amyloid-β antibodies [Raymond et al. 2008; Jordão et al. 2010]), which have a molecular weight of approximately 150 kDa, and stem cells (Park et al. 2011; Burgess et al. 2011b), which are approximately 7 to 10 μm in diameter. The majority of studies that have investigated the therapeutic effects of agents delivered via FUS BBBD have examined the effects on brain tumors. Treat et al. (2009) demonstrated increased survival in a 9L gliosarcoma mouse model. 1,3-Bis(2-chloroethyl)-1-nitrosourea (BCNU) (Liu et al. 2010a; Chen et al. 2010) and epirubicin (Liu et al. 2010b) have both been shown to improve survival and growth suppression in rat C6 gliomas when delivered using FUS. The effects were further enhanced by immobilizing the agents on magnetic nanoparticles and using a combination of FUS and magnetic targeting (Liu et al. 2010b; Chen et al. 2010). Recently, FUS combined with BCNU-loaded microbubbles was shown to produce a significant reduction in rat tumor volume (Ting et al. 2012).

One noncancer investigation of BBBD for targeted therapy has been for the treatment of Alzheimer's disease (Jordão et al. 2010). After treatment with FUS and amyloid-β antibodies, a statistically significant reduction in plaque volume and number was found in FUS-treated hemispheres compared with the contralateral hemisphere. In addition, the gene-loaded microbubbles have been used with FUS, resulting in increased gene expression in the sonicated tissue and demonstrating the possibility of future gene therapy in the brain (Huang et al. 2011).

FUS BBBD has not yet been tested in humans. However, its feasibility in nonhuman primates has been shown (McDannold et al. 2010b; Tung et al. 2011b; Marquet et al. 2011; McDannold et al. 2012). The first comprehensive investigation of BBBD in primates has recently been completed (McDannold et al. 2012). Promisingly, repeat disruption in multiple locations in the brain was performed in several primates with no discernable cognitive or functional impairment. This demonstrates for the first time that such a technique may be feasible in humans, and this also illustrates how close the field is to initial clinical trials.

10.4.7   Neuromodulation

Although the investigation of temporary functional changes in the brain induced by ultrasound began more than 50 years ago (Fry and Fry 1960), very little work has been performed in this area since. However, in the past 4 years, interest in neuromodulation has been revived by a small number of studies, suggesting that ultrasound may have useful stimulating effects in the brain (Tyler et al. 2008; Tufail et al. 2010; Yoo et al. 2011a,b), which could potentially offer a noninvasive alternative to deep brain stimulation. Beginning in 2008, Tyler et al. demonstrated in vitro that ultrasound had a stimulating effect on neurons in hippocampal slice cultures. After that, ultrasound-induced functional changes in mouse brain (Tufail et al. 2010) were reported, which are consistent with the early experiments in feline models (Fry and Fry 1960). A more recent study found that ultrasound stimulation of the thalamus reduces recovery time from anesthetic in rats (Yoo et al. 2011b). Although ultrasound neuromodulation has the potential for a wide range of applications in brain treatments, the mechanisms for the observed functional changes are poorly understood and much work remains to be conducted.

10.5   Discussion on the Future of Ultrasound in the Brain

It is an exciting time for ultrasound applications in the brain. The greatest research interest lies in the therapeutic applications of ultrasound in the brain, the extent of which we are only beginning to fully understand. The clinical investigations of completely noninvasive brain ablation at multiple centers worldwide have given transcranial FUS more broad recognition. BBBD is on the advent of clinical investigation. New research in sonothrombolysis and neuromodulation suggests that we have only realized a small fraction of the potential of transcranial FUS. The difficulties with delivery of ultrasound to the brain, which have previously deterred research in the field, have been addressed. The research direction should now be toward new applications of this technology, which has the potential to revolutionize neuromedicine.

Acknowledgments

The authors thank Dr. Alison Burgess for contributing Figure 10.6 and for her help in reviewing the manuscript. Support for this work was provided by the National Institutes of Health under grant nos. EB003268 and EB009032, as well as the Canada Research Chair Program.

References

Aarnio, J., Clement, G. T. & Hynynen, K. (2005), A new ultrasound method for determining the acoustic phase shifts caused by the skull bone, *Ultrasound in Medicine and Biology* 31(6), 771–780.

Aaslid, R., Markwalder, T. M. & Nornes, H. (1982), Noninvasive transcranial Doppler ultrasound recording of flow velocity in basal cerebral arteries, *Journal of Neurosurgery* 57(6), 769–774.

Alexandrov, A. V., Demchuk, A. M., Burgin, W. S., Robinson, D. J., Grotta, J. C. & Investigators, C. L. O. T. B. U. S. T. (2004), Ultrasound-enhanced thrombolysis for acute ischemic stroke: Phase I. Findings of the CLOTBUST trial, *Journal of Neuroimaging* 14(2), 113–117.

Chapter 10

Andrews, B. T., Mampalam, T. J., Omsberg, E. & Pitts, L. H. (1990), Intraoperative ultrasound imaging of the entire brain through unilateral exploratory burr holes after severe head injury: Technical note, *Surgical Neurology* 33(4), 291–294.

Aubry, J. F., Tanter, M., Pernot, M., Thomas, J. L. & Fink, M. (2003), Experimental demonstration of noninvasive transskull adaptive focusing based on prior computed tomography scans, *Journal of the Acoustical Society of America* 113(1), 84–93.

Aubry, J.-F., Marsac, L., Pernot, M., Robert, B., Boch, A.-L., Chauvet, D., Salameh, N., Souris, L., Darasse, L., Bittoun, J., Martin, Y., Cohen-Bacrie, C., Souquet, J., Fink, M. & Tanter, M. (2010), High intensity focused ultrasound for transcranial therapy of brain lesions and disorders [Ultrasons focalisés de forte intensité pour la thérapie transcrânienne du cerveau], *IRBM* 31(2), 87–91.

Auer, L. M. & van Velthoven, V. (1990), Intraoperative ultrasound (US) imaging. Comparison of pathomorphological findings in US and CT, *Acta Neurochirurgica* (Wien) 104(3–4), 84–95.

Azuma, T., Kawabata, K., Umemura, S., Ogihara, M., Kubota, J., Sasaki, A. & Furuhata, H. (2004), Schlieren observation of therapeutic field in water surrounded by cranium radiated from 500 kHz ultrasonic sector transducer, in *Proceedings of the IEEE Ultrasonics Symposium*, pp. 1001–1004.

Bakay, L., Ballantine, H., Hueter, T. & Sosa, D. (1956), Ultrasonically produced changes in the blood–brain barrier, *AMA Archives of Neurology and Psychiatry* 76(5), 457–467.

Ballantine, H. T., Bell, E. & Manlapaz, J. (1960), Progress and problems in the neurological applications of focused ultrasound, *Journal of Neurosurgery* 17, 858–876.

Barger, J. E. & Linzer, M., ed. (1979), Attenuation and dispersion of ultrasound in cancellous bone, in *Ultrasonic Tissue Characterization II*, U.S. Government Printing Office, Washington, DC, pp. 197–202.

Baron, C., Aubry, J.-F., Tanter, M., Meairs, S. & Fink, M. (2009), Simulation of intracranial acoustic fields in clinical trials of sonothrombolysis, *Ultrasound in Medicine and Biology* 35(7), 1148–1158.

Berland, L. L., Bryan, C. R., Sekar, B. C. & Moss, C. N. (1988), Sonographic examination of the adult brain, *Journal of Clinical Ultrasound* 16(5), 337–345.

Bing, K., Howles, G., Qi, Y., Palmeri, M. & Nightingale, K. (2009), Blood–brain barrier (BBB) disruption using a diagnostic ultrasound scanner and Definity® in mice, *Ultrasound in Medicine and Biology* 35(8), 1298–1308.

Blasberg, R. G., Patlak, C. & Fenstermacher, J. D. (1975), Intrathecal chemotherapy: Brain tissue profiles after ventriculocisternal perfusion, *Journal of Pharmacology and Experimental Therapeutics* 195(1), 73–83.

Bobo, R. H., Laske, D. W., Akbasak, A., Morrison, P. F., Dedrick, R. L. & Oldfield, E. H. (1994), Convection-enhanced delivery of macromolecules in the brain, *Proceedings of the National Academy of Sciences of the United States of America* 91(6), 2076–2080.

Borgnis, F. E. (1953), Acoustic radiation pressure of plane compressional waves, *Reviews of Modern Physics* 25(3), 653–664.

Burger, F. J. & Fuhrman, F. A. (1964), Evidence of injury by heat in mammalian tissues, *American Journal of Physiology* 206, 1057–1061.

Burgess, A., Huang, Y., Ganguly, M., Goertz, D. & Hynynen, K. H. (2011a), High intensity focused ultrasound (HIFU) treatment of acute stroke, in 87th Scientific Assembly and Annual Meeting of the Radiological Society of North America, Abstract LL-NRS-TU11A.

Burgess, A., Ayala-Grosso, C. A., Ganguly, M., Jordão, J. F., Aubert, I. & Hynynen, K. (2011b), Targeted delivery of neural stem cells to the brain using mri-guided focused ultrasound to disrupt the blood–brain barrier, *PLoS One* 6(11), e27877.

Burgin, W. S., Malkoff, M., Felberg, R. A., Demchuk, A. M., Christou, I., Grotta, J. C. & Alexandrov, A. V. (2000), Transcranial Doppler ultrasound criteria for recanalization after thrombolysis for middle cerebral artery stroke, *Stroke* 31(5), 1128–1132.

Caskey, C. F., Stieger, S. M., Qin, S., Dayton, P. A. & Ferrara, K. W. (2007), Direct observations of ultrasound microbubble contrast agent interaction with the microvessel wall, *Journal of the Acoustical Society of America* 122(2), 1191–1200.

Chen, P.-Y., Liu, H.-L., Hua, M.-Y., Yang, H.-W., Huang, C.-Y., Chu, P.-C., Lyu, L.-A., Tseng, I.-C., Feng, L.-Y., Tsai, H.-C., Chen, S.-M., Lu, Y.-J., Wang, J.-J., Yen, T.-C., Ma, Y.-H., Wu, T., Chen, J.-P., Chuang, J.-I., Shin, J.-W., Hsueh, C. & Wei, K.-C. (2010), Novel magneticultrasound focusing system enhances nanoparticle drug delivery for glioma treatment, *Neurooncology* 12(10), 1050–1060.

Cho, E., Drazic, J., Ganguly, M., Stefanovic, B. & Hynynen, K. (2011), Two-photon fluorescence microscopy study of cerebrovascular dynamics in ultrasound-induced blood–brain barrier opening, *Journal of Cerebral Blood Flow and Metabolism* 31, 1852–1862.

Choi, J., Feshitan, J., Baseri, B., Wang, S., Tung, Y.-S., Borden, M. & Konofagou, E. (2010), Microbubble-size dependence of focused ultrasound-induced blood–brain barrier opening in mice in vivo, *IEEE Transactions in Biomedical Engineering* 57(1), 145–154.

Choi, J. J., Selert, K., Vlachos, F., Wong, A. & Konofagou, E. E. (2011), Noninvasive and localized neuronal delivery using short ultrasonic pulses and microbubbles, *Proceedings of the National Academy of Sciences of the United States of America*.

Clement, G. T. & Hynynen, K. (2002a), Correlation of ultrasound phase with physical skull properties, *Ultrasound in Medicine and Biology* 28(5), 617–624.

Clement, G. T. & Hynynen, K. (2002b), A non-invasive method for focusing ultrasound through the human skull, *Physics in Medicine and Biology* 47(8), 1219–1236.

Clement, G. T., Sun, J., Giesecke, T. & Hynynen, K. (2000), A hemisphere array for non-invasive ultrasound brain therapy and surgery, *Physics in Medicine and Biology* 45(12), 3707–3719.

Clement, G. T., Sun, J. & Hynynen, K. (2001), The role of internal reflection in transskull phase distortion, *Ultrasonics* 39(2), 109–113.

Clement, G. T., White, P. J. & Hynynen, K. (2004), Enhanced ultrasound transmission through the human skull using shear mode conversion, *Journal of the Acoustical Society of America* 115(3), 1356–1364.

Cline, H. E., Schenck, J. F., Watkins, R. D., Hynynen, K. & Jolesz, F. A. (1993), Magnetic resonance-guided thermal surgery, *Magnetic Resonance in Medicine* 30(1), 98–106.

Cohen, Z. R., Zaubermann, J., Harnof, S., Mardor, Y., Nass, D., Zadicario, E., Hananel, A., Castel, D., Faibel, M. & Ram, Z. (2007), Magnetic resonance imaging-guided focused ultrasound for thermal ablation in the brain: A feasibility study in a swine model, *Neurosurgery* 60(4), 593–600, discussion 600.

Connor, C. W., Clement, G. T. & Hynynen, K. (2002), A unified model for the speed of sound in cranial bone based on genetic algorithm optimization, *Physics in Medicine and Biology* 47(22), 3925–3944.

Connor, C. W. & Hynynen, K. (2004), Patterns of thermal deposition in the skull during transcranial focused ultrasound surgery, *IEEE Transactions in Biomedical Engineering* 51(10), 1693–1706.

Culp, W. C., Flores, R., Brown, A. T., Lowery, J. D., Roberson, P. K., Hennings, L. J., Woods, S. D., Hatton, J. H., Culp, B. C., Skinner, R. D. & Borrelli, M. J. (2011), Successful microbubble sonothrombolysis without tissue-type plasminogen activator in a rabbit model of acute, ischemic stroke, *Stroke* 42, 2280–2285.

Daffertshofer, M., Gass, A., Ringleb, P., Sitzer, M., Sliwka, U., Els, T., Sedlaczek, O., Koroshetz, W. J. & Hennerici, M. G. (2005), Transcranial low-frequency ultrasound-mediated thrombolysis in brain ischemia: Increased risk of hemorrhage with combined ultrasound and tissue plasminogen activator: Results of a phase II clinical trial, *Stroke* 36(7), 1441–1446.

Deffieux, T. & Konofagou, E. E. (2010), Numerical study of a simple transcranial focused ultrasound system applied to blood–brain barrier opening, *IEEE Transactions on Ultrasonics, Ferroelectrics and Frequency Control* 57(12), 2637–2653.

Deng, J., Huang, Q., Wang, F., Liu, Y., Wang, Z., Wang, Z., Zhang, Q., Lei, B. & Cheng, Y. (2011), The role of caveolin-1 in blood–brain barrier disruption induced by focused ultrasound combined with microbubbles, *Journal of Molecular Neuroscience* 46(3), 677–687.

Dorme, C. & Fink, M. (1995), Focusing in transmit-receive mode through inhomogeneous media: The time reversal matched filter approach, *Journal of the Acoustical Society of America* 92(2), 1155–1161.

Duck, F. A. (1990), Acoustic properties of tissue at ultrasonic frequencies, in *Physical Properties of Tissue: A Comprehensive Reference Book*, Academic Press, London, pp. 102–111.

Edwards, M. K., Brown, D. L., Muller, J., Grossman, C. B. & Chua, G. T. (1981), Cribside neurosonography: Real-time sonography for intracranial investigation of the neonate, *American Journal of Roentgenology* 136(2), 271–275.

Elias, J. W., Huss, D., Khaled, M. A., Monteith, S. J., Frysinger, R., Loomba, J., Druzgal, J., Wylie, S., Voss, T., Harrison, M., Wooten, F. & Wintermark, M. (2011), MR-guided focused ultrasound lesioning for the treatment of essential tremor. A new paradigm for noninvasive lesioning and neuromodulation, Congress of Neurological Surgeons 2011 Annual Meeting, Abstract 966.

Fink, M. (1992), Time reversal of ultrasonic fields. I. Basic principles, *IEEE Transactions on Ultrasonics, Ferroelectrics and Frequency Control* 39(5), 555–566.

Francis, C. W., Onundarson, P. T., Carstensen, E. L., Blinc, A., Meltzer, R. S., Schwarz, K. & Marder, V. J. (1992), Enhancement of fibrinolysis in vitro by ultrasound, *Journal of Clinical Investigation* 90(5), 2063–2068.

Fry, F. J. & Barger, J. E. (1978), Acoustical properties of the human skull, *Journal of the Acoustical Society of America* 63(5), 1576–1590.

Fry, W. J. & Fry, F. J. (1960), Fundamental neurological research and human neurosurgery using intense ultrasound, *IRE Transactions on Medical Electronics* ME-7, 166–181.

Chapter 10

Gahn, G., Gerber, J., Hallmeyer, S., Hahn, G., Ackerman, R. H., Reichmann, H. & von Kummer, R. (2000), Contrast-enhanced transcranial color-coded duplexsonography in stroke patients with limited bone windows. *American Journal of Neuroradiology* 21, 509–514.

Gâteau, J., Marsac, L., Pernot, M., Aubry, J.-F., Tanter, M. & Fink, M. (2010), Transcranial ultrasonic therapy based on time reversal of acoustically induced cavitation bubble signature, *IEEE Transactions in Biomedical Engineering* 57(1), 134–144.

Goldman, D. E. & Hueter, T. F. (1956), Tabular data of the velocity and absorption of high-frequency sound in mammalian tissue, *Journal of the Acoustical Society of America* 28, 35–37.

Guthkelch, A. N., Carter, L. P., Cassady, J. R., Hynynen, K. H., Iacono, R. P., Johnson, P. C., Obbens, E. A., Roemer, R. B., Seeger, J. F. & Shimm, D. S. (1991), Treatment of malignant brain tumors with focused ultrasound hyperthermia and radiation: Results of a phase I trial, *Journal of Neurooncology* 10(3), 271–284.

Harris, A. B., Erickson, L., Kendig, J. H., Mingrino, S. & Goldring, S. (1962), Observations on selective brain heating in dogs, *Journal of Neurosurgery* 19, 514–521.

Hayner, M. & Hynynen, K. (2001), Numerical analysis of ultrasonic transmission and absorption of oblique plane waves through the human skull, *Journal of the Acoustical Society of America* 110(6), 3319–3330.

Heimburger, R. F. (1985), Ultrasound augmentation of central nervous system tumor therapy, *Indiana Medicine* 78(6), 469–476.

Holscher, T., Zadicario, E., Fisher, D. J. & Bradley, W. G. (2010), Transcranial clot lysis using high intensity focused ultrasound, *AIP Conference Proceedings* 12(15), 23–26.

Huang, Q., Deng, J., Wang, F., Chen, S., Liu, Y., Wang, Z., Wang, Z. & Cheng, Y. (2011), Targeted gene delivery to the mouse brain by MRI-guided focused ultrasound-induced blood–brain barrier disruption, *Experimental Neurology* 233(1), 350–356.

Hynynen, K., Clement, G. T., McDannold, N., Vykhodtseva, N., King, R., White, P. J., Vitek, S. & Jolesz, F. A. (2004), 500-element ultrasound phased array system for noninvasive focal surgery of the brain: A preliminary rabbit study with ex vivo human skulls, *Magnetic Resonance in Medicine* 52(1), 100–107.

Hynynen, K., Darkazanli, A., Damianou, C. A., Unger, E. & Schenck, J. F. (1993a), Tissue thermometry during ultrasound exposure, *European Urology* 23 Suppl 1, 12–16.

Hynynen, K., Damianou, C., Darkazanli, A., Unger, E. & Schenck, J. F. (1993b), The feasibility of using MRI to monitor and guide noninvasive ultrasound surgery, *Ultrasound in Medicine and Biology* 19(1), 91–92.

Hynynen, K. & Jolesz, F. A. (1998), Demonstration of potential noninvasive ultrasound brain therapy through an intact skull, *Ultrasound in Medicine and Biology* 24(2), 275–283.

Hynynen, K., McDannold, N., Clement, G., Jolesz, F. A., Zadicario, E., Killiany, R., Moore, T. & Rosen, D. (2006), Pre-clinical testing of a phased array ultrasound system for MRI-guided noninvasive surgery of the brain—A primate study, *European Journal of Radiology* 59(2), 149–156.

Hynynen, K., McDannold, N., Martin, H., Jolesz, F. & Vykhodtseva, N. (2003), The threshold for brain damage in rabbits induced by bursts of ultrasound in the presence of an ultrasound contrast agent (Optison®), *Ultrasound in Medicine and Biology* 29(3), 473–481.

Hynynen, K., McDannold, N., Vykhodtseva, N. & Jolesz, F. (2001), Noninvasive MR imaging-guided focal opening of the blood–brain barrier in rabbits, *Radiology* 220(3), 640–646.

Hynynen, K., Roemer, R., Anhalt, D., Johnson, C., Xu, Z. X., Swindell, W. & Cetas, T. (1987), A scanned, focused, multiple transducer ultrasonic system for localized hyperthermia treatments, *International Journal of Hyperthermia* 3(1), 21–35.

Hynynen, K. & Sun, J. (1999), Trans-skull ultrasound therapy: The feasibility of using image-derived skull thickness information to correct the phase distortion, *IEEE Transactions on Ultrasonics, Ferroelectrics and Frequency Control* 46(3), 752–755.

Hynynen, K., Vykhodtseva, N. I., Chung, A. H., Sorrentino, V., Colucci, V. & Jolesz, F. A. (1997), Thermal effects of focused ultrasound on the brain: Determination with MR imaging, *Radiology* 204(1), 247–253.

Ivancevich, N. M., Dahl, J. J. & Smith, S. W. (2009), Comparison of 3-D multi-lag cross-correlation and speckle brightness aberration correction algorithms on static and moving targets, *IEEE Transactions on Ultrasonics, Ferroelectrics and Frequency Control* 56(10), 2157–2166.

Ivancevich, N. M., Pinton, G. F., Nicoletto, H. A., Bennett, E., Laskowitz, D. T. & Smith, S. W. (2008), Real-time 3-D contrast-enhanced transcranial ultrasound and aberration correction, *Ultrasound in Medicine and Biology* 34(9), 1387–1395.

Jordão, J., Ayala-Grosso, C., Markham, K., Huang, Y., Chopra, R., McLaurin, J., Hynynen, K. & Aubert, I. (2010), Antibodies targeted to the brain with image-guided focused ultrasound reduces amyloid-beta plaque load in the TgCRND8 mouse model of Alzheimer's disease, *PLoS ONE* 5(5), e10549.

Jungehulsing, G., Brunecker, P., Nolte, C., Fiebach, J., Kunze, C., Doepp, F., Villringer, A. & Schreiber, S. (2008), Diagnostic transcranial ultrasound perfusion-imaging at 2.5 MHz does not affect the blood-brain barrier, *Ultrasound in Medicine and Biology* 34(1), 147–150.

Kinoshita, M., McDannold, N., Jolesz, F. & Hynynen, K. (2006a), Noninvasive localized delivery of Herceptin to the mouse brain by MRI-guided focused ultrasound-induced blood–brain barrier disruption, *Proceedings of the National Academy of Sciences of the United States of America* 103(31), 11719–11723.

Kinoshita, M., McDannold, N., Jolesz, F. & Hynynen, K. (2006b), Targeted delivery of antibodies through the blood–brain barrier by MRI-guided focused ultrasound, *Biochemical and Biophysical Research Communications* 340(4), 1085–1090.

Kirkham, F. J., Padayachee, T. S., Parsons, S., Seargeant, L. S., House, F. R. & Gosling, R. G. (1986), Transcranial measurement of blood velocities in the basal cerebral arteries using pulsed Doppler ultrasound: Velocity as an index of flow, *Ultrasound in Medicine and Biology* 12(1), 15–21.

Klotz, A. R., Lindvere, L., Stefanovic, B. & Hynynen, K. (2010), Temperature change near microbubbles within a capillary network during focused ultrasound, *Physics in Medicine and Biology* 55(6), 1549–1561.

Krasovitski, B. & Kimmel, E. (2004), Shear stress induced by a gas bubble pulsating in an ultrasonic field near a wall, *IEEE Transactions on Ultrasonics, Ferroelectrics and Frequency Control* 51(8), 973–979.

Kripfgans, O. D., Fowlkes, J. B., Miller, D. L., Eldevik, O. P. & Carson, P. L. (2000), Acoustic droplet vaporization for therapeutic and diagnostic applications, *Ultrasound in Medicine and Biology* 26(7), 1177–1189.

Kunwar, S., Chang, S., Westphal, M., Vogelbaum, M., Sampson, J., Barnett, G., Shaffrey, M., Ram, Z., Piepmeier, J., Prados, M., Croteau, D., Pedain, C., Leland, P., Husain, S. R., Joshi, B. H., Puri, R. K. & Group, P. R. E. C. I. S. E. S. (2010), Phase III randomized trial of CED of IL13-PE38QQR vs Gliadel wafers for recurrent glioblastoma, *Neurooncology* 12(8), 871–881.

Larkin, J. M. (1979), A clinical investigation of total-body hyperthermia as cancer therapy, *Cancer Research* 39(6 Pt 2), 2252–2254.

Lidar, Z., Mardor, Y., Jonas, T., Pfeffer, R., Faibel, M., Nass, D., Hadani, M. & Ram, Z. (2004), Convection-enhanced delivery of paclitaxel for the treatment of recurrent malignant glioma: A phase I/II clinical study, *Journal of Neurosurgery* 100(3), 472–479.

Lieberman, D. M., Laske, D. W., Morrison, P. F., Bankiewicz, K. S. & Oldfield, E. H. (1995), Convection-enhanced distribution of large molecules in gray matter during interstitial drug infusion, *Journal of Neurosurgery* 82(6), 1021–1029.

Lindsey, B. D., Light, E. D., Nicoletto, H. A., Bennett, E. R., Laskowitz, D. T. & Smith, S. W. (2011), The ultrasound brain helmet: New transducers and volume registration for in vivo simultaneous multi-transducer 3-D transcranial imaging, *IEEE Transactions on Ultrasonics, Ferroelectrics and Frequency Control* 58(6), 1189–1202.

Liu, H.-L., Hua, M.-Y., Chen, P.-Y., Chu, P.-C., Pan, C.-H., Yang, H.-W., Huang, C.-Y., Wang, J.-J., Yen, T.-C. & Wei, K.-C. (2010a), Blood–brain barrier disruption with focused ultrasound enhances delivery of chemotherapeutic drugs for glioblastoma treatment, *Radiology* 255(2), 415–425.

Liu, H.-L., Hua, M.-Y., Yang, H.-W., Huang, C.-Y., Chu, P.-N., Wu, J.-S., Tseng, I.-C., Wang, J.-J., Yen, T.-C., Chen, P.-Y. & Wei, K.-C. (2010b), Magnetic resonance monitoring of focused ultrasound magnetic nanoparticle targeting delivery of therapeutic agents to the brain, *Proceedings of the National Academy of Sciences of the United States of America* 107(34), 15205–15210.

Lynn, J. G. & Putnam, T. J. (1944), Histology of cerebral lesions produced by focused ultrasound, *American Journal of Pathology* 20(3), 637–649.

Lynn, J. G., Zwemer, R. L., Chick, A. J. & Miller, A. E. (1942), A new method for the generation and use of focused ultrasound in experimental biology, *Journal of General Physiology* 26(2), 179–193.

Mackay, J. & Mensah, G. A. (2004), *The Atlas of Heart Disease and Stroke*, World Health Organization.

Marmor, J. B. & Hahn, G. M. (1978), Ultrasound heating in previously irradiated sites, *International Journal of Radiation Oncology, Biology, Physics* 4(11–12), 1029–1032.

Marquet, F., Pernot, M., Aubry, J.-F., Montaldo, G., Tanter, M. & Fink, M. (2006), Non-invasive transcranial ultrasound therapy guided by CT-scans, *Conference Proceedings of the IEEE Engineering in Medicine and Biology Society* 1, 683–687.

Marquet, F., Tung, Y.-S., Teichert, T., Ferrera, V. & Konofagou, E. (2011), Noninvasive, transient and selective blood–brain barrier opening in non-human primates in vivo, *PLoS One* 6(7), e22598.

Martin, E., Jeanmonod, D., Morel, A., Zadicario, E. & Werner, B. (2009), High-intensity focused ultrasound for noninvasive functional neurosurgery, *Annals of Neurology* 66(6), 858–861.

Chapter 10

McDannold, N., Moss, M., Killiany, R., Rosene, D. L., King, R. L., Jolesz, F. A. & Hynynen, K. (2003), MRI-guided focused ultrasound surgery in the brain: Tests in a primate model, *Magnetic Resonance in Medicine* 49(6), 1188–1191.

McDannold, N., King, R. L. & Hynynen, K. (2004), MRI monitoring of heating produced by ultrasound absorption in the skull: In vivo study in pigs, *Magnetic Resonance in Medicine* 51(5), 1061–1065.

McDannold, N., Vykhodtseva, N. I. & Hynynen, K. (2006a), Microbubble contrast agent with focused ultrasound to create brain lesions at low power levels: MR imaging and histologic study in rabbits, *Radiology* 241(1), 95–106.

McDannold, N., Vykhodtseva, N. & Hynynen, K. (2006b), Targeted disruption of the blood–brain barrier with focused ultrasound association with cavitation activity, *Physics in Medicine and Biology* 51(4), 793–807.

McDannold, N., Vykhodtseva, N. & Hynynen, K. (2008), Blood–brain barrier disruption induced by focused ultrasound and circulating preformed microbubbles appears to be characterized by the Mechanical Index, *Ultrasound in Medicine and Biology* 34, 834–840.

McDannold, N., Clement, G. T., Black, P., Jolesz, F. & Hynynen, K. (2010a), Transcranial magnetic resonance imaging-guided focused ultrasound surgery of brain tumors: Initial findings in 3 patients, *Neurosurgery* 66(2), 323–332.

McDannold, N., Vykhodtseva, N., Arvanitis, C. & Livingstone, M. (2010b), BBB disruption in nonhuman primates using a clinical MRgFUS System: Preliminary results, *MR-Guided Focused Ultrasound 2010: 2nd International Symposium*.

McDannold, N., Arvanitis, C. D., Vykhodtseva, N. & Linvingstone, M. S. (2012), Temporary blood–brain barrier disruption via ultrasound and microbubbles: Evaluation of a noninvasive targeted drug delivery method in rhesus macaques, *Cancer Research* 72(14), 3652–3663.

Mei, J., Cheng, Y., Song, Y., Yang, Y., Wang, F., Liu, Y. & Wang, Z. (2009), Experimental study on targeted methotrexate delivery to the rabbit brain via magnetic resonance imaging-guided focused ultrasound, *Journal of Ultrasound Medicine* 28(7), 871–880.

Mesiwala, A., Farrell, L., Wenzel, H., Silbergeld, D., Crum, L., Winn, H. & Mourad, P. (2002), High-intensity focused ultrasound selectively disrupts the blood–brain barrier in vivo, *Ultrasound in Medicine and Biology* 28(3), 389–400.

Mitri, F. G., Greenleaf, J. F. & Fatemi, M. (2005), Chirp imaging vibro-acoustography for removing the ultrasound standing wave artifact, *IEEE Transactions on Medical Imaging* 24(10), 1249–1255.

Molina, C. A., Ribo, M., Rubiera, M., Montaner, J., Santamarina, E., Delgado-Mederos, R., Arenillas, J. F., Huertas, R., Purroy, F., Delgado, P. & Alvarez-Sabín, J. (2006), Microbubble administration accelerates clot lysis during continuous 2-MHz ultrasound monitoring in stroke patients treated with intravenous tissue plasminogen activator, *Stroke* 37(2), 425–429.

N.I.N.D.S. (1995), Tissue plasminogen activator for acute ischemic stroke, *New England Journal of Medicine* 333(24), 1582–1587.

Nock, L., Trahey, G. E. & Smith, S. W. (1989), Phase aberration correction in medical ultrasound using speckle brightness as a quality factor, *Journal of the Acoustical Society of America* 85(5), 1819–1833.

O'Donnell, M. & Flax, S. W. (1988), Phase-aberration correction using signals from point reflectors and diffuse scatterers: Measurements, *IEEE Transactions on Ultrasonics, Ferroelectrics and Frequency Control* 35(6), 768–774.

O'Reilly, M. A. & Hynynen, K. (2012), Blood–brain barrier: Real-time feedback-controlled focused ultrasound disruption by using an acoustic emissions-based controller, *Radiology* 263(1), 96–106.

O'Reilly, M. A., Waspe, A. C., Ganguly, M. & Hynynen, K. (2011), Focused-ultrasound disruption of the blood–brain barrier using closely-timed short pulses: Influence of sonication parameters and injection rate, *Ultrasound in Medicine and Biology* 37(4), 587–594.

Padilla, F., Bossy, E., Haiat, G., Jenson, F. & Laugier, P. (2006), Numerical simulation of wave propagation in cancellous bone, *Ultrasonics* 44 (Suppl 1), e239–e243.

Pardridge, W. M. (2005), The blood–brain barrier: Bottleneck in brain drug development, *NeuroRx* 2(1), 3–14.

Park, J., Zhang, Y. Z., Vkyhodtseva, N. & McDannold, N. (2011), Targeted delivery of neural stem cells in the rat brain via focused ultrasound-induced blood–brain barrier disruption, in *International Symposium on Therapeutic Ultrasound*, New York.

Park, J.-W., Jung, S., Jung, T.-Y. & Lee, M.-C. (2006), Focused ultrasound surgery for the treatment of recurrent anaplastic astrocytoma: A preliminary report, *AIP Conference Proceedings* 829, 238–240.

Patrick, J. T., Nolting, M. N., Goss, S. A., Dines, K. A., Clendenon, J. L., Rea, M. A. & Heimburger, R. F. (1990), Ultrasound and the blood–brain barrier, *Advances in Experimental Medicine and Biology* 267, 369–381.

Pernot, M., Aubry, J. F., Tanter, M., Marquet, F., Montaldo, G., Boch, A. L., Kujas, M., Seilhean, D. & Fink, M. (2007), High power phased array prototype for clinical high intensity focused ultrasound: Applications to transcostal and transcranial therapy, *Conference Proceedings of the IEEE Engineering in Medicine and Biology Society* 2007, 234–237.

Pernot, M., Aubry, J. F., Tanter, M., Thomas, J. L. & Fink, M. (2003), High power transcranial beam steering for ultrasonic brain therapy, *Physics in Medicine and Biology* 48(16), 2577–2589.

Pernot, M., Montaldo, G., Tanter, M. & Fink, M. (2006), "Ultrasonic stars" for time-reversal focusing using induced cavitation bubbles, *Applied Physics Letters* 88, 034102.

Pichardo, S., Sin, V. W. & Hynynen, K. (2011), Multi-frequency characterization of the speed of sound and attenuation coefficient for longitudinal transmission of freshly excised human skulls, *Physics in Medicine and Biology* 56(1), 219–250.

Pulkkinen, A., Huang, Y., Song, J. & Hynynen, K. (2011), Simulations and measurements of transcranial low-frequency ultrasound therapy: Skull-base heating and effective area of treatment, *Physics in Medicine and Biology* 56(15), 4661–4683.

Quesson, B., Vimeux, F., Salomir, R., de Zwart, J. A. & Moonen, C. T. W. (2002), Automatic control of hyperthermic therapy based on real-time Fourier analysis of MR temperature maps, *Magnetic Resonance in Medicine* 47(6), 1065–1072.

Ram, Z., Cohen, Z. R., Harnof, S., Tal, S., Faibel, M., Nass, D., Maier, S. E., Hadani, M. & Mardor, Y. (2006), Magnetic resonance imaging-guided, high-intensity focused ultrasound for brain tumor therapy, *Neurosurgery* 59(5), 949–955, discussion 955-6.

Raymond, S., Treat, L., Dewey, J., McDannold, N., Hynynen, K. & Bacskai, B. (2008), Ultrasound enhanced delivery of molecular imaging and therapeutic agents in Alzheimer's disease mouse models, *PLoS One* 3(5), e2175.

Ringelstein, E. B., Kahlscheuer, B., Niggemeyer, E. & Otis, S. M. (1990), Transcranial Doppler sonography: Anatomical landmarks and normal velocity values, *Ultrasound in Medicine and Biology* 16(8), 745–761.

Rubin, L. L. & Staddon, J. M. (1999), The cell biology of the blood–brain barrier, *Annual Review of Neuroscience* 22, 11–28.

Salahuddin, T. S., Johansson, B. B., Kalimo, H. & Olsson, Y. (1988), Structural changes in the rat brain after carotid infusions of hyperosmolar solutions: A light microscopic and immunohistochemical study, *Neuropathology and Applied Neurobiology* 14(6), 467–482.

Samiotaki, G., Vlachos, F., Tung, Y.-S. & Konofagou, E. E. (2011), A quantitative pressure and microbubble-size dependence study of focused ultrasound-induced blood–brain barrier opening reversibility in vivo using MRI, *Magnetic Resonance in Medicine* 67(3), 769–777.

Sauerbrei, E. E., Harrison, P. B., Ling, E. & Cooperberg, P. L. (1981), Neonatal intracranial pathology demonstrated by high-frequency linear array ultrasound, *Journal of Clinical Ultrasound* 9(1), 33–36.

Shang, X., Wang, P., Liu, Y., Zhang, Z. & Xue, Y. (2011), Mechanism of low-frequency ultrasound in opening blood–tumor barrier by tight junction, *Journal of Molecular Neuroscience* 43(3), 364–369.

Sheikov, N., McDannold, N., Sharma, S. & Hynynen, K. (2008), Effect of focused ultrasound applied with an ultrasound contrast agent on the tight junctional integrity of the brain microvascular endothelium, *Ultrasound in Medicine and Biology* 34(7), 1093–1104.

Sheikov, N., McDannold, N., Vykhodtseva, N., Jolesz, F. & Hynynen, K. (2004), Cellular mechanisms of the blood–brain barrier opening induced by ultrasound in presence of microbubbles, *Ultrasound in Medicine and Biology* 30(7), 979–989.

Shimm, D. S., Hynynen, K. H., Anhalt, D. P., Roemer, R. B. & Cassady, J. R. (1988), Scanned focussed ultrasound hyperthermia: Initial clinical results, *International Journal of Radiation Oncology, Biology, Physics* 15(5), 1203–1208.

Smith, N. B., Merrilees, N. K., Dahleh, M. & Hynynen, K. (2001), Control system for an MRI compatible intracavitary ultrasound array for thermal treatment of prostate disease, *International Journal of Hyperthermia* 17(3), 271–282.

Smith, S., Phillips, D. J. & von Ramm, O. T. (1977), Some advances in acoustic imaging through the skull, in D. G. Hazzard & M. L. Litz, ed., *Symposium on Biological Effects and Characterizations of Ultrasound Sources*.

Smith, S. W., Ivancevich, N. M., Lindsey, B. D., Whitman, J., Light, E., Fronheiser, M., Nicoletto, H. A. & Laskowitz, D. T. (2009), The ultrasound brain helmet: Feasibility study of multiple simultaneous 3D scans of cerebral vasculature, *Ultrasound in Medicine and Biology* 35(2), 329–338.

Smith, S. W., Trahey, G. E. & von Ramm, O. T. (1986), Phased array ultrasound imaging through planar tissue layers, *Ultrasound in Medicine and Biology* 12(3), 229–243.

Song, J. & Hynynen, K. (2010), Feasibility of using lateral mode coupling method for a large scale ultrasound phased array for noninvasive transcranial therapy, *IEEE Transactions in Biomedical Engineering* 57(1), 124–133.

Song, J., Pulkkinen, A., Huang, Y. & Hynynen, K. (2012), Investigation of standing-wave formation in a human skull for a clinical prototype of a large-aperture, transcranial MR-guided focused ultrasound (MRgFUS) phased array: An experimental and simulation study, *IEEE Transactions in Biomedical Engineering* 59(2), 435–444.

Spencer, M. P., Thomas, G. I., Nicholls, S. C. & Sauvage, L. R. (1990), Detection of middle cerebral artery emboli during carotid endarterectomy using transcranial Doppler ultrasonography, *Stroke* 21(3), 415–423.

Sun, J. & Hynynen, K. (1998), Focusing of therapeutic ultrasound through a human skull: A numerical study, *Journal of the Acoustical Society of America* 104(3 Pt 1), 1705–1715.

Tang, S. C. & Clement, G. T. (2009), Acoustic standing wave suppression using randomized phase-shift-keying excitations, *Journal of the Acoustical Society of America* 126(4), 1667–1670.

Tanter, M., Aubry, J. F., Gerber, J., Thomas, J. L. & Fink, M. (2001), Optimal focusing by spatio-temporal inverse filter. I. Basic principles, *Journal of the Acoustical Society of America* 110(1), 37–47.

Tanter, M., Thomas, J. L. & Fink, M. (2000), Time reversal and the inverse filter, *Journal of the Acoustical Society of America* 108(1), 223–234.

Thomas, J.-L. & Fink, M. A. (1996), Ultrasonic beam focusing through tissue inhomogeneities with a time reversal mirror: Application to transskull therapy, *IEEE Transactions on Ultrasonics, Ferroelectrics and Frequency Control* 43(6), 1122–1129.

Ting, C.-Y., Fan, C.-H., Liu, H.-L., Huang, C.-Y., Hsieh, H.-Y., Yen, T.-C., Wei, K.-C. & Yeh, C.-K. (2012), Concurrent blood–brain barrier opening and local drug delivery using drug-carrying microbubbles and focused ultrasound for brain glioma treatment, *Biomaterials* 33(2), 704–712.

Treat, L., Zhang, Y., McDannold, N. & Hynynen, K. (2009), MRI-guided focused ultrasound-enhanced chemotherapy of 9 l rat gliosarcoma: Survival study, *International Society of Magnetic Resonance in Medicine*, Annual Meeting May 2008, Toronto, Canada.

Trübestein, G., Engel, C., Etzel, F., Sobbe, A., Cremer, H. & Stumpff, U. (1976), Thrombolysis by ultrasound, *Clinical Science and Molecular Medicine* Suppl 3, 697s–698s.

Tsivgoulis, G., Eggers, J., Ribo, M., Perren, F., Saqqur, M., Rubiera, M., Sergentanis, T. N., Vadikolias, K., Larrue, V., Molina, C. A. & Alexandrov, A. V. (2010), Safety and efficacy of ultrasound-enhanced thrombolysis: A comprehensive review and meta-analysis of randomized and nonrandomized studies, *Stroke* 41(2), 280–287.

Tufail, Y., Matyushov, A., Baldwin, N., Tauchmann, M. L., Georges, J., Yoshihiro, A., Tillery, S. I. H. & Tyler, W. J. (2010), Transcranial pulsed ultrasound stimulates intact brain circuits, *Neuron* 66(5), 681–694.

Tung, Y.-S., Vlachos, F., Choi, J., Deffieux, T., Selert, K. & Konofagou, E. (2010), In vivo transcranial cavitation threshold detection during ultrasound-induced blood–brain barrier opening in mice, *Physics in Medicine and Biology* 55(20), 6141–6155.

Tung, Y.-S., Marquet, F., Teichert, T., Ferrera, V. & Konofagou, E. (2011a), Feasibility of noninvasive cavitation-guided blood–brain barrier opening using focused ultrasound and microbubbles in nonhuman primates, *Applied Physics Letters* 98(16), 163704.

Tung, Y.-S., Vlachos, F., Feshitan, J. A., Borden, M. A. & Konofagou, E. E. (2011b), The mechanism of interaction between focused ultrasound and microbubbles in blood–brain barrier opening in mice, *Journal of the Acoustical Society of America* 130(5), 3059.

Tyler, W. J., Tufail, Y., Finsterwald, M., Tauchmann, M. L., Olson, E. J. & Majestic, C. (2008), Remote excitation of neuronal circuits using low-intensity, low-frequency ultrasound, *PLoS One* 3(10), e3511.

van Velthoven, V. & Auer, L. M. (1990), Practical application of intraoperative ultrasound imaging, *Acta Neurochirurgica* (Wien) 105(1–2), 5–13.

Vignon, F., Aubry, J. F., Tanter, M., Margoum, A. & Fink, M. (2006), Adaptive focusing for transcranial ultrasound imaging using dual arrays, *Journal of the Acoustical Society of America* 120(5 Pt 1), 2737–2745.

Vlachos, F., Tung, Y.-S. & Konofagou, E. (2010), Permeability assessment of the focused ultrasound-induced blood–brain barrier opening using dynamic contrast-enhanced MRI, *Physics in Medicine and Biology* 55(18), 5451–5466.

Vlachos, F., Tung, Y.-S. & Konofagou, E. (2011), Permeability dependence study of the focused ultrasound-induced blood–brain barrier opening at distinct pressures and microbubble diameters using DCE-MRI, *Magnetic Resonance in Medicine* 66, 821–830.

von Ramm, O. T. & Smith, S. W. (1983), Beam steering with linear arrays, *IEEE Transactions in Biomedical Engineering* (8), 438–452.

Vykhodtseva, N., Sorrentino, V., Jolesz, F. A., Bronson, R. T. & Hynynen, K. (2000), MRI detection of the thermal effects of focused ultrasound on the brain, *Ultrasound in Medicine and Biology* 26(5), 871–880.

Vykhodtseva, N. I., Hynynen, K. & Damianou, C. (1995), Histologic effects of high intensity pulsed ultrasound exposure with subharmonic emission in rabbit brain in vivo, *Ultrasound in Medicine and Biology* 21(7), 969–979.

Wang, F., Cheng, Y., Mei, J., Song, Y., Yang, Y.-Q., Liu, Y. & Wang, Z. (2009), Focused ultrasound microbubble destruction-mediated changes in blood–brain barrier permeability assessed by contrast-enhanced magnetic resonance imaging, *Journal of Ultrasound Medicine* 28(11), 1501–1509.

Wang, J.-E., Liu, Y.-H., Liu, L.-B., Xia, C.-Y., Zhang, Z. & Xue, Y.-X. (2011), Effects of combining low frequency ultrasound irradiation with papaverine on the permeability of the blood–tumor barrier, *Journal of Neurooncology* 102(2), 213–224.

Wessels, T., Bozzato, A., Mull, M. & Klötzsch, C. (2004), Intracranial collateral pathways assessed by contrast-enhanced three-dimensional transcranial color-coded sonography, *Ultrasound in Medicine and Biology* 31(11), 1435–1440.

White, D. N., Clark, J. M., Chesebrough, J. N., White, M. N. & Campbell, J. K. (1968), Effect of the skull in degrading the display of echoencephalographic B and C scans, *Journal of the Acoustical Society of America* 44(5), 1339–1345.

White, J., Clement, G. T. & Hynynen, K. (2005), Transcranial ultrasound focus reconstruction with phase and amplitude correction, *IEEE Transactions on Ultrasonics, Ferroelectrics and Frequency Control* 52(9), 1518–1522.

White, P. J., Clement, G. T. & Hynynen, K. (2006), Longitudinal and shear mode ultrasound propagation in human skull bone, *Ultrasound in Medicine and Biology* 32(7), 1085–1096.

White, P. J., Whalen, S., Tang, S. C., Clement, G. T., Jolesz, F. & Golby, A. J. (2009), An intraoperative brain shift monitor using shear mode transcranial ultrasound: Preliminary results, *Journal of Ultrasound Medicine* 28(2), 191–203.

Wright, C., Hynynen, K. & Goertz, D. (2012a), In vitro and in vivo high-intensity focused ultrasound thrombolysis, *Investigative Radiology* 47(4), 217–225.

Wright, C. C., Hynynen, K. & Goertz, D. E. (2012b), Pulsed focused ultrasound-induced displacements in confined in vitro blood clots, *IEEE Transactions in Biomedical Engineering* 59(3), 842–851.

Wu, J. & Cubberley, F. (1997), Measurement of velocity and attenuation of shear waves in bovine compact bone using ultrasonic spectroscopy, *Ultrasound in Medicine and Biology* 23(1), 129–134.

Xia, C.-Y., Zhang, Z., Xue, Y.-X., Wang, P. & Liu, Y.-H. (2009), Mechanisms of the increase in the permeability of the blood–tumor barrier obtained by combining low-frequency ultrasound irradiation with small-dose bradykinin, *Journal of Neurooncology* 94(1), 41–50.

Xie, F., Boska, M., Lof, J., Uberti, M., Tsutsui, J. & Porter, T. (2008), Effects of transcranial ultrasound and intravenous microbubbles on blood brain barrier permeability in a large animal model, *Ultrasound in Medicine and Biology* 34(12), 2028–2034.

Yang, F.-Y., Liu, S.-H., Ho, F.-M. & Chang, C.-H. (2009), Effect of ultrasound contrast agent dose on the duration of focused-ultrasound-induced blood–brain barrier disruption, *Journal of the Acoustical Society of America* 126(6), 3344–3349.

Yoo, S.-S., Bystritsky, A., Lee, J.-H., Zhang, Y., Fischer, K., Min, B.-K., McDannold, N., Pascual-Leone, A. & Jolesz, F. A. (2011a), Focused ultrasound modulates region-specific brain activity, *Neuroimage* 56(3), 1267–1275.

Yoo, S.-S., Kim, H., Min, B.-K., Franck, E. & Park, S. (2011b), Transcranial focused ultrasound to the thalamus alters anesthesia time in rats, *Neuroreport* 22(15), 783–787.

Yousefi, A., Goertz, D. E. & Hynynen, K. (2009), Transcranial shear-mode ultrasound: Assessment of imaging performance and excitation techniques, *IEEE Transactions on Medical Imaging* 28(5), 763–774.

Zhang, Z., Xia, C., Xue, Y. & Liu, Y. (2009), Synergistic effect of low-frequency ultrasound and low-dose bradykinin on increasing permeability of the blood–tumor barrier by opening tight junction, *Journal of Neuroscience Research* 87(10), 2282–2289.

Zhang, Z., Xue, Y., Liu, Y. & Shang, X. (2010), Additive effect of low-frequency ultrasound and endothelial monocyte-activating polypeptide II on blood–tumor barrier in rats with brain glioma, *Neuroscience Letters* 481(1), 21–25.

Chapter 10

Index

Page numbers ending in "f" refer to figures. Page numbers ending in "t" refer to tables.

Printed and bound by CPI Group (UK) Ltd, Croydon, CR0 4YY

22/10/2024

01777614-0010